THE DERVENI PAPYRUS

This is the first comprehensive study of the Derveni Papyrus. The papyrus, found in 1962 near Thessalonica, is not only one of the oldest surviving Greek papyri but is also considered by scholars to be a document of primary importance for a better understanding of the religious and philosophical developments of the fifth and fourth centuries BC. Gábor Betegh aims to reconstruct and systematically to analyse the different strata of the text and their interrelation by exploring the archaeological context; the interpretation of rituals in the first columns of the text; the Orphic poem commented on by the author of the papyrus; and the cosmological and theological doctrines which emerge from the Derveni author's exegesis of the poem. Betegh discusses the place of the text in the context of late Presocratic philosophy and offers an important preliminary edition of the text of the papyrus with critical apparatus and English translation.

GÁBOR BETEGH is Assistant Professor of Philosophy at the Central European University, Budapest.

THE DERVENI PAPYRUS

Cosmology, Theology and Interpretation

GÁBOR BETEGH

CAMBRIDGE
UNIVERSITY PRESS

CAMBRIDGE UNIVERSITY PRESS
Cambridge, New York, Melbourne, Madrid, Cape Town, Singapore, São Paulo

Cambridge University Press
The Edinburgh Building, Cambridge CB2 8RU, UK

Published in the United States of America by Cambridge University Press, New York

www.cambridge.org
Information on this title: www.cambridge.org/9780521801089

First published 2004
Fourth printing 2006
This digitally printed version 2007

A catalogue record for this publication is available from the British Library

Library of Congress Cataloguing in Publication data
Betegh, Gábor.
The Derveni papyrus : cosmology, theology, and interpretation / Gábor Betegh.
p. cm.
Includes bibliographical references and index.
ISBN 0 521 80108 7
1. Derveni papyrus. 2. Greece – Religion. 3. Cosmology, Ancient.
I. Derveni papyrus. English & Greek. II. Title.
BL783.B48 2004
292.8 – dc22 2003055820

ISBN 978-0-521-80108-9 hardback
ISBN 978-0-521-04739-5 paperback

Contents

Preface

I first met the Derveni papyrus via a reference to the Heraclitus quotation in it. When I read the text in the anonymously published *ZPE* transcript, I found it astonishing and fascinating; I realised that I wished to know more about it even apart from its lesson on Heraclitus.

Presumably everyone interested in ancient philosophy, and especially in the earliest part of it, has the dream that by some archaeological miracle some new evidence will one day be found which can shed new light on the doctrines of individual philosophers or on the intellectual climate of the age. I felt I was reading such a document. Also, the commonplace of papyrology became personal experience: I found it outstandingly exciting to read a text which did not come down to us through the stemma of medieval manuscripts but had survived in a copy written by someone around the end of the classical age.

Yet, even though many of the phrases and ideas of the text had a familiar ring, I found the text beyond comprehension. The reasoning appeared contorted, I soon got lost in the jungle of allegorical identifications, and, on the whole, it seemed far from obvious what the author actually wanted to say, why he wanted to say it, and, more generally, what the purpose of the whole text was. As I turned for help to the secondary literature, numerous details became clearer, but I had the feeling that many of the basic questions were still open, or, for that matter, had not even been asked. Also, piece by piece, I gathered the details of the unfortunate editorial situation.

When I finally decided to work on the Derveni text more thoroughly, I first had the clear intention of concentrating solely on that aspect of it which is the most immediately connected to Presocratic philosophy: the cosmological views of the author. Sharing the assumption of most interpreters that the Orphic poem was no more than a pretext for the author, I believed I could examine the author's cosmology without bothering too much with the Orphic poem. Besides, my knowledge about early Orphism was scanty, and the little I knew admonished that it was safer to avoid this slippery field. However, I gradually came to realise that this approach is misguided. Not only does it conceal a genuinely interesting aspect of the text, but, in fact, it is impossible to assess the cosmological theory in isolation from the other constitutive elements of the text. I had to acknowledge that my original project was mistaken for the reason that the author's cosmogony is not a self-contained theory, but is intimately linked to his theological tenets and, possibly, to ideas which come up in other no less obscure texts of the age. Moreover, little by little, I came to the conclusion that these theological tenets may have connections, considerably stronger than usually assumed, with the ideas expressed in the poem. First of all, even if the author's explanations appear completely far-fetched, having nothing to do with the obvious, literary meaning of the verses quoted, there is a clear thematic connection: both the poem and the author's cosmological theory speak about the birth of the world, concentrating on the role of a supreme divine being. This led me to the hypothesis that, despite appearances, there may be more affinities between the poem and the author's ideas; or at least that this is a possibility worth exploring. The present work is an attempt to examine these ingredients of the Derveni text and their interrelation systematically and in a unified framework.

This synthetic approach notwithstanding, I found it essential first to separate the different strata of the text, analyse them one by one, and then consider the way they are related to and built upon each other. Hence the general structure of this book. In the first chapter I summarise the most important facts about the archaeological context of the find, especially those aspects of it that may be relevant for the evaluation of the text of the papyrus. It is also in the first chapter that I give a brief papyrological description of the document. In the second chapter I offer a preliminary discussion of the contents of the first six columns. The third chapter starts the analysis of the second part of the text, and deals with the reconstruction of the Orphic poem commented on by the Derveni author. Based on the results of this survey, I try to explore in chapter four the theological and cosmological ideas expressed in the poem. It is only in chapter five that I turn to the Derveni author's own doctrines, concentrating first on his system of identifications, the means by which he defines his cosmic god. Chapter six, in turn, focuses on the physical side of the cosmology, examining also the ontology underlying it. Up to this point, I give an internal reading of the papyrus, with the fewest possible references to other texts. In the remaining four chapters I try to take the reconstruction and interpretation one step further by analysing the text against the background of those figures and currents of thought of the Presocratic age that seem to have had the greatest impact on the author. Having presented a comparative analysis with Anaxagoras in chapter seven, and with Diogenes of Apollonia and Archelaus of Athens in chapter eight, I discuss the possible theological and cosmological connexions with Heraclitus and some of the gold plates in chapter nine. Finally, in chapter ten I examine the author's epistemology and hermeneutic approach with reference to Plato, Heraclitus, the medical writers and Empedocles. The methodology I have chosen – i.e. to provide a close reading of the text first, and to discuss it in its wider intellectual context only in a second step – naturally entails that I have to return to the same passages over and over again as I discuss them from different angles.

I present a text and translation before the interpretative essay, and give a complete *index verborum* at the end. In an appendix I discuss Richard Janko's recent hypothesis about the identity of the Derveni author.

This book is the outcome of a long project, and I incurred many intellectual and personal debts during this period – but I also made many friends through this work. The book in its present form is a revised and somewhat enlarged version of the doctoral thesis I wrote in a system of 'co-tutelle en alternance' at the École des Hautes Études en Sciences Sociales (Paris) and the Eötvös Loránd University (Budapest). I am deeply indebted to my two supervisors, Kornél Steiger, who gave me the first impetus to study the Presocratics during my undergraduate studies, and Jacques Brunschwig, from whom I have learned the most about Greek philosophy in general and research methodology in particular. Discussions with Jacques, be it about philosophy, music or the fine arts, are always outstandingly enriching. What I have learnt from him goes far beyond the scope of ancient philosophy. He is a model for me, and not only as a scholar.

My special thanks go to André Laks, with whom I had the most rewarding discussions on different aspects of the Derveni text and the appropriate hermeneutic approach to it. He has supported me in many other ways as well ever since we first met in Paris. During my study period in France, I also had the opportunity to discuss questions related to Orphism and allegorical interpretation with Luc Brisson. His valuable comments on the last draft have helped me to clarify my position at important points.

In Budapest, István M. Bodnár instructed me on cosmological issues and has helped me in all professional matters. I was fortunate to have him as a teacher, and now I am fortunate to have him as a colleague and a close friend.

Judit Horváth helped me more than once when I had special difficulties with the Derveni author's Greek. My thanks go to Andrea Nagy for helping me in checking the word index and to Ákos Brunner for his help in creating the indexes.

I wrote the larger part of my doctoral dissertation at Christ's College, Cambridge, during the academic year of 1997/98. My warmest gratitude is to David Sedley, my Cambridge supervisor. He corrected drafts of every chapter of the dissertation with great attentiveness and patience, and gave me much help also in transforming it into a book. His criticism and constructive suggestions have always been outstandingly helpful and stimulating. Everybody who has had the good fortune to have him as a reader can testify that he is the best possible critic: acute, patient, inventive, demanding, but always open to alternative views. The support and encouragement I have received from him ever since has been a constant source of inspiration.

I am also grateful to the members of the dissertation committee, Zsigmond Ritoók, László Tengelyi, György Karsai, Heinz Wissman and Glenn W. Most, all of whom made valuable comments that I tried to integrate into this book. Glenn W. Most also read chapters 2 and 10 in their penultimate version and his criticism helped me to formulate my views more clearly.

I did most of the revision during my fellowship at Harvard's Center for Hellenic Studies in Washington D.C. It was an extremely pleasant and rewarding working environment due to the kindness and helpfulness of the Director, Gregory Nagy, and the staff, the librarians and the Fellows. My special thanks to Sandy Blakely, Nikolay Grintzer, Jens Holzhausen, Roberto Polito, Gretchen Reydams-Schils and Sarolta Takács with whom I could discuss parts of the papyrus and my interpretation of it.

I am particularly grateful to Alberto Bernabé. He most helpfully showed me the relevant parts of his forthcoming edition of the Orphic fragments, sent me various important offprints, and has always been ready to answer my queries in correspondence. I am grateful also to Walter Burkert for having sent me his paper on theo-cosmogonies before publication. Let me also thank Muriel Hall and Michael Sharp of Cambridge University Press for their patience and dedication.

Finally, I also wish to thank all my friends and colleagues in Budapest, Paris and Cambridge. They all gave inspiration and helped my work in various ways. I am especially grateful to Nick Lumsden whose kind help arrived at a crucial moment and proved to be decisive for the realisation of this project.

My semesters in Paris were financed by a French governmental scholarship, administered by the Service Culturel, Scientifique et de la Coopération de l'Ambassade de France in Budapest. The Cambridge Overseas Trust, Christ's College and the Hungarian Soros Foundation financed the year I spent in Cambridge. For the final revision I received financial help from the Hungarian OTKA Foundation (FO32471).

I dedicate this book to my family. I suppose it is not easy to live with someone who spends much of his time in a different age and in a different land. One can understand that the wife of Hermotimus of Clazomenae got fed up after a while when the soul of her husband was travelling too much to distant places, while only his body stayed at home. My family, by contrast, have always been particularly patient and supportive. Let me express my gratitude first of all to my mother for all the help and encouragement she has given me over the years. Let me also thank my grandmother and mother-in-law for their support. But my thanks go, above all, to my wife, Ági, who has supported me unfailingly with her love, patience and care. I could not have accomplished this book without her help. Finally, let me also mention our daughter Juli who made me realise even more acutely how good it is to come back home after such spiritual journeys to remote lands and ages.

Abbreviations

DK	*Die Fragmente der Vorsokratiker*, 7th edn. by H. Diels, ed. with additions and Nachträge by W. Kranz. Berlin, 1954.
HGP	W. K. C. Guthrie, *A History of Greek Philosophy*, vols. 1–6. Cambridge, 1962–1981.
IG	*Inscriptiones Graecae*, eds. A. Kirchhoff et al. Berlin, 1873–.
KRS	G. S. Kirk, J. E. Raven and M. Schofield, *The Presocratic Philosophers*, 2nd edn. Cambridge, 1983.
L&M	*Studies on the Derveni Papyrus*, eds. A. Laks and G. W. Most. Oxford, 1997.
LIMC	*Lexicon iconographicum mythologiae classicae*. Zurich and Munich, 1981–99.
LSJ	Liddell and Scott, *A Greek-English Lexicon*, 9th edn., revised by H. Stuart Jones and R. McKenzie. Oxford, 1940.
OF	*Orphicorum Fragmenta*, ed. O. Kern. Berlin, 1922.
PCG	*Poetae Comici Graeci*, eds. R. Kassel and C. Austin. Berlin 1983–.
RE	*Paulys Realencyclopädie der classischen Altertumswissenschaft*, eds. G. Wissowa et al. Stuttgart and Munich, 1894–1978.
SVF	*Stoicorum Veterum Fragmenta*, ed. H. von Arnim. Leipzig, 1903–5.

Text and translation

The following text makes no claim to be a critical edition. Its main purpose is to show which solutions I have opted for in controversial cases, and to indicate in the apparatus the alternatives others have suggested. I do not, however, mention suggestions that have been superseded by more accurate readings of the papyrus. It has to be noted also that, because of the complex editorial situation and the constant exchange of information among scholars in informal ways, in some cases it is difficult to identify the ultimate origin of a supplement or emendation.

I was already about to submit the manuscript for publication when Richard Janko's 'interim text' was published (Janko (2002)). Without any doubt, it is the most important advance since 1982 when the first near-complete provisional transcript was published, and it should form the textual basis for the study of the papyrus until Prof. Tsantsanoglou's long-awaited edition becomes available. I have revised the following text taking into account Janko's new suggestions regarding both supplements and translation. It turns out that in some cases we are proposing the same new supplements, independently of each other.

The primary sources for the individual columns are as follows:

Coll. 1–7: Tsantsanoglou (1997) (I do not indicate the *ZPE* readings in the apparatus for they are unhelpful in view of Tsantsanoglou's new readings)

Col. 12: Tsantsanoglou *ap.* Brisson (1997)

Col. 20: Tsantsanoglou *ap.* Obbink (1997)

Coll. 17, 22, and 25: Tsantsanoglou's readings displayed at the exhibition of the Derveni papyrus in the Archaeological Museum of Thessaloniki (Apr.–Aug. 1998). I reinstate the consonant assimilations of the papyrus, however.

For the rest of the columns: the anonymous publication in *ZPE* 47 (1982) after p. 300, supplemented with further publicly available information. The sign 'Ts. *ap.* L&M' indicates Tsantsanoglou's remarks on

L&M's English translation. The sign 'Ts. *ap*. Bernabé' indicates information on Tsantsanoglou's new readings I gathered from Bernabé (2002) and (forthcoming). I have also made use of the supplements and emendations proposed by Janko (2001).

I had the opportunity of seeing the papyrus in June 1998 at the exhibition in the Archaeological Museum of Thessaloniki. The sign *pap. ut vid.* in the apparatus indicates modifications I propose on the basis of autopsy. It should be stressed, however, that – because of the unfavourable viewing conditions – I could not read all the fragments. I read all the fragments (with the possible exception of small bits containing a few letters) of coll. 10, 11 and 13–26; most fragments of coll. 7–9 and 12; and a few fragments of col. 4. I have included my readings only for the sake of information, and they should not in any case be treated as authoritative.

In preparing the translation I have made use of all the published translations, and above all the translation of Laks and Most as printed in L&M and the translations of Janko (2001) and (2002).

Text and translation

The Derveni Papyrus

COL. I.

]
]
]ι̣ν̣[
]ν ἕκαστον̣[ ]
]. α̣
]. δ. . . .
Ἐρι]ν̣ύ̣ων
]. . . .
]. . . .
].[. . . .]εια

COL. I.

. . . each one . . . of the [?Eri]nyes . . .

COL. 2.

[ ]
]ωι̣ [
[ ]. ρ̣ .[ ] Ἐ̣ριν[υ
. . . Ἐ̣ρινύω̣[ν]τιμῶcι.[
ψυ̣[χαί ε]ἰ̣cι[χ]ο̣αὶ cταγόcι .[.]εο.[
[ ]. ουc τιμὰc φ[έ]ρη[ι]
[ ]. ἑ̣κάcτο[ι]c ὀρ̣[ν]ί̣θ̣ε̣ιόν τι
κλ[αρμ]οcτο[υ]c τῆ̣ι̣ μ̣ουc̣[ι]κῆι
]. c̣τ̣[..]υτο̣[..]
].ε̣[
]
]
]ει̣

6. ὅταν τιc εἰc τοὺc νεκ]ρ̣οὺc Ts.

COL. 2.

. . . Erinyes . . . of the Erinyes . . . they honour . . . are so[uls] . . . funeral libations in droplets . . . brings honour . . . for each something birdlike . . . fitted to the music . . .

COL. 3.

[ ]
[ ]
±6].. ạịω̣c .[. . . .]cιμ̣. .[
δαίμ]ω̣γ γίνετα[ι ἑκά]cτωι ị .[
. . . .]. η ἐξώλεας [. . . .]ε̣τ̣.ị.. ε..[ ]δ̣ὲ
δ]α̣ίμονες οἱ κάτω[. . . . .]υ δε̣χ̣όμ̣[ενοι]
θ̣εῶν ὑπηρέται δ̣[ὲ κα]λ̣ο̣ῦνṭαị[ ]ι
εἰcὶν ὅπωcπερ ἄ[νδρες] ἄ̣δικοị θ̣.[ ]νοι,
αἰτίην [δ' ἔ]χουcị[
οἵουc. [..].[
.μ]υcτ[

4. ị.[Ts.: ịλ̣ [or ịα̣ [Ts. 5. οἳ] δ̣ὲ Battezzato *ap.* Janko: καλοῦνται] δ̣ὲ Janko 6. δ]α̣ίμονες ZPE || οἱ West *ap.* L&M: οἳ ZPE || κάτω[θεν Ts.: κάτω [εἰcίν ZPE || δε̣χ̣όμ̣[ενοι Betegh: ο]ὐ δέ̣χ̣ο̣υ̣[ται Janko: το]ῦδε̣ χο̣ῦ̣[Ts.: ο]ὐδ' ἔ̣χ̣ο̣υ̣[cι Ts. 8. θ̣α̣[νάτωι ζημιούμε]νοι West *ap.* Ts. 9. [δ' ἔ]χουcị[Burkert: [τ' ἔ]χουcị[Ts.: ἴc]χουcι ZPE 11. μ̣]υcτ[or ὕcτ[ερ Janko

COL. 3.

. . . daimon becomes to each . . . destroyed utterly . . . the daimones beneath . . . receive . . . and are called assistants of the gods . . . (they) are, like unjust men . . . and they are responsible . . . such as . . . initiate

COL. 4.

[.]ου ε .[ ]. ων[
ὁ κείμ[ενα] μ̣ετ̣α̣θ̣[±18]δ̣οῦναι
μᾶλλ[ον ἢ] ϲ̣ίνεται̣[±17 παρ]ὰ τῆϲ τύχηϲ γ[ὰρ
οὐκ εἴ̣[α λα]μμάνει̣[ν.] ἆρ᾽ οὐ τά̣[ξιν ἔχει ἐκ τῶ]ν̣δε κόϲμοϲ;
κατὰ̣ [ταῦτ]ὰ̣ Ἡράκλ̣[ε]ιτοϲ με̣[. . . . .] τ̣ὰ κοινὰ
κατ̣[αϲτρέ]φ̣ει τὰ ἴδ̣[ι]α̣· ὅϲπερ ἴκελ̣[α ἱερο]λόγωι λέγων [ἔφη]·
‾ἥλι̣[οϲ . . .]τ̣ου κατὰ φύϲιν ἀ̣νθρω[πηΐου] ε̣ὖροϲ ποδόϲ [ἐϲτι]
τοὺ̣[ϲ οὔρου]ϲ̣ οὐχ ὑπε̣ρβάλλων· ει .[. . . . .]ρουϲε[
[ἐ]κ̣[βήϲεται]ι̣, Ἐρινύε[ϲ] νιν ἐξευρήϲου̣[ϲι, Δίκηϲ ἐπίκουροι.]
[±12 ὑπερ]βατόν ποῆι κ[
]α̣θυο̣.[
]α δίκηϲ[
]μηνιτακ[
]. . . . ι̣ϲ̣ .[

1. τ]οῦ Ts.–Parássoglou: τ]οῦ ἑα̣[υτοῦ Lebedev *ap.* Ts.–Parássoglou 2. μ̣ετ̣α̣θ̣[έμενοϲ ὠφέλειαν θέλει Ts.: μ̣ετ̣ὰ̣ θ̣[εῶν Ts.–Parássoglou 3. μᾶλλ[ον Ts.–Parássoglou || παρ]ὰ Ts.–Parássoglou (tentatively):]ὰ Ts.: τ]ὰ Janko: ἀλλ]ὰ Bremmer *ap.* Janko || γ[Ts.: π̣[άθη Janko 4. οὐκ εἴ̣[α Ts.–Parássoglou: οὐκέτ[ι Burkert: οὐκ <ἂν> εἴ̣[η Lebedev || λα]μμάνει̣[Ts.–Parássoglou || ἆρ᾽ οὐ Ts.–Parássoglou || τά̣[ξιν ἔχει ἐκ Ts.: τά̣[ξιν ἔχει ὁ Burkert: τά̣[ϲϲεται διὰ Janko || τῶ]ν̣δε Lebedev *ap.* Ts.–Parássoglou: τό]ν̣δε or τή]ν̣δε Ts.–Parássoglou 5. με̣[ταϲκευάζων] or μα̣[ρτυρόμονοϲ Ts.–Parássoglou: με̣[γάλα νομίζων Janko 6. ἴδ̣[ι]α Ts.–Parássoglou || ὅϲπερ ἴκελ̣[α ἱερο]λόγωι λέγων [ἔφη]· Sider: ἰκελ̣[οῖ μυθο]λόγωι λέγων [ὧδε]· Ts.: || ἴκελ̣[οϲ Janko: ἴκελ̣[ωϲ Ts.–Parássoglou || θεο]λόγωι or ἀϲτρο]λόγωι or ἱερῶι] λόγωι Ts.–Parássoglou || [ἔφη] Ts.–Parássoglou 7. ‾ἥλι̣[οϲ *paragr. pap. ut vid.* || ἥλι̣[οϲ . . .]τ̣ου Ts.–Parássoglou: ἥλι̣[οϲ ἑωυ]τ̣οῦ Ts. 8. τοὺ̣[ϲ Ts.–Parássoglou || οὔρου]ϲ̣ Ts.: ὅρου]ϲ̣ Ts.–Parássoglou || ει .[Schönbeck: εἰ γά̣[ρ τι εὔ]ρουϲ Ts.–Parássoglou || ἑ[ωυτοῦ] Ts. 9. [ἐ]κ̣[βήϲεται]ι̣ or [ὑ]π̣[ερβαλε]ῖ̣ Ts.–Parássoglou || Δίκηϲ ἐπίκουροι Ts.–Parássoglou: ἐπικουρήϲουϲι Sider: καὶ φυλάξουϲι Ts.–Parássoglou 10. γὰρ Δίκηι ἐάν τι Sider: ἄτοπα, ἢμ μὴ τιϲ Burkert: οὕτω δ᾽ ἔφη ἵνα Lebedev *ap.* Ts.–Parássoglou || κ[Ts.: κ[αὶ (ἀπο- / κατα-)ϲβέϲουϲι Sider: κ[αὶ ποδὸϲ οὔρωι Burkert: κ[αὶ ἀαφῆ τὸν λόγον Lebedev *ap.* Ts.–Parássoglou 11. πόπαν]α̣ Lebedev *ap.* Ts.–Parássoglou || θύο̣υ̣[ϲι Lebedev *ap.* Ts.–Parássoglou 12. Δίκηϲ [ἐπικουροι Janko 13. μηνι or δυϲ]μήνιτα or ὀξυ]μήνιτα Ts. || τα[κτῶι Ts.: τὰ [κοινά Janko

COL. 4.

who (?alters) what lays . . . to give rather than that harms (*verb*) . . . for [. . .] does not allow to obtain (?it) from chance. Is it not on account of these that the cosmos possesses order? In the same way, Heraclitus . . . the common . . . overturns what is private; he who speaking as someone telling holy discourses said:

> The sun . . . according to nature is a human foot in width, not transgressing its boundaries. If . . . oversteps, the Erinyes, the guardians of Justice, will find it out.

. . . would make a transgression . . . of justice . . .

COL. 5.

]ῃ δει̣ν̣[ὰ
χρη[στη]ριαζομ[ ].ο̣ι̣. ε̣[
χ̣ρ̣ης̣[τ]ηριάζον[ται].[.]. . . . . .[. .]ι
αὐ̣τοῖc πάριμεν̣ [εἰc τὸ μα]ν̣τεῖον ἐπερ̣[ω]τ̣ής̣[οντες,]
τῶν μαντευομένω̣ν̣ [ἕν]εκεν, εἰ θεμι[. . .]. . . . .[ ]
ἐ̣ν̣ Ἅιδου δεινὰ τί ἀπ̣ιστοῦcι; οὐ γινώς̣[κοντες ἐ]ν̣ύπνια
ο̣ὐδὲ τῶν ἄλλων πρ̣αγμάτων ἕκαστ̣[ον], δ̣ιὰ ποίω̣ν ἂν
π̣α̣ρ̣α̣δειγμάτων π̣[ι]στεύοιεν; ὑπό τ[ῆc τε]ἁ̣μαρτ<ί>ης̣
κ̣αὶ [τ]ῆc ἄλλης ἡδον[ῆ]c νενικημέν̣[οι, οὐ] μ̣α̣ν̣θ̣[άνο]υ̣cιν
[οὐδὲ]π̣ιστεύουcι. ἀπ̣[ι]cτίη δὲ κἀμα̣[θίη τὸ αὐτό· ἢν γὰρ]
[μὴ μα]ν̣θάνωcι μη̣[δ]ὲ γινώ[c]κ̣ωc[ιν, οὐκ ἔcτιν ὅπωc]
[πιcτεύcου]ς̣ι̣ν καὶ ὁρ[ῶντεc]
[τ]ὴν ἀπιcτί[ην]
]φαίνεται̣ [

1.]ῃ Ts.: τὰ ἐν Ἅι]δο̣υ Janko || δ̣ε̣ι̣ν̣[ὰ Janko: δε. [Ts. 2. χρη[στη]ριαζομ[Ts.: χρη[στη]ριαζόμ[ενοι or χρη[στη]ριαζόμ[εθα Janko 3. χ̣ρ̣ης̣[τ]ηριάζον[ται Janko || χ̣ρ̣ης̣-[τ]ηρ̣ιάζον[ται Ts. || . . . [. .]ι Ts.: ἔ̣π̣[ειc]ι Burkert 4. τὸ μα]ν̣τεῖον Ts.: τι μα]ν̣τεῖον Janko || ἐπερ̣[ω]τ̣ής̣[οντεc] Ts.: ἐπερ̣[ω]τ̣ής̣[ουcι Burkert 5. θεμι[. . .]. . . . [ ] Ts.: θέμι[c ἀπ]ι̣c̣τ̣ε̣ῖ̣ν̣ [τὰ Janko 6. δεινὰ. τί Janko 7. πρ̣αγμάτων Ts.: παραδειγμάτων Griffiths *ap.* Janko: ὁραμάτων Janko || ἕκαστ̣[ον]Ts.: ἕκαστ̣[α] Janko 8. ὑπό τ[ῆc τε] Ts. *ap.* Bernabé: ὑπό τ̣[ῆc τε] Burkert: ὑπό [τε γὰρ] Ts. || ἁ̣μαρτ<ί>ηc Ts.: ἁ̣μαρτη[ματων ZPE 9. ἡδον[ῆ]c Ts. *ap.* Bernabé: ἡδον[ῆ]c̣ ZPE 10. [οὐδὲ] Ts. || τὸ αὐτό Janko: ταὐτόν Ts. || ἀπ̣[ι]cτίη Janko: ἀ[πι]cτίη Ts. 11. γινώ[c]κ̣ωc[ιν Ts.: γινώ[c]κ̣ωc[ι Janko || οὐκ ἔcτιν ὅπωc Ts. 12. [ἐνύπνια] Janko

COL. 5.

. . . terrors . . . consult an oracle . . . they consult an oracle . . . for them we go into the oracular shrine to inquire for oracular answers, whether it is right . . . the terrors of Hades, why do they disbelieve? Not understanding dreams, nor any of the facts, on the basis of what kind of warning would they believe? Overcome by fault and by pleasure as well, they neither learn, nor believe. Disbelief and lack of understanding [? are the same thing]. For if they neither understand, nor do they learn, [it is not possible that they believe] even when they see . . . disbelief . . . appears . . .

COL. 6.

[±8 εὐ]χ̣αὶ καὶ θυς[ί]α̣ι μ[ειλ]ί̣ς̣ς̣ο̣υςι τ̣ὰ̣[ς ψυχάς,] ἐπ̣[ωιδὴ δ]ὲ̣ μάγων δύν[α]ται δ̣αίμονας ἐμ[ποδὼν γι̣[νομένο]υς μεθιστάναι· δαίμονες ἐμπο[δὼν ὄντες εἰςὶ ψ[υχαὶ τιμω]ρ̣οί. τὴν θυς[ία]ν̣ τούτου ἕνεκε[μ] π̣[οιοῦς]ι̣[ν οἱ μ̣ά̣[γο]ι̣, ὡ̣ςπερεὶ ποινὴν̣ ἀποδιδόντες. τοῖ<ς> δὲ ἱεροῖ[ς] ἐπιςπένδουςιν ὕ[δω]ρ καὶ γάλα, ἐξ ὧνπερ καὶ τὰς χοὰς ποιοῦςι. ἀνάριθμα̣ [κα]ὶ̣ πολυόμφαλα τὰ πόπανα θύουςιν, ὅτι καὶ αἱ ψυχα[ὶ ἀν]ά̣ριθμοί̣ ε̣ἰςι. μύςται Εὐμενίςι προθύουςι κ[ατὰ τὰ] α̣ὐτὰ μ̣ά̣γοις· Εὐμενίδες γὰρ ψυχαί εἰςιν. ὧν ἕνεκ[εν τὸν μέλλοντ]α θεοῖς θύειν ὀ̣[ρ]ν̣ί̣θ[ε]ιον πρότερον [±11].ι̣ς̣π̣ο̣τ̣ε̣[. .]ται [. . .]ω[.]τε καὶ τὸ κα̣[ ]ου . . . [. .]. ι. εἰςὶ δὲ [. . .].ι . . . [ ].τουτο .[ὅςαι δὲ [ ]ων ἀλλ̣[φ̣ο̣ρ̣ου[ ] . . .[

1. εὐ]χ̣α̣ὶ Ts. *ap.* Bernabé: εὐ]χ̣α̣ὶ Ts. 3. γι̣[νομένο]υς Ts. *ap.* Bernabé: γι̣[νομένου]ς Ts. || ἐμπο[δὼν Ts.: ἐμπο[δίζουςιν Henrichs || ὄντες εἰςὶ Janko: εἰςὶ Ts. 4. ψ[υχαὶ τιμω]ρ̣οί Ts.: ψ[υχαῖς ἐχθ]ρ̣οί Ts. || θυς[ία]ν̣ Ts.: θυς[ίη]ν̣ Janko 5. μ̣ά̣[γο]ι Ts. *ap.* Bernabé: μά̣[γο]ι Ts. || τοῖ<ς> δὲ Ts.: τοῖζε Janko: τοιδ̣ε ZPE 12. ω[.]τε Ts.: ὥ[ς]τε Janko 13. [. . .].ι Ts.: [ψυχ]α̣ὶ Janko 14. ἀλλ̣[Ts.: ἀλλ̣[ὰ or ἄλλ̣[ων Janko 15. φ̣ο̣ρ̣ου[Ts.: φ̣ο̣ρ̣οῦ[ςι Janko

COL. 6.

. . . prayers and sacrifices appease the souls, and the enchanting song of the magi is able to remove the daimones when they impede. Impeding daimones are avenging souls. This is why the magi perform the sacrifice, as if they were paying a penalty. On the offerings they pour water and milk, from which they make the libations, too. They sacrifice innumerable and many-knobbed cakes, because the souls, too, are innumerable. Initiates make the preliminary sacrifice to the Eumenides, in the same way as the magi. For the Eumenides are souls. On account of these, he who is going to sacrifice to the gods, first birdlike . . . and the . . . (they) are . . . as many as . . .

COL. 7.

. . . (.)]οcε̣[
. . ὕ]μνον̣ [ὑγ]ι̣ῆ καὶ θεμ[ι]τ̣ὰ λέγο[ντα˙ . . . .]τ̣ο γὰρ
[τῆ]ι ποήcει. [κ]αὶ εἰπεῖν οὐχ οἷόν τ[ε ἅμα τὴν ὀ]νομάτων
[θέ]cιγ καὶ τ[ὰ] {ῥ}ῥηθέντα. ἔcτι δὲ ξ[ένη τις ἡ] πόηcιc
[κ]α̣ὶ ἀνθρώ[ποιc] αἰνι̣[γμ]ατώδηc. [ὁ δ]ὲ̣ [Ὀρφεὺ]c̣ αὐτο̣[ῖc
[ἐ]ρίcτ' αἰν[ίγμα]τα οὐ̣κ ἤ̣θελε λέγειν, [ἐν αἰν]ίγμαc̣[ι]ν δὲ
[μεγ]άλα. ἱερ[ολογ]ε̣ῖ̣ται μὲν οὖγ καὶ ἀ̣[πὸ το]ῦ πρώτου
[καὶ] μέχρι <τ>οῦ̣ [τελε]υτ̣α̣ίου ῥήματοc. ὡ̣[c δηλοῖ] καὶ ἐν τῶι
[εὐκ]ρινήτω[ι ἔπει· θ]ύ̣ρ̣αc γὰρ ἐπιθέ[cθαι κελ]εύcαc τοῖ̣[c]
[ὠcὶ]ν αὐτ[οὺc οὔτι νομο]θ̣ε̣τ̣εῖμ φη[ςιν τοῖc]πολλοῖc̣
τοὺc τὴ]ν ἀκοὴν [ἁγνεύο]νταc κατ̣[ὰ]
]c̣ειτ̣[. .].
]ωι τ[. .]εγ .[. . . .]. . [
ἐν δ]ὲ τῶι ἐχομ[ένωι
].τ.. ε̣ι̣γ .[

2. ἱερουργεῖ]τ̣ο or ἠ̣ινίζε]τ̣ο or Ts.: [ἱερολογεῖ]τ̣ο Janko 3. [τῆ]ι Ts.: τ]ῆ̣ι Janko || [κ]αὶ Ts.: κ̣αὶ Janko || τ[ε ἅμα τὴν ὀ]νομάτων Betegh: τ[ε τὴν τῶν ὀ]νομάτων Ts.: τ[ῆν Janko 4. θέ]cιγ Janko and Betegh: [λύ]cιγ Ts.: [φύ]cιγ Ts. *ap.* L&M || καὶ τ[ὰ] Janko: καίτ[οι] Ts. || ξ[ένη Ts. *ap.* Bernabé: ξ[ένη Ts.: ξ[ενική West *ap.* Ts. || τις ἡ West *ap.* Ts. 5. [κ]α̣ὶ Ts.: καὶ Janko || αὐτο̣[ῖc Janko: αὐτ[ῆι] Ts. 6. [ἐ]ρίcτ' Ts.: [ἀ]πίcτ' or [ὕβ]ριcτ' Janko: [ἀό]ριcτ' F. Ferrari *ap.* Janko: [ἄ]ριcτ' ZPE || ἤ̣θελε Ts.: ἔ̣θελε Janko 7. [μεγ]άλα Ts.: [με]γάλα Janko || οὖγ *pap. ut vid.*: οὖν Ts. 8. [καὶ] Betegh: [ἀεὶ] Ts. || <τ>οῦ̣ Ts. 9. [εὐκ]ρινήτω[ι Ts.: [εὐθ]ρυλήτω[ι Janko || ἔπει Ts.: ἔπει Burkert || τοῖ̣[c] Ts.: τοῖc̣ Janko 10. φη[ciν τοῖc] Ts.: φη[ci τοῖc̣] Janko 11. τοὺc τὴ]ν Ts.: ἀλλὰ διδάcκειν τοὺc τὴ]ν Janko: διδάcκειν τοὺc τὴ]ν Burkert 14. ἐχομ[ένωι Ts.: ἐχομ[ένωι λέγει· Janko

COL. 7.

. . . hymn saying sound and lawful things. For . . . by his poetry. For it is not possible to state what way the words are used and at the same time the text itself. His poetry is something strange and riddling for people. But Orpheus did not intend to tell them captious riddles, but momentous things in riddles. Indeed, he is telling a holy discourse from the first and up to his last word. As he also makes clear in the well-chosen verse: for having ordered them to put doors to their ears he says that he is [? not legislating] for the many . . . [? but only for] those pure in hearing . . . according . . . in the next verse . . .

COL. 8.

[. . . . . δ]ε̣δήλω[ται ἐν τῶι]δ̣ε τῶι ἔπ̣[ει]˙
[ο]ἳ Διὸϲ ἐξεγ̣ένοντο [ὑπερμεν]έοϲ βαϲιλῆοϲ
‾ὅπωϲ δ᾽ ἄρχεται ἐν τῶ[ιδε δη]λοῖ˙
Ζεὺϲ μὲν ἐπεὶ δὴ π̣α̣[τρὸϲ ἑο]ῦ πάρα θέ[ϲ]φατον ἀρχὴν
[ἀ]λκήν τ᾽ ἐγ χείρεϲϲι ἔ[λ]αβ[εγ κ]α̣ὶ̣ δαίμου̣[α] κυδρόν
[τα]ῦτα τὰ ἔπη ὑπερβατὰ ἐό[ν]τ̣α λανθά̣[νει],
[ἔϲ]τιν δὲ ὧδ᾽ ἔχοντα· Ζεὺϲ μὲν ἐπε̣ὶ τ[ὴν ἀλ]κὴν
[πα]ρὰ πατρὸϲ ἑοῦ ἔλαβεγ καὶ δαίμονα̣ [κυδρ]ό̣ν.
[οὕτω] δ᾽ ἔχοντα οὐκ ἀκούειν τὸν Ζᾶ̣[να ἀμφιϲβη]τεῖ
τοῦ πατρ]ὸ̣ϲ ἀλλὰ τὴν ἀλκὴν λαμβ̣[άνειν παρ᾽ αὐτο]ῦ̣.
ἄλλωϲ δ᾽ ἔ]χοντα π̣αρὰ̣ θέϲφατα δ[όξειεν ἂν λαβεῖ]ν̣
[τὴν ἀλκήν . .].κ̣α̣ὶ̣ γὰρ̣ τούτωι δ̣ο̣[κεῖ
[. . . . . .ἀ]νάγκην νομίζο̣ι̣τ᾽ [ἂν
[. . . . .] καὶ μαθὼν ο̣ ο . . . [

1. [. . . ὡϲ . . .] Janko || δ]ε̣δήλω[ται ZPE: ε̣δήλω[ϲεν Ts. *ap.* Bernabé || τῶι]δ̣ε Ts. *ap.* Bernabé: τῶιδ]ε ZPE 2. ἐξεγ̣ένοντο Ts. *ap.* Bernabé: ἐξεγ̣ένοντο ZPE || [ὑπερμεν]έοϲ ZPE and West: [περιϲθεν]έοϲ Janko 3. τῶ[ιδε Ts. *ap.* Bernabé: το̣[ῖϲδε ZPE 4. δὴ π̣α̣[τρὸϲ Ts. *ap.* Bernabé: δὴ̣ [πατρὸϲ ZPE || θέ[ϲ]φατον *pap. ut vid.*: θ̣έ̣[ϲ]φατον ZPE 5. χείρεϲϲι ἔ[λ]αβ[εγ ZPE: χείρεϲϲ{ι} ἔ[λ]αβ[εγ West: χείρεϲϲι {ε}[λ]άβ[εγ Rusten 6. λανθά̣[νει] *pap. ut vid.*: λανθ̣ά̣[νει] ZPE 7. [ἔϲ]τιν Ts. *ap.* L&M: [. .].ιν ZPE 8. δαίμονα̣ *pap. ut vid.*: δαίμο̣ν̣α̣ ZPE 9. [οὕτω] δ᾽ Ts. *ap.* L&M: [χρὴ ὧ]δ᾽ ZPE || ἀμφιϲβη]τεῖ Betegh: ἐπικρα]τεῖ Ts. *ap.* L&M: ὅπωϲ κρα]τεῖ Janko: τὴν δυναϲ]τεί|αν πατρ]ὸ̣ϲ Janko 10. [τοῦ πατρ]ὸ̣ϲ Ts. *ap.* L&M: .]ε̣ϲ ZPE || λαμβ̣[*pap. ut vid.*: λαμ̣β̣[ZPE || λαμ̣β̣[άνει παρ᾽ αὐτο]ῦ̣ Ts. *ap.* L&M 11. ἄλλωϲ δ᾽ ἔ]χοντα Ts. *ap.* L&M: ταύτην ἔ]χοντα Janko: . . .]οντα ZPE || παρὰ̣ *pap. ut vid.*: π̣αρὰ̣ ZPE || δ[όξειεν ἂν λαβεῖ]ν̣ Ts. *ap.* L&M: δ[όξει ἀπογαίνει]ν̣ Burkert 12. [τὴν ἀλκήν . .] Ts. *ap.* L&M: ἄρχοντα αὐτόν Burkert || δ̣ο̣[κεῖ Ts. *ap.* L&M: μα[ZPE 13. νομιζο̣ι̣τ᾽ [ἂν Ts. *ap.* L&M: νομιζομ[ZPE

COL. 8.

. . . has been made clear in this verse:

Who were born from Zeus the mighty king

And how he (sc. Zeus) begins, he makes clear in this:

And when Zeus took from his father the prophesied rule
And the strength in his hands and the glorious daimon

They ignore the fact that these words are transposed. They are to be taken as follows: 'Zeus when he took the strength from his father and the glorious daimon.' Since this is so he (sc. Orpheus) does not maintain that Zeus hears [from his father] but that he takes the strength [from him]. If it is taken the other way, [it would seem] that he [took it] contrary to the divine decrees . . . for seems to this . . . might be believed necessary . . . and knowing . . .

COL. 9.

ε̣ἶ̣ν̣α̣ι̣. τ̣ὴ̣[ν ἀλ]κ̣ὴν οὖν τοῦ ἰ̣cχυρ̣[ο]τάτου ἐπόη̣[cεν]
εἶναι ὡς̣[περ]ε̣ὶ παῖδα πατρ̣ό̣c. οἱ δὲ οὐ γινώcκον[τεc]
τὰ λεγό[μεν]α δοκοῦcι̣ τ̣ὸ̣ν̣ Ζᾶνα παρὰ τοῦ αὑτο[ῦ]
πατρὸc [τὴν] ἀλκήν [τε καὶ] τὸν δαίμονα λαμμά[νειν].
γιγνώcκ[ων] οὖ̣ν τὸ πῦρ [ἅτε] μ̣εμειγμένον τοῖc
ἄλλοιc ὅτι ταρ̣άccοι καὶ κ[ωλ]ύ̣οι τὰ ὄντα cυνίcταcθαι
διὰ τὴν θάλψιν, ἐξή̣λλαξ[ε ὅc]ο̣ν̣ γε ἱκανόν ἐcτιν
ἐξαλλαχθὲμ μὴ κωλύ[ειν τὰ] ὄντα cυμπαγῆναι.
ὅcα δ᾽ ἅ[ν] ἁ̣φθῆι ἐπικρα[τεῖται, ἐπικ]ρ̣ατηθὲν δὲ μίcγεται
τοῖc ἄλ̣[λο]ιc. ὅτι δ᾽ ἐγ χείρ[εccιν ἔλαβ]ε̣ν ἠινίζετο
ὥcπε[ρ τ]ἄλλα τὰ π̣[±14]. ομεν [. . . .]α
[βε]βαιότατα νοε̣ι̣[±18]ν ἰcχυρὸc
ἔφη τὸν Ζᾶνα τ̣.[±16]ν δαίμονα
[ὡ]ς̣περεὶ ε[ἰcχ]υ̣ρ̣ο̣ῦ̣

1. ε̣ἶ̣ν̣α̣ι̣ Ts. *ap.* L&M || τ̣ὴ̣[ν̣ ἀ̣λ̣]κ̣ὴν Betegh: τ̣ὴ̣[ν̣ ἀ̣ρ̣]χ̣ὴν Ts. *ap.* L&M 2. ὡς̣[περ]ε̣ὶ Rusten 5. τὸ πῦρ Ts. *ap.* L&M: τὸ [. . . . .] ZPE: [θερμὸν μὴ] Rusten || [ἅτε] Betegh: [cύμ-] Ts. *ap.* L&M 6. ὄντα *pap. ut vid.*: ὄ̣ν̣τ̣α ZPE 7. ἐξή̣λλαξ[ε Betegh: ἐξή̣λλαξ[εν Burkert: ἐξα̣λλας̣[cει] Rusten: ἐξα̣λλας̣[c . .]. τ̣ε ZPE || ὅc]ο̣ν̣ γε Betegh *pap. ut vid.* above the space between the Ο̣ and the Ν̣ there seems to be a small letter, perhaps a Θ, Σ or E, possibly a correction by the scribe. The letter before Γ̣ is almost certainly an N.: ὅc]ο̣ν τ̣ε Ts.-Parássoglou: .]. .τ̣ε ZPE: ὥ̣ς̣τ̣ε Rusten 8. cυμπαγῆναι Ts.-Parássoglou: cυμπαγῆνα[ι] ZPE 9. ἐπικ]ρ̣ατηθὲν ZPE: ἐπικ]ρ̣ατηθέν<τα> Janko 10. χείρ[εccιν ZPE: χείρ[εcc᾽ {ι} Janko 11. .]α ZPE: τ]ὰ Janko 12. νοε̣ι̣[. . . ZPE: νοε̣ῖ̣[ται Janko 14. . .]ν ZPE: τὸ]ν Janko

COL. 9.

being. He made the strength to belong to the strongest, just as a son belongs to the father. Those who do not understand the meaning of the things said think that Zeus takes the strength and the daimon from his father. Now, knowing that fire, [in as much as] it is mixed with the others, agitates the things that are and hinders them from getting set together because of fomenting, he removed it to an adequate distance, so that once it is removed, it does not hinder the things that are from coagulating. For whatever is kindled is dominated (sc. by fire), and when dominated, it mixes with the other things. But (as to the expression) that 'he took in his hand', he gave a riddling meaning just like the other [. . .] knows firmest [. . .] strong, he said that Zeus [. . .] the daimon . . . just as . . . . strong.

COL. 10.

καὶ λέγẹι̣[ν· οὐ γ]ὰρ λẹ́[γ]ειν οἷόν τẹ μ̣ὴ φωνοῦντ[α].
ẹ̓νόμιζε δ̣ẹ̀ τὸ αὐτὸν εἶναι τὸ λ̣έγẹιν τε καὶ φωνẹῖν.
λέγειν δὲ κ̣αὶ διδάσκειν τὸ αὐτὸ δ[ύ]ναται· οὐ γὰρ
οἷόν τε δ̣ι̣[δ]άσκειν ἄνευ τοῦ λέγειν ὅσα διὰ λόγων
διδάσκετα̣[ι]. νομίζεται δὲ τὸ διδ̣άσκειν ἐν τῶι
λέγειν ẹἶ̣ṿ[αι]. οὐ τοίνυν τὸ μὲν διδάσκειν ἐκ τ̣ọῦ̣
λέγειν ἐχ[ωρί]σθη τὸ δὲ λέγειν ẹ̓κ τοῦ φωνεῖν,
τὸ δ' αὐτ̣ọ̀[ν δύνα]ται φωνεῖ̣ν καὶ λέγειν καὶ διδάς̣[κειν].
οὕτως [οὐδὲν κωλ]ύει πανομφεύουσαν καὶ πά̣ν̣[τα]
διδά[σκουσαν τὸ αὐ]τ̣ὸ εἶναι.
¯τροφ[ὸν δὲ λέγων αὐ]τ̣ὴ̣ν̣ α̣ἰ̣ν̣ί̣[ζε]ται ὅτι [ἅ]σσα
ὁ ἥλι̣[ος θερμαίνων δι]α̣λύει ταῦτα ἡ νὺξ ψύ[χουσα
συ[νίστησιν ±18]ἅσσα ὁ ἥλιος ẹ̓θ̣ẹ́ρ̣[μ . . .]
]τα[

1. φωνοῦντ[α] ZPE: φωνοῦντ[ι] Janko 7. ἐχ[ωρί]σθη Ts. *ap.* L&M: ẹ̓[χωρί]σθη Rusten: ẹ̓[φημί]σθη ZPE 9. οὕτως Ts. *ap.* Bernabé: οὕτω[ς ZPE || κωλ]ύει Ts. *ap.* Bernabé: κωλ]ύ̣ει ZPE || πά̣ν̣[τα] Ts. *ap.* Bernabé: [πάντα] ZPE 10. αὐ]τ̣ὸ Ts. *ap.* Bernabé: αὐτ]ὸ ZPE 11. τροφ[ὸν δὲ λέγων Ts. *ap.* L&M: τροφ[ZPE: τροφ[ὸς δὲ θέων West: τροφ[ὸς δὲ πάντων: τροφ[ὸν δὲ φήσας Janko || αὐ]τ̣ὴ̣ν Ts. *ap.* L&M:]. . ZPE: ἡ ν]ύ̣ξ West and Rusten || α̣ἰ̣ν̣ί̣[ζε]ται Ts. *ap.* L&M: . . . . ι[. .].αι ZPE: ν̣ο̣μ̣ί[ζε]τ̣αι Rusten 12. ἥλι̣[ος Ts. *ap.* L&M: ἥ̣λ̣ι̣[ος ZPE || θερμαίνων δι]α̣λύει ἥλι̣[ος Ts. *ap.* L&M:]αύει ZPE: ξηραίνει καὶ ἐξ]αύει Rusten || ψύ[χουσα Ts. *ap.* L&M: τ .[ZPE: τρ[έφει Rusten 13. συ[νίστησιν or συ[μπήγνυσιν Ts. *ap.* L&M: ς̣[ZPE || καὶ] ἅσσα Janko || ἥλιος Ts. *ap.* L&M: ἥλ[ι]ος ZPE || ẹ̓θ̣ẹ́ρ̣[μ . . .] Ts. *ap.* L&M: .[. .]ρ̣[ZPE 14. τα[ZPE: τα[ῦτα Ts. *ap.* L&M

COL. 10.

. . . and to say. For it is not possible to say without uttering; and he (sc. Orpheus) considered 'to say' and 'to utter' to be the same; and 'to say' and 'to teach' mean the same (*or*: have the same power): for it is not possible to teach without saying whatever is taught through discourses. Teaching is considered to reside in saying. Accordingly, 'to teach' was not distinguished from 'to say' on the one hand, and 'to say' from 'to utter' on the other, but 'to utter', 'to say' and 'to teach' mean the same. Thus nothing prevents 'all-pronouncing' and 'teaching all things' from being the same thing.

By saying that she is 'nurse', he (sc. Orpheus) expresses in riddling form that whatever the sun dissolves by heating, the night unites by cooling . . . those things which the sun heated . . .

COL. II.

[τ]ῆϲ Νυκτόϲ. ἐξ ἀ̣[δύτοι]ο δ' αὐτὴν [λέγει] χρῆϲαι γνώμην ποιού[με]νοϲ ἄδυτ̣ον ε̣ἶναι τὸ βάθοϲ τῆϲ νυκτόϲ· οὐ γ[ὰρ] δύνει ὥ[ϲ]περ̣ τὸ φῶϲ, ἀλλά νιν ἐν τῶι αὐτῶι μέ[νο]ν̣ αὐγὴ κ̣ατ̣α̣[λ]αμμάνει. χρῆϲαι δὲ καὶ ἀρκέϲαι ταὐτ̣ὸ̣ [δύ]ναται. ϲκέψαϲθαι δὲ χρὴ ἐφ' ὧι κεῖτα̣[ι τὸ] ἀρκέϲαι κ̣α̣ὶ τὸ χρῆϲαι.

‾χρᾶν τὸνδε τὸν θεὸν νομίζον̣[τεϲ ἔρ]χονται [π]ευϲόμενοι ἅϲϲα ποῶϲι· τὰ δ' [ἐπὶ τούτ]ω̣ι λ̣έ̣γει· [ἣ δὲ] ἔχρηϲεν ἅπαντα τά οἱ θέ̣[μιϲ ἦν ἀνύϲαϲ]θ̣αι [ἐν τούτ]ο̣ιϲ ἐδή̣λ̣ωϲεν ὅ[τ]ι ο[[±12 τὰ] ἐ̣όντ̣α ι̣[

1. [τ]ῆϲ Ts. *ap*. Bernabé: [τ]ῆ̣ϲ ZPE || Νυκτόϲ Ts. *ap*. Bernabé: Ν̣υκτόϲ ZPE || χρῆϲαι Ts. *ap*. Bernabé: χρῆϲ̣αι ZPE 2. ποιού[με]νοϲ Ts. *ap*. Bernabé: ποιοῡ́[με]νοϲ ZPE || ε̣ἶναι Ts. *ap*. Bernabé: ε̣[ἶ]ναι ZPE 3. τὸ Janko: τ̣ὸ ZPE 5. ταὐτ̣ὸ̣ [ZPE: ταὐτ̣ὸ̣[ν Kapsomenos 9. τα .[ZPE: τὰ δ' Ts. *ap*. Bernabé: τὰδ' Janko || [ἐπὶ τούτ]ω̣ι Betegh:]ω̣ι ZPE: 'after this' Ts. *ap*. L&M: [ἐν ἐχομέν]ω̣ι Janko || λέ̣γει Ts. *ap*. Bernabé: λ̣έ̣γει ZPE 10. [ἣ δὲ] ZPE: [ἣ οἱ] West || ἦν Ts. *ap*. Bernabé: ἦεν West || ἀνύϲαϲ]θ̣αι Ts. *ap*. Bernabé: ἀκοῦ]ϲ̣αι Janko: ἀνύϲϲ]αι West: .]αι ZPE 11. [ἐν τούτ]ο̣ιϲ Rusten: [. . . . .].ιϲ ZPE 12. τὰ] ἐ̣όντ̣α Ts. *ap*. Bernabé:]. οντ̣αι̣[ZPE 11–12. beside (*or:* in spite of) that things that are . . . possible to . . . Ts. *ap*. L&M

COL. II.

. . . of Night. He says that she proclaims the oracle out of the innermost shrine (ἐξ ἀ[δύτοι]ο), his view being that the depth of the night is 'never setting' (ἄδυτον); for it does not set (δύνει) as the light does, but the sunlight overtakes it as it remains stationary. Now, 'furnish/proclaim oracles' (χρῆσαι) and 'assist' mean the same. But one has to examine to what 'assist' and 'furnish/proclaim oracles' apply (as a name).

In the belief that this god proclaims oracles, they come inquiring what they should do. After this he says:

[? She] proclaimed an oracle about all that was right to him to hear.

. . . he made clear that . . . the things that are . . .

COL. 12.

καὶ ἀφα̣[ιρεῖ]ν̣. τὸ δ' ἐχόμε̣[νον ἔ]π̣οϲ ὧδ' ἔχει·
ὡϲ ἄρ̣ξ[ηι κα]τὰ καλὸν ἕδ̣οϲ νιφόεντοϲ Ὀλύμπου.
Ὄλυμπ[οϲ καὶ χ]ρ̣όνοϲ τὸ αὐτόν. οἱ δὲ δοκοῦντεϲ
Ὄλυμπ̣[ον καὶ]ο̣ὐρανὸν [τ]αὐτὸ εἶναι ἐξαμαρ-
τάν[ουϲ]ι̣[ν οὐ γ]ινώϲκοντ̣ε̣ϲ ὅτι οὐρανὸν οὐχ οἷόν τ̣ε̣
μακ[ρό]τ̣ερον ἢ εὐρύτε[ρο]ν εἶναι, χρόνον δὲ μακρὸν
εἴ τιϲ [ὀνομ]άζο[ι] ο̣ὐ̣κ ἂ̣[ν ἐξα]μαρτάνοι· ὁ δὲ ὅπου μὲν
οὐρανὸν θέ̣[λοι λέγειν, τὴμ] προϲθήκην εὐρὺν
ἐποε̣ῖτο, ὅπου̣ [δὲ Ὄλυμπον, το]ὐ̣ν̣α̣ν̣τίον, εὐρὺμ μὲν
οὐδέποτε, μα[κρὸν δέ· νιφό]ε̣ν̣τ̣α̣ δὲ φήϲαϲ εἶνα̣ι
τῆι [δ]υνάμει ε[±14]ι̣ νιφετῶδε
[. . .] νιφετω[δ ±14 λ]ε̣υ̣κ̣ὸν ε̣[ ]
[. . .]λ̣αμπ̣[ρ]π̣ολιὸν δ' α[. .].
].ια καὶ τα̣ [
]. . τ̣ο̣δ̣ε[

1–15. Ts. *ap*. Brisson 1. ἀφα̣[ιρεῖ]ν Janko: ἀφα̣[Ts. *ap*. Brisson: ἀφ. [. ZPE || ἔ]π̣οϲ Ts. *ap*. Brisson: ἔ̣]π̣οϲ Burkert 2. ἄρ̣ξ[ηι κα]τὰ Janko: ἂν̣ ἔ̣[χοι κά]τα Ts. *ap*. Brisson: ἄρ̣χ[ηι κα]τὰ Janko: ἄ[ρξαι κα]τὰ West: α[. . . ZPE 4. Ὄλυμπ̣[ον καὶ ο]ὐρανὸν ZPE 5. τάν[ουϲ]ι̣[ν Ts. *ap*. Brisson: τάν[ουϲι]ν ZPE: τάν[ουϲ]ι̣[Janko 9. [. . . Ὄλυμπον . . .] ZPE: [. . . χρόνον . . .] Janko 11. ε[ZPE: ε[ἰκάζει Ts. *ap*. Brisson: ἐ[χρήϲατο Janko || νιφετῶδε Ts. *ap*. Brisson: νιφετῶδει̣ [ ] ZPE 12. τὸ δὲ] νιφετῶ[δεϲ ψυχρόν τε καὶ Ts. *ap*. Brisson || λ]ε̣υ̣κ̣ὸν ε̣[ ] Ts:] . . . ονε̣[ZPE 13. λ̣αμπ̣[ρ Ts. *ap*. Brisson: λ̣άμπ̣[ει Janko: . .]. αμ[ZPE || δα[. .]. Ts. *ap*. Brisson: δ' ἀ[έρ]α̣ Brisson 14–15. not in ZPE

COL. 12.

. . . and to take away (sc. his rule). And the next line goes like this:

so that he may rule on the lovely abode of snowcapped Olympus

Olympus and time are the same. Those who think that Olympus and the heaven are the same are entirely mistaken, for they do not know that the heaven cannot be longer rather than wider; but if someone were to call time long, he would not be wrong at all. And whenever he (sc. Orpheus) wanted to speak about heaven, he added the epithet 'wide', whereas whenever (he wanted to talk) about Olympus, on the contrary, he never (added the epithet) 'wide', but 'long'. By saying that it is 'snow-capped', the power . . . snowy . . . snowy . . . white . . . bright . . . grey . . . and . . .

COL. 13.

Ζεὺc μὲν ἐπεὶ δὴ̣ π̣ατρ̣ὸc ἑ̣οῦ πάρα̣ [θ]έ̣cφατ' ἀκούcα[c]·
οὔτε γὰρ τότ̣ε ἤκουcεν, ἀλλὰ δεδήλωτ̣αι ὅπωc
ἤκουcεν, οὔτε ἡ Νὺξ κελεύει. ἀλλὰ δηλοῖ ὧδε λέγων·
‾αἰδοῖογ κατ̣έπινεν, ὃc αἰθέρα ἔχθορε πρῶτοc.
‾ὅτι μὲμ πᾶ̣cαν τὴμ πόηcιν περὶ τῶν πραγμάτων
αἰνίζεται κ[α]θ' ἔποc ἕκαcτον ἀνάγκη λέγειν.
‾ἐν τοῖc α[ἰδοίο]ιc ὁρῶν τὴγ γένεcιν τοὺc ἀνθρώπου[c]
νομίζον̣[ταc ε]ἶν̣αι τούτωι ἐχρήcατο, ἄνευ δὲ τῶν̣
αἰδοίων [οὐ γίν]εcθαι, αἰδοίωι εἰκάcαc τὸν ἥλιο[ν]
ἄνε̣[υ γὰρ τοῦ ἡ̣λίου] τὰ ὄντα τοιαῦτα οὐχ οἷ̣ό̣ν̣ [τ' ἦν]
γε̣ν̣[έcθαι ±9]ενων τῶν ἐ̣φόντων [
πρ[±14]τ̣ὸν ἥλιο[μ] π̣άντα ο[
]. . .[

1. [θ]έ̣cφατ' Ts. *ap*. Bernabé: [θέ]cφατ' ZPE || ἀκούcα[c] Ts. *ap*. Bernabé: ἀκούcα̣[c] ZPE 2. τότ̣ε Betegh and Janko: τόδ̣ε Ts. *ap*. Bernabé: τό . ε ZPE || δεδήλωτ̣αι Ts. *ap*. Bernabé: δεδήλω[τ]αι ZPE 3. δηλοῖ Ts. *ap*. Bernabé: δηλ̣οῖ ZPE 4. κατ̣έπινεν Ts. *ap*. Bernabé: κα[τ]έπινεν ZPE || αἰθέρα ZPE: αἰθέρο̲c̲ Lamberton *ap*. L&M 5. πᾶ̣cαν Ts. *ap*. Bernabé: πᾶ[c]αν ZPE || τῶν Ts. *ap*. Bernabé: τῶμ ZPE 6. κ[α]θ' Ts. *ap*. Bernabé: κ̣[α]θ' ZPE: κ[α]τ̲' Janko || ἔποc Burkert: ἔ̆ποc ZPE 7. α[ἰδοίο]ιc Ts. *ap*. Bernabé: α[ἰδοίο]ι̣c ZPE 8. νομίζον̣[ταc ε]ἶν̣αι Ts. *ap*. Bernabé: νομίζο[νταc εἶ]ν̣αι ZPE 9. γίν]εcθαι Ts. *ap*. Bernabé: γίν]ε̣cθαι ZPE 10. [τ' ἦν] Janko: [τε] ZPE 10–13. possible for the things that are to become such . . . things that are . . . calm . . . the sun all . . . nor to the things that are . . . to encompass . . . Ts. *ap*. L&M 11. γε̣ν̣[έcθαι Ts: γ. . [ZPE ||] ἐνων ZPE: καὶ γενομ]ένων Janko 12. πρ[ῶτον χωριcθέντα Janko

COL. 13.

Zeus when he heard the prophecies from his father

For neither did he hear this time – but it has been made clear in what sense he heard – nor does Night command (this time). But he makes this clear by saying as follows:

He swallowed the phallus of [. . .], who sprang from the aither first.

Since in his whole poetry he speaks about facts enigmatically, one has to speak about each word in turn. Seeing that people consider that generation is dependent upon the genitalia, and that without the genitals there is no becoming, he used this (word), likening the sun to a phallus. For without the sun the things that are could not have become such . . . things that are . . . the sun everything . . .

COL. 14.

ἐ]{κ̣}χθο̣ρηι τὸ{ν} λαμπροτάτον τε [καὶ θε]ρ̣μ̣ό̣[τ]ατον
χωριcθὲν ἀφ' ἑωυτ̣οῦ. τοῦτον οὖν τὸγ Κρόνον
γενέcθαι φηcὶν ἐκ τ̣οῦ ἡλίου τῆι γῆι, ὅτι αἰτίαν ἔcχε
διὰ τὸν ἥλιον κρούεcθαι πρὸc ἄλλ̣ηλα.
διὰ τοῦτο λέγει· ὃc μέγ' ἔρεξεν. τὸ δ̣' ἐπὶ τούτωι·
¯Οὐρανὸc Εὐφρονίδηc, ὃc πρώτιcτο̣c βαcίλευcεν,
¯κρούντα τὸν Νοῦμ πρὸc ἄλληλ[α] Κ̣ρόνον ὀνομάcαc
μέγα ῥ̣έξαι φηcὶ τὸν Οὐρανόν· ἀφ[αι]ρ̣εθῆ̣ναι γὰρ
τὴμ βαcιλείαν αὐτόγ. Κρόνο̣ν̣ δὲ ὠ̣ν̣ό̣μαcεν ἀπὸ τοῦ
ἔ̣[ρ]γ̣ου αὐτὸν καὶ τἆλλα κατὰ [τὸν αὐτὸν λ]ό̣γον.
[τῶν ἐ]όντωγ γὰρ ἁπάντ[ω]ν̣ [ ]νων
[ ] ὡ̣c ὁρ[ ]ν φύcιν [ ]ν
[ ]c ἀφαιρ[εῖ]cθαι δ' αὐ̣[τὸμ τὴμ βαcιλ]ε̣ίαν
[ ]μ̣ένων τ̣[ῶν] ἐ[ό]ν̣τ̣[ων ±12]ντα

1. ἐ]{κ̣}χθο̣ρηι Ts. *ap.* Bernabé: ἐ̣χθο̣ρηι ZPE ‖ τὸ{ν} Rusten: τὸν ZPE ‖ λαμπροτάτον Ts. *ap.* Bernabé: λαμ̣π̣ρ̣οτάτον ZPE ‖ [καὶ θε]ρ̣μ̣ό̣[τ]ατον Ts. *ap.* Bernabé: [καὶ λ]ε̣υ̣κ̣ό̣[τ]ατον ZPE 2. οὖν Ts. *ap.* Bernabé: οὖ̣ν ZPE 4. ἄλληλα Ts. *ap.* Bernabé: ἄλλ̣ηλ̣α ZPE 5. ἐπὶ Ts. *ap.* Bernabé: ἐ̣πὶ ZPE 6. Οὐρανὸc Εὐφρονίδηc ZPE: Οὐρανὸ̲ν̲ Εὐφρονίδη̲ν̲ West 7. τὸν Ts. *ap.* Bernabé: τ̣ὸν ZPE ‖ Κ̣ρόνον *pap. ut vid.*: Κ̣ρόνον Ts. *ap.* Bernabé: Κ̣ρ̣όν̣ον ZPE 8. μέγα Ts. *ap.* Bernabé: μ̣έγα ZPE ‖ ἀφ[αι]ρ̣εθῆ̣ναι Ts. *ap.* Bernabé: ἀ[φα]ι̣ρ̣ε̣θῆ̣ναι ZPE 10. ἔ̣[ρ]γου ('activity') Ts. *ap.* L&M: .[.].ου ZPE 11. [τῶν ἐ]όντωγ Ts. *ap.* L&M: [ἕκαcτ]ον· τῶγ ZPE ‖ ἁπάντ[ω]ν̣ Ts. *ap.* L&M: ἁπαντ[ZPE [οὔπω κρουομέ]νων̣ Ts. *ap.* L&M:]ναν ZPE 12. [ὁ Νοῦc] ὡ̣c ὁρ[ίζω]ν Ts. *ap.* L&M: [. . . . .]cο̣ρ̣[. . .]ν ZPE: [. . . .]ὡ̣c ὁρ[ᾶι τὴ]ν Janko ‖ [τὴν ἐπωνυμίαν ἔcχε]ν Ts. *ap.* L&M:]ν ZPE 13. [Οὐρανό]c Ts. *ap.* L&M: .]c ZPE ‖ δ' αὐ̣[τὸμ Ts. *ap.* L&M: δ' α[ὐτὸν ZPE ‖ φηcι τὴμ βαcιλ]ε̣ίαν Ts. *ap.* L&M: τὴμ βαcιλε]ίαν ZPE 14. [κρουο]μ̣ένων Ts. *ap.* L&M: [. . . . . .].ενων ZPE ‖ τ̣[ῶν] ἐ[ό]ν̣τ̣[ων Ts. *ap.* L&M ‖]ντα ZPE: τὰ ἐό]ντα or τὰ ὄ]ντα Janko

COL. 14.

[?he made] to leap the brightest and hottest having separated it from himself. He says that this Kronos was born from the sun to the earth because he became the cause through the sun that they were struck against each other. For this reason he says: '(He) who did a great deed.' And after this:

Ouranos son of Night, who first of all ruled

Naming Mind that strikes (? the beings) against each other Kronos, he says that it did a great deed to Ouranos; for (he says that) he (sc. Ouranos) was deprived of his kingdom. He named Kronos himself from his action and the others too according to the same principle. For of all the things that are . . . nature . . . that he got deprived of his kingdom . . . the things that are . . .

COL. 15.

κρ̣[ο]ύε<ι>ν αὐτὰ πρὸ̣[ϲ ἄλλ]ηλα, κἀ̣[μ] π̣οήϲηι τὸ[ν ἥλι]ον
χω̣ριϲθέντα, δ̣ιαϲ̣τῆναι δίχ' ἀλλήλων τὰ ἐό̣ν̣τα·
χωρ̣[ι]ζομένου γὰ̣ρ τοῦ ἡλίου καὶ ἀπολαμβανομένου
ἐμ μ̣έϲωι πήξαϲ ἴϲχει καὶ τἄνωθε τοῦ ἡλίου
καὶ τὰ κάτωθεν. ἐχόμενον δὲ ἔποϲ·
¯ἐκ τοῦ δὴ Κρόνοϲ α̣ὖτιϲ, ἔπειτα δ̣ὲ μητίετα Ζεύϲ,
¯λέγει <ὅ>τι ἐκ τοῦδε̣ [ἀ]ρ̣χή ἐϲτιν ἐξ ὅϲου βαϲιλεύει. ἥδε
ἀρχὴ̣ διηγεῖται ὅ̣[τι τὰ] ἐ̣όντα κρούωμ πρὸϲ ἄλληλα
διαϲτήϲαϲ ἔ̣ϲ̣[χε εἰϲ τὴ]ν̣ νῦμ μετάϲταϲιν οὐκ ἐξ ἑτέρ̣[ων]
ἕτερ' ἀλλ' ἑτε[ρ ±9]
¯τὸ δ' ἔπειτ̣[α δὲ μητίετα Ζεὺ]ϲ, ὅτι μὲν οὐχ ἕτερ[οϲ]
ἀ̣λ̣λὰ ὁ αὐ̣[τὸϲ δηλοῖ. ϲημαίν]ε̣ι δὲ [τ]όδ̣ε̣·
¯μῆτιγ κα[ὶ μακάρων κατέχ]ω̣μ βαϲιληίδα τιμ̣[ήν]
εϲμ̣[±20]τ̣αι ἶναϲ̣ απα̣[
ει̣[

1. κρ̣[ο]ύε<ι>ν Rusten: κρ̣[ο]ύεν ZPE || κἀ̣[μ] Betegh: κα̣[ὶ] ZPE || τὸ[ν ἥλι]ον Betegh: τὸ [πρῶτ]ον Ts. *ap.* L&M: τὸ [λοιπ]ὸν Burkert: το[. . . .]ον ZPE 6. α̣ὖτιϲ Ts. *ap.* Bernabé: [α]ὖτιϲ ZPE 7. τοῦδε̣ ZPE: τοῦδ' ἡ̱ Janko || [ἀ]ρ̣χη Ts. *ap.* Bernabé: [ἀ]ρχη ZPE || βαϲιλεύει. Rusten: βαϲιλεύει ZPE || ἥδε Ts. *ap.* Bernabé: ηδε ZPE: ἡ δὲ Burkert 8. ἀρχὴ̣ ZPE: ἀρχή. L&M || ὅ̣[τι Rusten: . [. . ZPE || τὰ]ἐ̣όντα Rusten: . . .]. οντα ZPE: τ]ὰ ὄντα Burkert 9. διαϲτήϲαϲ Rusten: διαϲτηϲαϲ ZPE || ἔ̣ϲ̣[χε εἰϲ τὴ]ν̣ Betegh: . . [. . . . . τὴ]ν̣ ZPE: ἔ̣χ̣[ει εἰϲ τὴ]ν̣ Rusten: τ̣' ἐ̣[ποίει τὴ]ν Burkert: ἐ̣π̣[όηϲε τὴ]ν Janko 10. ἕτερ' ἀλλ' ἑτε[ρ ZPE: ἕτερ' ἀλλ' ἑτέ[ρωϲ τὰ αὐτά Burkert: ἕτε[ρ' ἐκ τῶν αὐτῶν Janko 11. οὐχ Ts. *ap.* Bernabé: ο̣ὐ̣χ ZPE 12. δηλοῖ. Janko: δῆλον Ts. || ϲημαίν]ε̣ι Ts. *ap.* Bernabé: λέγε]ι Janko: κα]ὶ West:]ι ZPE 13. μῆτιγ Betegh: Μῆτιγ ZPE || κα .[Ts. *ap.* Bernabé: κα̣ὶ [μακάρων κατέχ]ω̣μ West: κατ̣[πίνογ καὶ Burkert: κα̣ὶ [κατέπινεν ἑλ]ὼ̣μ or εἶχ]ε̣μ or κατέχ]ε̣μ Janko: κα[±14]ω̣μ ZPE 14. εϲμ̣[Ts. *ap.* Bernabé: ἐϲ μ̣[Janko ||]τ̣αι ἶναϲ̣ απα̣[Ts. *ap.* Bernabé:]αι ιν. . . . . [ZPE:]τ̣αι ἶναϲ̣ ἀπά̣[ϲαϲ Janko 15. ει̣[Ts. *ap.* Bernabé

COL. 15.

. . . striking them against each other and, if he made the [su]n separate, (the result is that) the things which are stood apart from one another. For as the sun got separated and encircled, he coagulated and held fast both the things that are above and those which are below the sun. And in the next verse,

From him in turn Kronos, and then wise Zeus

he says that there has been this rule since he has been the king. But his rule gets the explanation that it is by striking the things that are against each other that he separated them into their current reconfiguration, not different (?ones) from different ones, but [. . .]. And the (expression) 'and then wise Zeus' makes it clear that (it is) not another one, but the same one. He indicates this:

Holding wisdom and royal honour over the blessed gods

COL. 16.

[αἰδοῖ]ον τὸν ἥλιον ἔφ[η]ςεν εἶναι δε[δήλ]ωται. ὅτι δὲ ἐκ τῶν ὑπαρχόντων τὰ νῦν ὄντα γίνεται λέγει· πρωτογόνου βασιλέως αἰδοίου, τῶι δ' ἄρα πάντες ἀθάνατοι προσέφυμ μάκαρες θεοὶ ἠδὲ θέαιναι καὶ ποταμοὶ καὶ κρῆναι ἐπήρατοι ἄλλα τε πάντα, ἅσσα τότ' ἦγ γεγαῶτ', αὐτὸς δὲ ἄρα μοῦνος ἔγεντο. [ἐ]ν τούτοις σημαίνει ὅτι τὰ ὄντα ὑπῆ[ρ]χεν ἀεί, τὰ δὲ ν[ῦ]ν ἐόντα ἐκ τῶν ὑπαρχόντων γίν[ετ]αι. τὸ δὲ [αὐ]τὸς δὲ ἄρα μοῦνος ἔγεντο, τοῦτο δὲ [λ]έγων δηλοῖ [αὐ]τὸν Νοῦμ πάντων ἄξιον εἶναι μόν[ο]ν ἐόντα [ὥσπερ]εὶ μηδὲν τἆλλα εἴη· οὐ γὰρ [οἷόν τε τα]ῦτα εἶναι [τὰ ὑπάρχ]οντα ἄ[νε]υ τοῦ Νοῦ. [καὶ ἐν τῶι ἐχ]ομένωι [ἔπει τούτ]ου ἄξιον πάντων [τὸν Νοῦν ἔφησεν ε]ἶναι· [νῦν δ' ἐστὶ]ν βασιλεὺ[c] πάντ[ων, καί τ' ἔσσετ' ἔπ]ειτα [. . . . . .] Νοῦς καὶ .[±20].τον

1. δε[δήλ]ωται Ts. *ap*. Bernabé: δε[δήλω]ται ZPE 2. ἐκ Ts. *ap*. Bernabé: ἐκ ZPE 3. τῶι West: τοῦ ZPE 4. ἀθάνατοι Ts. *ap*. Bernabé: ἀθάνατοι ZPE 5. ἄλλα Ts. *ap*. Bernabé: ἄλλα ZPE 6. ἅσσα Ts. *ap*. Bernabé: [ὅ]σσα ZPE 9. δὲ ἄρα *pap. ut vid.*: δ' ἄρα ZPE: δ' {ε} ἄρα Janko 10. [αὐ]τὸν Ts. *ap*. Bernabé: [αὐ]τὸν ZPE || <τὸν> Janko [δῆλον ὡς] Νοῦς καὶ β[ασιλεὺς πάντων τὸ α]ὐτὸν 12. [τὰ ὑπάρχ]οντα Janko: [τὰ. . . . .]οντα ZPE: εἰ ἦν τὰ ἐ]όντα Ts. *ap*. L&M || [καὶ ἐν τῶι ἐχ]ομένωι Ts. *ap*. Bernabé:]ομένωι ZPE 13. [ἔπει τούτ]ου Ts. *ap*. Bernabé:]ου ZPE || [τὸν Νοῦν ἔφησεν ε]ἶναι· Ts. *ap*. Bernabé:].ναι ZPE: [δηλοῖ τὸν Νοῦν ε]ἶναι Janko 14. [νῦν δ' ἐστὶ]ν βασιλεὺ[c] πάντ[ων, καὶ τ' ἔσσετ' ἔπ]ειτα Ts. *ap*. Bernabé with supplements by West: [. . . . . .]ν βασιλεὺ[c] πάντ[ων ±14]ειτα ZPE 15. It is clear that Mind and king of all is the same thing. Ts. *ap*. L&M [δῆλον ὡς] Νοῦς καὶ β[ασιλεὺς πάντων τὸ α]ὐτὸν

COL. 16.

It has been made clear above [that] he called the sun a phallus. Since the beings that are now come to be from the already subsistent he says:

[with?] the phallus of the first-born king, onto which all
The immortals grew (*or:* clung fast), blessed gods and goddesses
And rivers and lovely springs and everything else
That had been born then; and he himself became solitary.

In these (verses) he indicates that the beings always subsisted, and the beings that are now come to be from (*or:* out of) subsisting things. And as to (the phrase): 'and he himself became solitary', by saying this, he makes clear that the Mind itself, being alone, is worth everything, as if the others were nothing. For it would not be possible for the subsisting things to be such without the Mind. And in the following verse after this he said that the Mind is worth everything:

Now he is king of all and will always be

. . . Mind and . . .

COL. 17.

π[ρ]ọτερον ἦν πρ[ὶν ὀνο]μ̣α̣ς̣θῆνα̣ι̣· ἔ̣π̣[ει]τα ὠνομάςθ̣[η]·
ἦγ γὰρ καὶ πρόcθεν ʽ[ἐ]ὼνʼ ἢ τὰ νῦν ἐόντα cυ̣cταθῆναι
ἀὴρ καὶ ἔcται ἀεί· οὐ γὰρ ἐγένετο, ἀλλὰ̣ ἦν· δι' ὅ τι δὲ
ἀὴρ̣ ἐκλήθη δεδήλωται ἐν τοῖc προτέροιc. γενέcθαι δὲ
ἐνομίcθη ἐπείτ' ὠνομάcθη Ζεύc, ὡcπερεὶ π̣ρότερον
μὴ ἐών. καὶ ὕcτατον ἔφηcεν ἔcεcθαι τοῦτον, ἔπειτ'
ὠνομάcθη Ζεὺc κ̣αὶ τοῦτο αὐτῶι διατελεῖ ὄνομα ὄν,
μέχρι εἰc τὸ αὐτὸ ε̣ἶ̣δ̣οc τὰ νῦν ἐόντα cυνεcτάθη
ἐν ὧιπερ πρόcθεν̣ ἐόντα ἠιωρεῖτο. τὰ δ' ἐόντα δ̣[ηλοῖ]
γενέcθαι τοιαῦτ̣[α] δ̣ιὰ τοῦτον καὶ γενόμενα ε̣[ἶναι
ἐν τούτωι̣ π[άντα. cη]μαίνει δ' ἐν τοῖc ἔπεcι τọ[ῖcδε]·
¯Ζεὺc κεφα[λή, Ζεὺc μέc]c̣α, Διὸc δ' ἐκ̣ [π]άντα τέτ[υκται]·
¯κ̣εφαλὴ̣ν̣ [±14]ντ' αἰν̣[ί]ζεται ọ[
κ̣εφαλη̣[±15] ἀρχὴ γίνεται c̣υ̣[
δ̣[±16 cυcτ]αθῆναι ν̣[

1. π[ρ]ọτερον Ts.: π[ρό]τερον ZPE || πρ[ὶν ὀνο]μ̣α̣ς̣θῆνα̣ι Ts.: πρ̣[ὶν ὀνο]μ[α]c̣θῆνα̣[ι] ZPE 2. πρόcθεν [ ]ων ἢ *pap. ut vid.*: πρόcθεν ὢν ἢ Ts. The letters ωΝ are superscript between line 1 and 2, probably as a correction by the scribe. As the ω is above the space between the H and the T and the N is above the space between the T and the A, whereas the edge of the fragment is an oblique stroke (/), right before the ω, touching both the horizontal stroke at the bottom and the rounded upper part of it, there must have been one more letter before ων. Especially so, as it would be unnatural to write the word ὢν above ἢ and τά to be inserted between πρόcθεν and ἢ. 3. ἀὴρ̣ Ts. *ap*. Bernabé: ἀὴρ ZPE || δεδήλωται Ts. *ap*. Bernabé: δεδ̣ήλωται ZPE 6. ἐπείτ' Ts.: ἔπειτ. ZPE 7. ὄν Ts.: ὄν̣ ZPE 8. ε̣ἶ̣δ̣οc Ts.: εἶ̣δ̣οc̣ ZPE || cυνεcτάθη ZPE: cυ{νε}cταθῆ<ι> Betegh 9. πρόcθεν̣ ἐόντα Ts.: πρόcθεν [ἐ]όντα̣ ZPE || τὰ δ' ἐόντα δ̣[ηλοῖ] Ts.: τὰ δὲ ὄντα [ ] ZPE: τὰ δ' ἐόντα φ̣[ηcὶ Janko 10. τọιαῦτ[α] Ts.: τοιαῦτ̣[α] ZPE || ε̣[ἶναι Ts. *ap*. Bernabé: π̣[άλιν]Ts.: π̣[άντα Janko: [εἶναι Janko: [ ] ZPE 11. τούτωι̣ Ts. *ap*. Bernabé: τούτω[ι ZPE || π[άντα Ts. *ap*. Bernabé: [πάντα Janko: μένειν or κεῖcθαι West: [. . . . ZPE || τọ[ῖcδε] Ts. *ap*. Bernabé: τ[οῖcδε] ZPE 12. κεφα[λή Ts. *ap*. Bernabé: κεφα̣[λή ZPE || μέc]c̣α̣ Ts. *ap*. Bernabé: μέcc]α̣ ZPE || ἐκ Ts.: ἐκ̣ ZPE || τέτ[υκται] Ts. *ap*. Bernabé: τέτ̣[υκται] ZPE 13. κ̣εφαλὴ̣ν̣ Ts. *ap*. Bernabé: [κ]εφαλη̣[ZPE || [λέγων τὸν ἀέρα τὰ ἐό]ντ' Ts. *ap*. Bernabé: [φήcαc ἔχειν τὰ ἐό]ντ' Janko || ọ[ἷαπερ Ts. *ap*. Bernabé: ọ̔[τι Janko: ọ[ZPE 14. κ̣εφαλὴ̣[ν Ts. *ap*. Bernabé: κ̣εφαλη̣[ZPE: ἡ | κ̣εφαλὴ̣ Janko || ἔχοντα Ts. *ap*. Bernabé || οὗτοc Ts. *ap*. Bernabé: αὕτη Janko || γὰρ Ts. *ap*. Bernabé || c̣υ̣[cτάcεωc Ts. *ap*. Bernabé: c̣ώ̣[ματοc Burkert: c̣. [ZPE 15. δ̣[Ts.: . [ZPE || ν̣[Ts.: . [ZPE

COL. 17.

. . . it existed before it was named. Then it was named. For air existed even before the things that are now were set together and always will exist. For it was not born, but existed. And the reason why air received its name has been made clear above. But it was thought that it was born, because it got the name Zeus, just as if it did not exist previously. And he said that this will be 'the last', because it was named Zeus, and it will continue to be his name until the things that are now have got set together into the same form in which they were floating as they were before. And it is clear that they became such because of this, and having come to be, they are all in it. . . . he indicates in the following words:

Zeus the head, Zeus the middle, and from Zeus all things have their being

Head . . . expressed in a riddling form . . . head . . . beginning (*or:* rule) comes about . . . setting together . . .

COL. 18.

καὶ τὰ κατα[φερό]μενα ʽ[. . .]α φάμενος [εἶπε]ṿ' τὴν δ[ὲ γὴν] καὶ τἆλλα πάν[τ]α εἶναι
ἐν τῶι ἀέρịι π̣[νε]ῦμα ἐόν. τοῦτ' οὖν τὸ πνεῦμα Ὀρφεὺς
ὠνόμαcεμ Μοῖραν. οἱ δ' ἄλλοι ἄνθρωποι κατὰ φάτιμ Μọῖραν
ἐπικλῶcαι φαcὶν cφίcιγ καὶ ἔcεcθαι ταῦθ' ἅccα Μοῖρα
ἐπέκλωcεν, λέγοντεc μὲν ὀρθῶc οὐκ εἰδότεc δὲ
οὔτε τὴμ Μοῖραν ὅ τί ἐcτιν οὔτε τὸ ἐπικλῶcαι. Ὀρφεὺc γὰρ
τὴμ φρόνηc[ι]μ Μοῖραν ἐκάλεcεν· ἐφαίνετο γὰρ αὐτῶι
τοῦτο προcφερέcτατον ε[ἶ]ναịι ἐ̣ξ ὧν ἅπαντεc ἄνθρωποι
ὠνόμαcαμ. πρὶμ μὲγ γὰρ κληθῆ̣ναι Ζῆνα, ἦμ Μοῖρα
φρόνηcιc τοῦ θεοῦ ἀεί τε καị [δ]ιὰ παντόc· ἐπεὶ δ' ἐκλήθη
Ζεύc, γενέcθαι αὐτὸν ἐ̣[νομ]ί[cθ]η̣, ὄντ̣α μὲγ καὶ πρόcθεν,
[ὀ]νομαζόμ[ε]νον δ' ο[ὔ. διὰ τοῦτο λέ]γẹι Ζεὺc πρ̣ῶτοc̣
[γένετο, πρῶ]τογ γ' ἐ̣ọ́ντα [±10]ἔπειτ̣[. .] . . . . [ ].ο
[. . . . . . . ἄ]ṿθρωπ̣[οι οὐ γιγνώcκοντ]εc τὰ λεγόμενα [ ]
]τὸν Ζῆνα [ ]

1. κατα[φερό]μενα Ts. *ap.* L&M: κατω̣[. . .]μενα ZPE || [. . .]α ZPE: ἃ Janko || [εἶπε]ṿ or [ἔλεγε]ṿ Janko: [. . . .]ṿ ZPE || τὴν δ[ὲ γὴν] Janko: τὴν δ[ίνην] Burkert: τήνδ[ε Ts. *ap.* L&M: τὴν δ[. . . .] ZPE || πάν[τ]α Ts. *ap.* Bernabé: πάν[τ]α̣ ZPE 2. τῶι Ts. *ap.* Bernabé: τῷι ZPE || ἐόν Ts.: ἐόν ZPE || π̣[νε]ύμα<τι> ἐόν<τι> Janko 8. ἐ̣ξ ὧν ZPE: ἐ̣ξ ὧ Merkelbach 11. ἐ̣[νομ]ί[cθ]η̣ Ts. *ap.* Bernabé: δ̣[οκοῦc]ι ZPE || ὄντ̣α Ts. *ap.* Bernabé: ὄṿτ̣α ZPE 12. δ' ο[ὔ Ts. *ap.* L&M: δ' ὅ[τε. ZPE: δ' ο[ὔπω Janko || διὰ τοῦτο λέ]γẹι Ts. *ap.* Bernabé: .]. . ZPE || πρῶτοc Ts. *ap.* Bernabé: πρ̣ῶ̣τοc̣ ZPE 13. [γένετο Ts. *ap.* L&M || πρῶ]τογ Ts. *ap.* L&M || γ' ἐ̣ọ́ντα Janko: γ. . ντα[ZPE: γ̣ὰρ̣ Ts. *ap.* L&M || ἔπειτ̣[α Ts. *ap.* L&M: ἐπείτ̣[ε Janko 14. οἱ Ts. *ap.* L&M || δ' Janko || ἄ]ṿθ̣ρωπ̣[οι Ts. *ap.* L&M: ἀ]ṿθ̣ρώπ̣[ων ZPE || οὐ γιγνώcκοντ]εc Ts. *ap.* L&M: οἱ μὴ cυνιέcαντ]εc ZPE 15. νομίζουcι πρωτόγονον εἶναι Ts. *ap.* L&M

COL. 18.

. . . and those moving downwards. But speaking about [. . .] he means that the [? earth] and all the other things are in the air, it being breath. Now Orpheus named this breath Moira. But all other men according to the common usage say that Moira spun for them and that those things which the Moira has spun will be, on the one hand speaking correctly, but on the other hand not knowing either what Moira is or what spinning is. For Orpheus called wisdom Moira. This seemed to him to be the most suitable out of the names that all men have given. For before Zeus received his name, Moira was the wisdom of the god always and through everything. But since Zeus received his name, they think that he was born, even though he existed even before, but was not named. For this reason he says 'Zeus was born first', as he was first . . . then . . . men [? not understand]ing what is said (τὰ λεγόμενα) . . . Zeus . . .

COL. 19.

εκ[. . . .]ṭạ̀ ἐọ́ντα ἓν [ἕκ]αστογ κέκ[λητ]αι ἀπὸ τοῦ ἐπ̣ικρατοῦντος, Ζεὺ[ϲ] πάντα κατὰ τὸν αὐτὸν λόγον ἐκλήθη· πάντῳγ γὰρ ὁ ἀὴρ ἐπικρατεῖ τοcοῦτον ὅcομ βούλẹται. Μοῖραṿ δ' ἐπικλῶcαι λέγοντες τοῦ Διὸc τὴμ φρόνηcιν ἐπικυρῶcαι λέγουcιν τὰ ἐόντα κα̣ὶ τὰ γινόμενα καὶ τὰ μέλλοντα, ὅπωc χρὴ γενέcθαι τε καὶ εἶναι κα[ὶ]παύcαcθαι. βαcι̣λ̣εῖ δὲ αὐτὸν εἰκάζει (τοῦτο γάρ οἱ προcφέρειν ἐφα[ί]νετο ἐκ τῶν λεγομένων ὀνομάτων) λέγων ὧδε· Ζεὺ̣ϲ βαcιλεὺc, Ζεὺc δ' ἀρχὸc ἁπάντων ἀργικέραυνοc. [βαcιλέ]α ἔφη εἶναι ὅτι πολλ[ῶν . . . .]. ωμ μία [ἀρχὴ κ]ρατεῖ καὶ πάντα ṭελεῖ [. . . . .]. οὐδενὶ [. . . . .]χ̣ιν τẹ[λ]ẹcαι [[. . . . .] ἀρχὸν δε[[. . . . . . . ἄρ]χεται διạ[

1. εκ[ZPE: ἐπ̣[εὶ δὲ Kapsomenos: ἐκ̣[τότε or ἐπ̣[εὶ Ts. *ap.* L&M: ἕκαστα Janko || ṭạ̀ ἐọ́ντα Ts. *ap.* L&M: τὰ ἐό]ντα Kapsomenos: .] . . . ṿτα ZPE 3. πάντῳγ Ts. *ap.* L&M: πάντạc̣ Kapsomenos and ZPE 4. βούλεται Kapsomenos and Ts. *ap.* Bernabé: βούλẹṭαι ZPE 6. γε̱νόμενα Burkert 7. βαcι̣λ̣εῖ Ts. *ap.* Bernabé: βαc[ι]λεῖ ZPE 11–15. .]. ωμ μία ZPE: ἐόντων ἀρ]χ̣ῶμ 12. [ἀρχὴ κ]ρατεῖ ZPE || [ἅcca βροτῶ]ι Ts. *ap.* L&M:]. ZPE 13. [ἄλλωι ἔξεc]ṭιν Ts. *ap.* L&M:]χ̣ιν ZPE 14. δε[ZPE: δ' ἔ[φη ἁπάντων ὅτι ἅπαντα Ts. *ap.* L&M 15. διạ[ZPE: διạ̀ Ts. *ap.* L&M || [τοῦτον or ạ[ὐτὸν Janko

COL. 19.

. . . existing things have been called each single name by reason of what dominates (them); all things were called Zeus according to the same principle. For the air dominates all as far as it wishes. And when they say that the Moira spun they say that the wisdom of Zeus ordains how the things that are and the things that come to be and the things that are going to be must come to be and be and cease. And he likens him to a king (for this seemed to him the most fitting of all the names that are said) saying as follows:

Zeus the king, Zeus who rules all with the bright bolt

He said that he is king because many . . . one rule prevails and accomplishes all . . . no one . . . to accomplish . . . ruler . . . is ruled (*or:* starts) . . .

COL. 20.

ἀνθρώπω[ν ἐμ]πόλεσιν ἐπιτ̣ελέσαντες [τὰ ἱε]ρὰ εἶδον, ἔλασσον σφᾶς θαυμάζω μὴ γ̣ινώσκειν. οὐ γὰρ οἷόν τε ἀκοῦσαι ὁμοῦ καὶ μαθεῖν τὰ λ̣εγόμενα. ὅσοι δὲ παρὰ τοῦ τέχνημ ποιουμένου τὰ ἱερὰ, οὗτοι ἄξιοι θαυμάζεσθαι καὶ οἰκτε[ί]ρεσθαι, θαυμάζεσθαι μ̣ὲν ὅτι δ̣οκοῦντες πρότερον ἢ ἐπιτελέσαι εἰδήσειν ἀπέρχονται ἐπιτελέσαντες πρὶν εἰδέναι οὐδ' ἐπανερόμενοι ὥσπερ ὡς εἰδότες τ̣ι̣ ὧν εἶδον ἢ ἤκουσαν ἢ ἔμαθον· [οἰ]κτε<ί>ρεσθαι δὲ ὅτι οὐκ ἀρκεῖ̣ σφιν τὴν δαπάνην προανηλῶσθαι ἀλλὰ καὶ τῆς̣ γνώμης στερόμενοι προσαπέρχονται.

—πρὶμ μὲν τὰ [ἱ]ε̣ρὰ ἐπιτελέσαι ἐλπίζον[τε]ς εἰδήσειν ἐπ[ιτελέσ]αντ̣[ες] δὲ στερηθέντες κα̣[ὶ τῆς] ἐλπί[δος] ἀπέρχονται. τω̣[±10].ουντ[. . .] λ̣ογος̣ . . [. . .]ται̣ .[. .].να .[ ]. ι τῆι ἑαυ̣τ̣οῦ ο . . [μ]ητρὶ μὲν [ ] δ' ἀδελφη[ ]ως εἶδε] [

2. γ̣ινώσκειν Ts. *ap.* Obbink: γ[ι]νώσκειν ZPE 5. ὅτι Ts. *ap.* Obbink: ὅτι̣ ZPE 7. ἐπανε<ι>ρόμενοι Rusten 8. εἰδότες Ts. *ap.* Obbink: εἰδότ̣ε̣ς ZPE || τ̣ι̣ Ts. *ap.* Obbink: τ[ι] ZPE || [οἰ]κτε<ί>ρεσθαι Ts. *ap.* Obbink: [οἰ]κ̣τε<ί>ρεσθαι ZPE 9. ἀρκεῖ̣ Ts. *ap.* Obbink: ἀρκε[ῖ] ZPE 10. τῆς̣ Ts. *ap.* Bernabé: τῆ̣ς̣ ZPE || προσαπέρχονται ZPE: πρὸς ἀπέρχονται Ts. *ap.* Bernabé 11. [ἱ]ε̣ρὰ Ts. *ap.* Obbink: [ἱε]ρὰ ZPE || ἐλπίζον[τε]ς Ts. *ap.* Obbink: ἐλπίζο̣ν̣[τε]ς ZPE: ἐλπίζο<u>υσιν</u> Janko || εἰδήσειν Ts. *ap.* Obbink: εἰδ̣ήσειν ZPE 12. ἐπ[ιτελέσ]αντ̣[ες] Ts. *ap.* Obbink: ἐπ̣[ιτελέσ]αν[τες] ZPE || δὲ Ts. *ap.* Obbink: δ̣ὲ ZPE || στερηθέντες Ts. *ap.* Obbink: στερηθ̣έ̣ντες ZPE || κα̣[ὶ Ts. *ap.* Obbink: κ̣α̣[ὶ ZPE 13. τω̣[Ts. *ap.* Obbink: τ. [ZPE: τῶ̣ [ι Janko || ταῦτα or γὰρ or δὲ τὰ ἔπη Janko ||]. ουντ[. . .]λ̣ογος ZPE:].υόντ[ων ὁ] λ̣όγος̣ Janko || . .[. . .]ται̣ .[. .]. να ZPE: φα̣[ίνε]ται̣ τ̣ὸ̣[ν Ζ]ᾶ̣να Janko 14. . .]. ι Ts. *ap.* Obbink || λ̣[έγειν τῆι μητ]ρ̣ὶ or θυγατ]ρ̣ὶ or Janko || μ]ητρὶ ZPE: τῆι μ]ητρὶ Janko: Δήμ]ητρι 15. ὁμιλῆσαι or τῆι ἑαυτοῦ Janko ||] δ' ἀδελφη[Ts. *ap.* Obbink: τῆι] δ' ἀδελφῆ[ι Janko

COL. 20.

. . . those men who, while performing the rites in the cities, have seen the holy things, I wonder less that they do not have knowledge. For it is not possible to hear and at the same time to understand (*or*: learn) what is being said (τὰ λεγόμενα). But all those who (hope to acquire knowledge?) from someone who makes craft of the holy rites deserve to be wondered at and pitied. Wondered at because, thinking that they will know before they perform the rites, they go away after having performed them before they have attained knowledge, without even asking further questions, as though they knew anything of what they have seen or heard or learned; and pitied because it is not enough for them to have spent their money in advance, but they also go off deprived even of their judgement. Hoping before performing the holy rites that they will attain knowledge, they go away after having performed them deprived of hope too. . . . by his own . . . mother . . . sister . . .

COL. 21.

οὔτε τὸ ψυχ[ρὸν] τῶι ψυχρῶι. θόρνηι δὲ λέγ[ων] ḍηλοῖ ὅτι ἐν τῶι ἀέρι κατὰ μικρὰ μεμερισμένα ἐκινεῖτο καὶ ἐθόρνυτο, θορνύμενα δ' ἕκα<c>τα συνεστάθη πρὸς ἄλληλα. μέχρι δὲ τούτου ἐθόρνυτο, μέχρι ἕκαστον ἦλθεν εἰς τὸ σύνηθες. Ἀφροδίτη Οὐρανία καὶ Ζεὺς καὶ ἀφροδισιάζειν ḳαὶ θόρνυσθαι καὶ Πειθὼ καὶ Ἁρμονία τῶι αὐτῶι θεῶι ὄνομα κεῖται. ἀνὴρ γυναικὶ μισγόμενος ἀφροδισιάζειν λέγεται κατὰ φάτιν. τῶγ γὰ[ρ] νῦν ἐόντωμ μιχθέντων ἀλḷ[ή]λοις Ἀφροδίτη ὠν[ο]μάσθη. Πειθὼ δ' ὅτι εἶξεν τὰ ἐ[ό]ντα ἀλλήḷọ[ι]ṣịṿ. ε[ἴ]κειν δὲ καὶ πείθειν τὸ αὐτόṿ. [Ἁ]ρμονί' δὲ ὅτι πο[λλὰ. . . . . ἡ]ρμοσε τῶν ẹόντων ẹḳάστω[ι]· ἦμ μὲγ γ[ὰρ καὶ πρ]όσθεν, ὠṿọμάσθη δὲ γενέσ[θαι] ἐπεὶ διεκρίθ[η· τὸ δὲ δι]ạκριθῆṿ[α]ị δηλοῖ ὅτ[ι].[. . .]. ẹịṣ εδι. . .[. . . ἐκ]ράτει ὥστε διεḳ[ρίθησαν] [±15].[. . .]ν.[ ]ṿυν

1. θόρνηι ZPE: θορν<ύ>ηι Janko: θορ{ν}ῆι Janko: θόρ{ν}ηι Ts. *ap.* L&M: θορνῆι West 2. ἀέρι Ts. *ap.* Bernabé: ἀẹ́ṛι ZPE || ἐκινεῖτο Ts. *ap.* Bernabé: ẹ̓κινεῖτο ZPE 3. δ'ἕκα<c>τα συνεστάθη Ts. *ap.* L&M: δὲ κατασυνεστάθη Kapsomenos and ZPE 5. ἦλθεν *pap. ut vid.*: ἦḷθεν ZPE 7. θεῶι ὄνομα Ts. *ap.* Bernabé: θεῶị ọ̈νομα ZPE || <καὶ ἀφροδισιάζειν καὶ θόρνυσθαι> ἀνὴρ Ricciardelli Apicella || ἀνὴρ <γὰρ> Janko 8. μισγόμενος Ts. *ap.* Bernabé: μισγọ́μενος ZPE || ἀφροδισιάζειν <καὶ θόρνυσθαι> Janko || λέγεται Kapsomenos and Ts. *ap.* Bernabé: λέγετạι ZPE 8–9. κατὰ | φάτιν τῶγ *pap. ut vid.*: κατὰ φάτιν | τῶγ ZPE || κατὰ φάτιν <καὶ ἀφροδισιάζειν καὶ θόρνυσθαι> Merkelbach 11. ἀλλήḷọ[ι]ṣịν ZPE: ἀλλήḷ[οι]ṣ Janko || αὐτόṿ ZPE: αὐτό Janko || [Ἁ]ρμονί'α *pap. ut vid.*: [Ἁ]ρμονία ZPE 12. ἡ]ρμοσε ZPE: συνή]ρμοσε Kapsomenos || ἑḳάστω[ι] Ts. *ap.* L&M: ẹḳάστω[ι] ZPE 13. ὠṿọμάσθη ZPE: ε̣̲νọμί̲σθη Janko 14. τὸ ZPE: τῶι Janko || .[. . .]. ẹịς ZPE: ṭ[ὰς μεί]ξẹịς Janko 15. εδι. . . ZPE: ἐδίωḳẹ Janko || καὶ ἐκ]ṛάτει Janko: .]ατει ZPE || ὥσṭε Ts. *ap.* L&M: ωσ[ZPE διεḳ[ρίθησαν Janko: διε .[ZPE

COL. 21.

. . . nor the cold to the cold. By saying 'by mating', he (sc. Orpheus) makes clear that divided up into small (pieces) they were moving and mating in the air, and as they were mating, they all got put together with each other. They were mating until each came to its like. Aphrodite Ourania, and Zeus, and to aphrodise, and to mate, and Peitho, and Harmonia are given as name to the same god. A man mingling with a woman is said by common usage to aphrodise. For (this reason) as the things that are now got mixed with one another, (this god) has got the name Aphrodite. (It has got the name) Peitho, because the beings yielded to one another; and to yield and to persuade is the same. (It has got the name) Harmonia, because many of the beings got fitted to one another. For they existed even before, but were spoken of as 'being born' since they were separated out . . . to be separated out makes clear that . . . ruled so that they got separated out . . .

COL. 22.

πάν̣[τ' οὖ]ν ὁμοίω[ς ὠ]νόμασεν ὡς κ̣άλλιστα ἠ[δύ]ν̣ατο γινώσκων τῶν ἀνθρώπων τὴμ φύσιν, ὅτι οὐ̣ πάντες ὁμοίαν ἔχουσιν οὐδὲ θέλουσιμ πάντες ταὐτά· κρατιστεύοντες λέγουσι ὅ τι ἂν αὐτῶν ἑκάστωι ἐπὶ θυμὸν ἔλθηι, ἅπερ ἂν θέλοντες τυγχάνωσι, οὐδαμὰ ταὐτά, ὑπὸ πλεονεξίας, τὰ δὲ καὶ ὑπ' ἀμαθίας. Γῆ δὲ καὶ Μήτηρ καὶ Ῥέα καὶ Ἥρη ἡ αὐτή. ἐκλήθη δὲ Γῆ μὲν νόμωι, Μή̣τηρ δ' ὅτι ἐκ ταύτης πάντα γ[ίν]εται. Γῆ καὶ Γαῖα κατὰ [γ]λ̣ῶσσαν ἑκάστοις. Δημήτηρ̣ [δὲ] ὠνομάσθη ὥσπερ̣ ἡ Γῆ Μήτηρ, ἐξ ἀμφοτέρων ἓ[ν] ὄνομα· τὸ αὐτὸ γὰρ ἦν. – ἔστι δὲ καὶ ἐν τοῖς Ὕμνοις εἰρ̣[η]μένον·

‾Δη̣μήτηρ [Ῥ]έα Γῆ Μή̣τ̣ηρ Ἑστία Δηι̣ώι. καλε[ῖτ]αι γὰρ̣ καὶ Δηιὼ ὅτι ἐδηι[ώθ]η ἐν τῆι μείξει. δηλώσει δ̣ὲ [ὅτ]αν κ̣ατὰ τὰ ἔπη̣ γέν̣[ητα]ι. Ῥέα δ' ὅτι πολλὰ καὶ̣ . . .[ζῶια ἔφυ [±11]ἐ̣ξ αὐτῆς. Ῥέα καὶ̣ [κα̣ [±17]. η δ' ἐκ[

1. πάν̣[τ' Ts.: πά[ντ' ZPE || ὡς Ts.: ὡ̣ς ZPE || ἠ[δύ]ν̣ατο Ts.: ἠ[δύν]ατο̣ ZPE 2. γινώσκων Ts.: γινώσκ̣ων ZPE || ἀνθρώπων Ts.: ἀ̣νθρώπων ZPE || οὐ̣ πάντες Ts.: οὐ π̣άντες ZPE 3. οὐδὲ θέλουσιμ Ts.: οὐδ' ἐθέλουσιμ Janko || πάντες Ts.: π̣άντες ZPE 6. ὑπὸ Ts.: ὑ̣πὸ ZPE 7. Γῆ Ts.: Γῆ̣ ZPE || Ἥρη Ts.: Ἥρ̣η ZPE 8. δ<έ> Janko 9. [γ]λ̣ῶσσαν Ts.: [γ]λῶσσαν ZPE || Δημήτηρ̣ [δὲ] Ts.: Δημήτη[ρ δὲ] Kapsomenos and ZPE 10. ὥσπερ̣ Ts.: ὥσπε[ρ] Kapsomenos and ZPE || ὄνομα Ts.: ὄ̣νομα Kapsomenos and ZPE 11. εἰρ̣[η]μένον̣ Ts.: ε̣ἰ̣[ρη]μένον ZPE 12. Μή̣τ̣ηρ Ts.: Μή[τ]ηρ ZPE || καλε[ῖτ]αι Ts.: καλε̣[ῖτ]αι ZPE || γὰρ̣ Ts.: γὰρ ZPE 13. μείξει Ts.: μ̣είξει ZPE || δ̣ὲ, [ὅτ]αν Janko: δ̣ὲ [λί]αν Ts.: δ. [. .]α̣ν ZPE 14. κ̣ατὰ τὰ ἔπη̣ Ts.: κ[α]τὰ τὰ επ[.] || ZPE γέν̣[ητα]ι Janko: γεν̣[νᾶν] Ts.: γε̣. [. . .] ZPE || Ῥέα Ts.: Ῥέα̣ ZPE || δ<έ> Janko || . . . [ZPE: π̣α̣ν̣[τοῖα] Ts. *ap.* L&M 15. ζῶια Ts. and Janko: ζῷ̣ια ZPE || κα̣ὶ̣ [Ts.: κα̣ [ZPE || ἔφυ [ῥυέντα Ts. *ap.* L&M: ἔφυ [ῥέοντα Burkert: ἔφυ [ῥαιδίως Janko || Ῥέα ZPE: ῥέα Janko || καὶ̣ [Ῥείη Ts.: καλ̣[ῶν Burkert 16. κατ̣[ὰ γλῶσσαν ἑκάστοις Ts. *ap.* L&M: κα[ZPE || Ἥ]ρ̣η δ' ἐκ[λήθη Ts. *ap.* L&M:] ηοεκ̣[ZPE

COL. 22.

So he (sc. Orpheus) named all things in the same way as finely as he could, knowing the nature of men, that not all of them have a similar nature nor do all want the same things. When they have the power, they say anything that occurs to each one's heart, whatever they happen to want, never the same things, through greed (*or*: arrogance), sometimes also through lack of understanding. Earth (Ge), Mother (Meter), Rhea and Hera is the same (*or*: are one and the same). She/it was called Earth (Ge) by convention; Mother, because all things are born from her (*or:* from this one). Ge and Gaia according to each one's dialect. And (she/it) was called Demeter as the Mother Earth (Ge Meter), one name from the two; for it was the same.

And it is said in the *Hymns* too: 'Demeter Rhea Ge Meter Hestia Deio'. For (she/it) is also called Deio because she/it was torn (*or:* ravaged: ἐδηιώθη) in the mixing/sexual intercourse. He will make it clear when, according to the verses, she is born. . . . And (she/it) is called Rhea because many and . . . animals were born . . . from her. Rhea and . . .

COL. 23.

τ̣ο̣ῦ̣το τὸ ἔπος πα̣[ρα]γωγὸμ πεπόηται καὶ το[ῖc μ]ὲν
π̣ο̣λλ̣οῖc ἄδηλόν ἐcτι τοῖc δ̣ὲ ὀρθῶc γινώcκο̣υcι
εὔδηλον ὅτι Ὠκεανόc ἐcτιν ὁ ἀήρ, ἀὴρ δὲ Ζεύc·
οὔκουν ἐμήcατο τὸ̣ν Ζᾶνα ἕτεροc Ζεὺc ἀλλ' αὐτὸc
αὑτῶι cθένοc μέγα. οἱ δ' οὐ γινώcκοντεc τ̣ὸ̣ν
Ὠκεανὸν ποταμὸν δοκοῦcιν εἶναι ὅτι εὐρὺ ῥέοντα
προcέθηκεν. – ὁ δὲ cημαίνει τὴν αὑτοῦ γνώμην
‾ἐν τοῖc λεγομέν[ο]ιc καὶ νομιζομένοιc ῥή̣μ̣αcι.
καὶ γὰρ τῶν ἀν[θ]ρώπων τοὺc μέγα̣ δ̣υ̣νατ̣[οῦ]νταc
μεγάλουc φαcὶ ῥυῆναι. τὸ δ' ἐχόμενον·
‾ἶναc δ' ἐγκα̣[τέλε]ξ' Ἀχελώιου ἀργυ̣[ρ]οδίνε̣[ω]
‾τῶ[ι] ὕδα[τι οὐ τίθη]c̣ι Ἀχελῶιον ὄνομ[α. .].δ̣ .[ ]
τὰ[c]δ' ἶνα[c ἐγκαταλ]έξαι ἐcτ[ὶν τ]ὸ̣ ἐγγε̣[νέcθ]αι
τη .[ ]. .[±11].ων αυ[±13]. .
ἑκαc̣[τ ±12]δεβουλ[
. .ν[±13]ο̣ντο̣[

1. πεπόηται ZPE: πεποίηται Kapsomenos 2. ἐcτι Ts. *ap*. Bernabé: ἐ̣c̣τ̣ι̣ν ZPE: ἐcτιν Kapsomenos || δ̣ὲ ZPE: δὲ Kapsomenos || γινώcκουcιν Kapsomenos and Ts. *ap*. Bernabé: γινώcκο̣υcι ZPE 5. τ̣ὸ̣ν ZPE: τὸ̣ν Kapsomenos 6. δοκοῦcιν Kapsomenos and Ts. *ap*. Bernabé: δ̣οκοῦcιν ZPE 8. λεγομέν[ο]ιc ZPE: λεγομένοιc Kapsomenos || ῥή̣μ̣αcι ZPE: ῥήμαcι Kapsomenos 9. δ̣υ̣νατ̣[οῦ]νταc ZPE: δυνατ̣ο̣ῦ̣νταc Kapsomenos 11. ἐγκα̣[τέλε]ξ' Ts. *ap*. Bernabé: ἐγ α̣[. . . c]c' ZPE: ἐγκ̣α̣[τέλαc]c' West || ἀργυ̣[ρ]οδίνε̣[ω] Janko: ἀργυ̣[ρ]οδίν̣ε̣[ω] ZPE: ἀργυ̣[ρ]οδίνο̣[υ] West 12. οὐ Betegh: ὡc Burkert *ap*. Bernabé: ὅτι Janko || τίθη]c̣ι Burkert *ap*. Bernabé: δίδω]c̣ι Ts. *ap*. Bernabé . .].δ̣. [ ] ZPE: δ]ῆ̣ λ̣ο̣[ν Janko 13. ἐγκαταλ]έξαι Ts. *ap*. L&M: ἐγκατελά]ccαι West: ἐγ̣ α. . .]ccαι ZPE || ἐcτ[ὶν ZPE: ἐcτ[ὶ Janko || ἐγγε̣[νέcθ]αι Janko: ἐγγ̣ [. . . .]αι ZPE 14. τη̣ [ ]. .[ZPE: τὴν̣ ἀ̣ρ[χήν Janko 15.]δεβουλ[ZPE:]δ' ἐβουλ[or]δὲ βουλ[Janko

COL. 23.

This verse has been made misleading and it is unclear to the many, but to those who understand correctly it is clear that Okeanos is the air and air is Zeus. It is not the case that another Zeus contrived Zeus, but that the same one (contrived) for himself great strength. But those who do not understand think that Okeanos is a river because he (sc. Orpheus) added the epithet 'broadly flowing'. But he indicates his meaning in current and customary expressions. For they say that the very powerful among men 'flowed great'. And the next verse:

he placed in it the sinews of the silver-eddying Achelous

He does [not give] the name Achelous to water. [? The meaning of the phrase] that he placed in the sinews is that the . . . appears in . . . each . . .

COL. 24.

ἴϲα ἐϲτὶν ἐκ τοῦ [μέ]ϲου μετρούμενα, ὅϲα δ[ὲ μ]ὴ κυκλοειδέα οὐχ οἷόν τε ἰϲομελῆ εἶναι. δηλοῖ δὲ τόδε·

‾ ἣ πολλοῖϲ φαίνει μερόπεϲϲι ἐπ' ἀπείρονα γαῖαν.

‾ τοῦτο τὸ ἔποϲ δόξειεν ἄν τιϲ ἄλλωϲ ἐρῆϲ[θ]αι, ὅτι ἣν ὑπερβάληι μ̣ᾶλλον τὰ ἐόντα φαί̣ν̣ε̣τ̣αι ἢ̣ πρὶν ὑπερβάλλειν. ὁ δὲ οὐ τοῦτο λέγει φαίνειν αὐτήν, εἰ γὰρ τοῦτο ἔλεγε, οὐκ ἂμ πολλοῖϲ ἔφη φαίνειν αὐτὴν, ἀλλὰ πᾶϲιν ἅμα τ̣οῖϲ τε τὴγ γῆν ἐργαζομένοιϲ καὶ τ̣οῖϲ̣ ναυτιλλομ̣ένοιϲ ὁπότε χρὴ πλεῖν, τούτοιϲ τὴν ὥραν. εἰ γὰρ μ̣ὴ ἦν ϲελήνη, οὐκ ἂν ἐξηύρ[ι]ϲκον οἱ ἄνθρωποι τὸν ἀριθμὸν οὔτ̣ε τῶν ὡρέων ο[ὔ]τε τῶν ἀνέμω[ν. . . . .]καὶ τἆλλα π̣ά̣ντα [. . . . .]ην

εκ[±18]ϲαεν̣[±13].

[ ]

[ ]ϲ

1. [μέ]ϲου Burkert and ZPE: [πο]ϲοῦ Kapsomenos and Ricciardelli Apicella 3. μερόπεϲϲ'{ι} Merkelbach 4. ἐρῆϲ[θ]αι ZPE: ε<ἰ>ρῆϲ[θ]αι Ricciardelli Apicella 5. ὑπερβάληι ZPE: ὑπερβάλ<λ>ηι Merkelbach || μ̣ᾶλλον ZPE: μᾶλλον Kapsomenos || φαί̣ν̣ε̣τ̣αι ZPE: φ̣α̣ί̣ν̣εται Kapsomenos 6. <φήϲαϲ> φαίνειν Janko 8. ἅμα Kapsomenos and Ts. *ap*. L&M: ἀλλὰ Merkelbach and ZPE || τ̣οῖϲ ZPE: τοῖϲ Kapsomenos 9. τ̣οῖϲ̣ ZPE: τ]οῖϲ Kapsomenos || ναυτιλλομ̣ένοιϲ ZPE: ναυτιλλομένοιϲ Kapsomenos || τούτοιϲ <ἢ ἀροῦν> Merkelbach 10. ὥραν ZPE: θ']ὥραν Burkert: χ]ώραν Kapsomenos || μὴ Ts. *ap*. Bernabé: μ̣ὴ ZPE: μ[ὴ Kapsomenos || ἐξηύρ[ι]ϲκον ZPE: ἐξηύριϲκον Kapsomenos 11. ἀριθμὸν Ts. *ap*. Bernabé: ἀ̣ρ̣ι̣θμὸν ZPE || οὔτ̣ε ZPE: οὔτε Kapsomenos

COL. 24.

. . . are equal measured from the centre, but those which are not round-shaped cannot be of equal limbs. This (verse) makes it clear:

which shines for many articulate-speaking humans on the boundless earth.

Someone might think that this verse is said wrongly, namely that when she is at her utmost, the things that are show up more than before she is at her utmost. But he does not mean this (by saying that) she shows, for if he had meant this, he would not have said that she shows for many, but that for all at the same time, both for those who work the land, and for those who sail when they have to sail, and for the former the seasons. For if there were no moon, people could not have discovered the counting either of the seasons, or of the winds . . . and all the others . . .

COL. 25.

ḳαὶ λαμπρό[τ]ῃτα· τὰ δ' ἐξ ὧν ἡ σελήνη [λ]ẹυκότατα μẹν τῶν ἄλλωγ κατὰ τὸν αὐτὸν λόγον μεμερισμένα θερμὰ δ' οὔḳ ἐστι. ἔστι δὲ καὶ ἄλλα νῦν ẹ̓ν τῶι ἀέρι ἑκὰς ἀλλήλων α[ἰ]ωρούμεν', ἀλλὰ τῆς μὲν ἡμέρης ἄδηλ' ἐστὶν ὑ[π]ọ̀ τοῦ ἡλίου ἐπικρατούμενα τῆς δ̣ὲ νυκτὸς ἐόντα δῆλά ἐστιν. ἐπικρατεῖται δὲ διὰ σμικ[ρ]ότητα. αἰωρεῖται δ' αὐτῶν ἕκαστα ἐν ἀνάγκηι, ὡς ἂμ μὴ συνίηι πρὸς ἄλληλα· εἰ γὰρ μή, συνέλθοι <ἂν> ἁλέα ὅσα τὴν αὐτὴν δύναμιν ἔχει, ἐξ ὧν ὁ ἥλιος συνεστάθη. τὰ νῦν ἐόντα ὁ θεὸς εἰ μὴ ἤθẹλεν εἶναι, οὐκ ἂν ἐπόησεν ἥλιον. ἐποίησε δὲ τοιοῦτογ καὶ τ[ο]σοῦτον γινόμενον οἷος ἐν ἀρχῆι τοῦ λόγου διηγεῖ̣τ̣α̣ι. τὰ δ̣' ẹ̓πὶ τούτοις ἐπίπροσθε π[ο]ιεῖται [οὐ β]ου[λό]μενο[ς] πάντας γιν[ώ]σκẹ[ι]ṿ. ἐν δὲ [τ]ῶιδε σημαί[ν]ε[ι]·

[αὐτ]ἀ̣ρ̣ [ἐ]π̣εὶ δ[ὴ πάν]τα Διὸ[ς φρὴμ μή]σατ̣[ο ἔ]ρ̣γα
[ ].τ̣ρονη .[
[ ]πηγ. . [
[ ]ων .[

1. λαμπρό[τ]ῃτα ZPE: λαμπρό[τ]ατα Janko || ὧν Ts.: ᾧ̣ν ZPE || σελήνη Ts.: σελήνῃ ZPE || [λ]ẹυκότατα Ts.: [λε]υκότατα ZPE: [λε]υκότερα Janko 2. ἄλλωγ κατὰ Ts.: ἄλλωṿ ḳατὰ ZPE: ἄλλωγ <καὶ> ḳατὰ Janko 3. θερμὰ δ' οὔḳ Ts.: θερμạ̀ δ' οὔ[κ] ZPE 4. ἐστὶν Ts.: ἐ̣σ̣τὶν ZPE || α[ἰ]ωρούμεν<α> Janko || ἡμέρης ZPE: ἡμέρης Janko 5. ὑ[π]ọ̀ Ts.: ụ̔[πὸ] ZPE || δ̣ὲ Ts.: δὲ ZPE 6. ἐπικρατεῖται Ts.: ẹ̓πικρατεῖται ZPE 7. ἀνάγκηι Ts.: ἀνάγκῃι ZPE 8. ἄλληλα Ts.: ἄλληλ̣α ZPE || <ἂν> West 11. καὶ τ[ο]σοῦτον Ts.: ḳαị̀ τ̣[ο]σοῦτον ZPE || ἐν Ts.: ἐṿ ZPE 12. διηγεῖ̣τ̣α̣ι Ts. *ap.* Bernabé: διηγεῖ[τ]α̣ι ZPE || δ̣' ẹ̓πὶ Ts. *ap.* Bernabé: δ' [ἐ]πὶ ZPE || τούτοις Ts. *ap.* Bernabé: τούτοις̣ ZPE || ἐπίπροσθε Ts. *ap.* Bernabé: ẹ̓π̣ίπ̣ροσθε ZPE || π[ο]ιεῖται Ts. *ap.* Bernabé: π̣[ο]ιεῖται ZPE 13. β]ου[λό]μενο[ς] Ts. *ap.* Bernabé: β]ου[λό]μενọ[ς] ZPE: β]ου[λο]μένο[υ] Janko || δὲ Ts.: δ̣ὲ ZPE || σημαί[ν]ε[ι] Ts.: σημαί[νει] ZPE 14. [αὐτ]ἀ̣ρ̣ [ἐ]π̣εὶ δ[ὴ Ts.: [. . .]. .[.]μει[ZPE || πάν]τα Ts.:]ω̣ς̣τ̣α ZPE: π]ạ́ṿτα Janko || Διὸ[ς φρὴμ μή]σατ̣[ο ἔ]ρ̣γα Ts.: . . . διο[. . . . . . .]σα .[. . .]. ZPE: νόο]ς αị̓[γιόχ]ọιọ | [μήσατο Janko 15–17. Ts.: not in ZPE

COL. 25.

. . . and brightness. Those things out of which the moon is (composed) are the whitest of all, divided according to the same principle (*or:* measure), but they are not hot. And there are other things now in the air, floating far away from each other. But during the day they are invisible being dominated by the sun, whereas during the night it is visible that they exist. They are dominated because of their smallness. Each of them floats in necessity, so that they do not come together with one another; for otherwise all those which have the same characteristics as those out of which the sun was set together would come together in one mass. If the god had not wished that the things which are now should exist, he would not have made the sun. But he made it of such a sort and of such a size as is explained in the beginning of the account. Those (words) which come after these he puts before (as a screen) not wishing all men to understand. In this verse he indicates:

But once [? the heart] of Zeus devised all deeds

COL. 26.

μη[τρ]òc μὲν ὅτι μήτηρ ὁ Νọῦ̣c ἔcτιν τῶν ἄλλωṿ
ἑᾶc δὲ ὅτι ἀγαθῆc. δηλοῖ δẹ̀ καὶ ἐν τοῖcδẹ τοῖc ἔπẹcιν
¯ὅτι ἀγαθηṇν cημαίνει.
¯Ἑρμῆ Μαιάδοc υἱὲ διάκτορε δῶτορ ἐᾴων.
¯δηλοῖ δὲ καὶ ἐν τ[ῶ]ιδε·
¯δοιοὶ γάρ τε πίθοι κατακήαται ἐν Διὸc οὐδεῖ
δώρων οἷα διδοῦcι, κακῶν, ἕτεροc δέ τ' ἐάων.
¯οἱ δὲ τὸ {ῥ}ῥῆμα οὐ γινώcκοντεc δοκοῦ̣cιν εἶναι
μητρὸc ἑαυτοῦ· ὁ δ' εἴπερ ἤθελεν ἑαυτοῦ μητρὸc
ἐμ φιλότητι ἀποδẹῖ̣ξαι θέλοντα μιχθῆ̣ναι τὸν
θεόν, ἐ̣ξῆν αὐτῶι γράμματα παρακλίναντι̲
μητρὸc ἑοῖο εἰπε[ῖ]ν· οὕτω γ[ὰ]ρ̣ ἂν ἑαυτọῦ̣ γίνοιτο
[υἱὸc δ']ạὐτῆc ἂν ε[ἴη. . . . . δ]ῆ̣λον ὅτι ṿ[. . . . . . .][ ]
[. . . . .]εν τῆι cυ .[. . . . .]ἀμφοτερ̣[
[. . . . . . ἀ]γαθη .[. . . . .].α[
[. . . . . .].εναι̣[

1. μη[τρ]òc Ts. *ap*. Bernabé: μη[τρ]ọ̀c ZPE || μήτηρ Ts. *ap*. Bernabé: μ̣ήτηρ ZPE || Νọῦ̣c Ts. *ap*. Bernabé: Ν[οῦ]c ZPE || τῶν Ts. *ap*. Bernabé: τῷν ZPE 2. ἑᾶc ZPE: ἑᾶc Kapsomenos || ἀγαθῆc Ts. *ap*. Bernabé: ἀγαθῆc̣ Kapsomenos and ZPE || δẹ̀ ZPE: δ]ὲ Kapsomenos: δὲ Janko || ἔπεcιν Ts. *ap*. Bernabé: ἔπẹ[cιν] ZPE: ἔ[πεcιν Kapsomenos: ἔπẹcι Janko 3. cημαίνει ZPE: cημαί̣νει̣ Kapsomenos 4. ἐᾴων ZPE: ἐάων Kapsomenos 5. τ[ῶ]ιδε ZPE: τῷ̣ιδε Kapsomenos 8. δοκοῦ̣cιν ZPE: δοκοῦcιν Kapsomenos 10. ἀποδẹῖ̣ξαι Ts. *ap*. Bernabé: ἀπọδẹῖ̣ξαι ZPE: ἀπọδẹῖξαι Kapsomenos: ἀποδεῖξαι Janko || μιχθῆ̣ναι ZPE: μιχθῆναι Kapsomenos 11. ἐ̣ξῆν ZPE: ἐξῆν Kapsomenos || <β> γράμματα Kapsomenos || παρακλίναντι̲ Janko: παρακλίναντα ZPE 12. μητρὸc ZPE: μ̣ῃ̣τ̣ρọ̀c Kapsomenos || εἰπε[ῖ]ν ZPE: εἰπε̣ῖ̣ν Kapsomenos 13. [υἱὸc δ']ạὐτῆc ἂν ε[ἴη Ts. *ap*. L&M:]ạυτῆc ανε[ZPE || ṿ[Janko: ṿ[ἱὸc Janko: Ζ̣[εὺc Ts. *ap*. L&M 14. cυ .[ZPE: cυμ̣[μείξει or cυṿ[ουcίαι Burkert

COL. 26.

. . . '[of?] mother' because the Mind is the mother of the others; and 'of his own (ἑᾶς)' because she is good. He makes clear in these verses too that it means good:

Hermes Diaktoros son of Maia, giver of goods (ἐάων).

He makes it clear in these as well:

For two urns are placed on Zeus' threshold,
Of gifts such as they give: of evils, and the other one of goods (ἐάων)

Those who do not understand the term think that it is '[of?] his own mother'. But if he had wished to show the god wishing to mingle in love of his own (ἑαυτοῦ) mother, it would have been possible for him by altering some letters to say of his own (ἑοῖο) mother. For thus it would have become his own (ἑαυτοῦ), and he would be her son . . . . her . . . clear that . . . both . . . good . . .

CHAPTER I

The find

On 15 January, 1962, during the widening of the national road leading from Thessalonica to Kavala, a large unlooted cist grave was discovered at Derveni, some 10 km to the north of Thessalonica. The following day another similar but even larger cist grave came to light 4.5 metres away. The Greek Archaeological Service undertook a systematic excavation of the graves and the surrounding area between January and August 1962, under the direction of C. I. Makaronas and under the supervision of Petris Themelis. This survey led to the discovery of five more burials: the looted cist grave Γ, the cist graves Δ, E and H, and the pit grave Z. The tombs and their grave goods proved to be of outstanding importance in many respects, but the two most notable objects are the stunning bronze krater with a Dionysiac scene found in tomb B,[1] and the remains of a charred papyrus scroll which was discovered, through the attentiveness of P. Themelis, among the remains of the funeral pyre on top of the covering slabs of tomb A.

Thanks to the recent full publication of the Derveni tombs by P. Themelis and J. Touratsoglou, we can now get a clear idea of the site.[2] As the archaeological context might be relevant for the understanding of the papyrus, let me summarise the most important points on the basis of their description.

The site was located on the *chora* of the ancient settlement of Lete. It is not clear, however, whether or not the burial ground was directly connected with the city.[3] The tombs were built in the pass of Derveni, along the road known as the *via Egnatia*. Tombs A, B and Δ, located close to each other, are similar in architecture and size, and seem to form a group on their own. The dimensions of the graves are as follows. Tomb A: 2.07 × 0.90 m, 1.08 m

[1] For the Derveni krater, see Giouri (1978).

[2] Themelis and Touratsoglou (1997). The first and for a long time only report provided by Makaronas (1963) was not only very brief but also misleading in some respects, notably about the funeral pyre of tomb A (see n. 8 below). See now also Sideris (2000).

[3] Tzanavari (1997) thinks that the burial ground was the necropolis of Lete.

deep; tomb B 3.06 × 1.53 m, 1.62 m deep; tomb Δ: 2.40 × 1.80 m, 1.50 m deep. The walls of the tombs are constructed of large blocks of poros laid in isodomic masonry. Tombs A and B were each covered with four similar large blocks. Tomb Δ was covered only with a wooden ceiling which subsequently collapsed into the tomb and crushed most of the grave goods. The walls of the chambers in all three tombs were coated with plaster and decorated with painting. The chamber of tomb A had a garland of blue leaves and berries on a red ground, bordered by a yellow and a blue band. The most colourful is tomb B, where the lower half of the walls was painted red and decorated with a guilloche of branches with blue-red leaves and black berries (perhaps olive). The white plaster of tomb Δ was decorated only with a blue band.

Tombs A and B were remarkably rich in grave goods. According to the excavators, tomb A 'was suffocatingly full of bronze and clay pots, vessels, jewellery and various small objects'.[4] Particularly notable in this tomb is the large bronze krater that contained the remains of the cremation and those of two wreaths, one of them of gold, forming oak leaves, and the other of gilded bronze with gilded clay berries. Among other vases of clay and bronze (but none in silver or gold)[5] found in tomb A, there were three *oinochoiai* and other drinking vessels. Other objects included a bronze lamp, many alabastra, some items in coloured glass, ivory figurines, bronze rings, several dozen strigils and many knucklebones.

Among the remains of the funeral pyre, found on top of the covering slabs, there were spearheads, a pair of greaves, the remains of a horse's harness and of a gilded wreath, various other small objects (including a large number of knucklebones, glass objects and two ivory eyes), fragments of at least six clay column capitals, and, in the north-west corner of the ruins, the carbonised remains of a scroll, which would become famous as the 'Derveni papyrus'.[6]

Tomb B, even richer and more carefully constructed than tomb A, contained the magnificently decorated 'Derveni krater' made of bronze with ornate cast volute handles and a scene depicting Dionysos with Ariadne, dancing Maenads, Silenus and other unidentified figures. The krater was standing on a poros block and contained the remains of the cremated dead, gold jewellery, the remains of a gilded wreath and a gold quarter stater of Philip II. Another gold wreath was found on the lid covering the krater.

4 Themelis and Touratsoglou (1997) 193.

5 Silver vases were found only in tomb B, with the exception of one, found in Z (Z12).

6 For the full catalogue of the objects found in and on tomb A, see Themelis and Touratsoglou (1997) 28–59.

A large number of different bronze and silver vessels surrounded the base of the krater, while another group of clay vessels was placed at the middle of the tomb. Spears, a sword, a knife and a pair of greaves were lying at the north-west corner of the tomb, along the wall. The tomb also contained a group of alabastra and various smaller objects, like boxes, nails and spatulas, and stems of wreaths.

The grave goods in tomb Δ were similar in nature to those found in the first two graves, but somewhat less numerous and of a less refined quality. They also included drinking and other vessels of bronze and clay, alabastra, wreaths (two with gilded bronze leaves, another with a solid gold myrtle branch), spearheads and swords, and pieces of jewellery, and an eight-stater of Philip II. The uneven arrangement of the goods in tomb Δ suggested to the excavators that the offerings of a first burial were later pushed aside to make room for a second, inhumed body.

On the basis of the bones and the nature of the grave goods, the archaeologists have concluded that tomb A contained the cremation of a man, tomb B the cremation of a man and a woman,[7] and tomb Δ the inhumation of a man and a woman. The careful construction of the tombs and the abundance of the grave goods, although not unusual in Macedonian tombs of the same period, show the relatively high economic and social status of those buried there. Moreover, the weapons, armour and especially the horse harnesses in tombs A and B indicate that these men belonged to the upper military class.

The remains of the pyres of tombs A and B and the objects contained in them can offer us some idea of the ritual. The corpse was cremated a few metres away from the grave, on top of a richly decorated couch; some of the objects found in the remains of the pyre of tomb A seem to have decorated the couch. According to the reconstruction of Themelis and Touratsoglou, the clay column capitals found in the ashes of the pyre formed part of an elaborate structure on which the couch was laid. They have also suggested that the ivory eyes found among the ashes 'probably come from clay (or wooden) masks-images of the dead (?) or of chthonic demons (?) which were burned in the pyre' (p. 207). For parallels, they quote near contemporary examples from Salamis in Cyprus and from the late archaic tombs at Vergina. Other objects found among the remains of the pyre on top of tomb A indicate that some offerings were burnt on the pyre with the body. After cremation the bones were wrapped in cloth and put into the krater, which was then placed in the grave. The grave was

[7] On this point, see the detailed study in Musgrave (1990).

finally closed in with the covering slabs, and the remains of the pyre thrown on top of the slabs.[8]

On the basis of the coins, metal vessels and pottery, Themelis and Touratsoglou maintain that 'all the evidence favours a date for the burials in the late fourth to early third century'.[9]

THE SCRIPT

The discovery of the papyrus scroll among the remains of the pyre of tomb A was, without exaggeration, sensational. To begin with, it was the first papyrus found in mainland Greece. In contrast to the sand of Egypt and the volcanic tufa in Herculaneum, the humidity of the Greek soil is unfavourable to the conservation of papyri. In this case, however, the fire of the pyre evaporated all the humidity from the fibres, and the resultant carbonisation saved the roll from putrefaction. The age of the roll only added to the excitement raised by the discovery, as papyrologists agreed that it might be the oldest Greek papyrus found so far.[10]

Some words, such as μύϲται, θεοῖϲ θύειν and χοάϲ (apparently all from col. 6), were readily legible on the loose parts of the roll, and they already suggested the religious nature and high interest of the text. The roll was immediately transferred from the site to the Archaeological Museum of Thessalonica, where, on the invitation of C. Makaronas, S. G. Kapsomenos offered a first expert evaluation and dating of the text. Makaronas also offered the rights of publication to Kapsomenos.

Anton Fackelmann, conservator of the Vienna National Library collection of papyri, executed the extremely difficult task of the unrolling. He first rendered the papyrus less fragile by the application of the juice of fresh papyrus plant, and then peeled off fragments from the side of the roll, using static electricity to move the scraps. This method resulted in a collection of some 200 fragments, some larger, some very small. The fragments were then arranged in 9 groups and put between glass plates for conservation. The grouping does not correspond to the original order of the fragments, and in some cases was apparently based merely on the size of the individual fragments. The establishment of the original order of the fragments and

8 Makaronas (1963) implied that the pyre was originally burnt where its remains were found, that is, on the covering slabs of the tomb. This, as noted already by Ginouvès (1994) 187, cannot be correct, because the krater containing the ashes had to be put into the grave before the covering slabs were placed on the top. (Ginouvès, however, mistakenly assigns the papyrus to tomb B.)

9 Themelis and Touratsoglou (1997) 221.

10 This title might now go to the badly preserved roll found in 1982 in a tomb in Athens and dated to the fifth century. But, as far as I know, this roll has not been made legible as yet.

the decipherment of the text were carried out with the help of high quality photographs, some of which were on display together with the papyrus at the Archaeological Museum of Thessalonica in the spring of 1998.[11]

The bottom part of the papyrus was consumed by the flames, and only 7–8 cm of the upper half escaped. This means that we have 15–17 lines of writing in the better preserved columns, of which the upper 10–11 lines have an almost continuous text, whereas usually only a few letters are legible in the bottom lines.[12] Only small fragments of 9–10 lines, some with hardly any legible letters, remain from the badly damaged first columns.

Kapsomenos, with the help of K. Tsantsanoglou, first reconstructed 22 columns, with a number of unidentified fragments remaining.[13] Tsantsanoglou (1997) has established, however, that we possess the remains of 26 columns in total, a fact which has also provided a definitive column numbering. The first columns, which made up the outer layers of the roll, were more exposed to the destructive effects of fire and time, and are accordingly in a much worse state of preservation. The remaining upper parts of the last columns, on the other hand, contain very few lacunae. In accordance with an established ancient practice, a relatively large block of space (about 17 cm) is left blank after the last column in order to make the use of the roll easier.

It is not clear how much of the original roll has reached us. According to a survey made by Eric Turner, the height of early literary papyri varies between 12.7 cm (comic fragment, Brit. Lib. pap. 1824. P. Hib. I 6., early third century BC) and 21.7 cm (Eur. *Antiope*, Brit. Lib. pap. 485. P. Petrie I 1–2, mid third century BC).[14] On the other hand, the number of lines per column is normally between 21 and 31, with the unusually high number 36–7 in the case of the highest roll. On the basis of these data, it seems that we have about half of the lines of the individual columns in the Derveni papyrus. It is even more difficult to guess how many columns are lost before our column I.[15] Moreover, it is possible that the text continued on a second roll. The available evidence, however, does not allow us to verify or falsify this hypothesis.[16]

The usual length of the individual lines varies between 30 and 45 letters, which number, however, can occasionally be considerably lower in lines

[11] For the description of the conservation process and for a primary papyrological description, see Kapsomenos (1964). For some published photographs, see Turner and Parsons (1987) plate 51; Tsantsanoglou and Parássoglou (1988) tav. IV; and the dust cover of L&M. For a full list, see the entry 'Illustrations' in the Bibliography by M. S. Funghi in L&M.

[12] It is mostly in these bottom lines that we may expect new readings with the allocation of smaller, hitherto unplaced fragments.

[13] Cf. Kapsomenos (1963) and [Anon.] (1982).

[14] Cf. Turner (1980b) 37.

[15] See however p. 328 n. 9 below.

[16] See however p. 130 below.

preceding a quotation, or, in some cases, at the end of a period. The length of the lines was apparently set in order to accommodate a hexameter.[17] The scribe usually avoids separating words, the only surviving exceptions to this rule being ἐξαμαρ|τάνουcιν in col. 12.4–5 and ἐπι|τελέcαντες in col. 20.6–7.

Although the papyrus is completely carbonised, the writing is easily legible when light is shed on it at the appropriate angle. The bilinear script shows a very neat, skilled book hand with small characters, using a sharp kalamos. The square, separate letters are comparable to the lapidary style of fourth-century inscriptions. The writing, however, is 'fast'; particularly notable are the sigma and omega written in one sequence.[18]

Paragraphoi usually, but apparently not in all cases, precede and follow the lemmata, and occasionally indicate punctuation with or without a dash in the line.[19] Iota adscript and consonant assimilation are regular features. There are a few supralinear corrections, apparently by the first hand.

Most papyrologists agree on the basis of its palaeographic features that the manuscript must be dated to the second half of the fourth century BC.[20] Tsantsanoglou and Parássoglou give a more precise dating, and assign the papyrus to the period between 340 and 320.[21] It should be noted, however, that the extreme scarcity of comparative evidence makes the dating difficult and insecure. Moreover, one cannot completely exclude other factors that might blur the picture. It is not inconceivable, for example, that the nature of the text prompted the reproduction of a more antiquated handwriting. Naturally, the date of the manuscript need not be identical with the date of the text itself.

The dialect used in the text is mixed. Most commentators consider it basically Attic with an Ionic overlay, whereas West and Janko maintain

17 So, e.g., Obbink (1997) 44 n. 9. The length of a hexameter, roughly 36 characters, was regularly used in papyri as a stichometric unit even for prose texts. It does not mean, however, that the actual lines in papyri were of this width. Typically the lines contained 18 characters, which means that two papyrus lines made up one 'line' taken as a unit. Cf. Sedley (1998a) 119–20 with reference to Cavallo (1983) 20–2. Nevertheless, in the absence of other documents, we cannot know much about fourth-century BC practice, and the possibility cannot be excluded that early papyri used longer lines. Cf. Parsons in Turner and Parsons (1987) 151 n. 113, with reference to J. Irigoin, who raises the possibility that the Derveni papyrus might be taken as evidence for the origin of the latter stichometric practice. See also p. 95 below.

18 For a detailed palaeographic and papyrological description, see Kapsomenos (1963); Turner and Parsons (1987) 92; and, for col. 4, Tsantsanoglou and Parássoglou (1988).

19 See also p. 95–6 below.

20 Alternative views have been put forward, however. Turner, for example, first preferred the dating between 325 and 275 BC, but, as Parsons indicates in Turner and Parsons (1987) 151 n. 111, he later settled for the fourth century BC. Irigoin (1972) 547, on the other hand, proposes an earlier date, possibly as early as the first half of the fourth century.

21 Tsantsanoglou and Parássoglou (1988) 125.

that it is fundamentally Ionic with Attic features which, according to West but with Janko's dissent, might be due to the transmission of the text.[22] Notably, sometimes the same word appears in different dialectal forms (e.g., σμικρο-, μικρο-; ὄντα, ἐόντα).[23] Moreover, Janko has drawn attention to the sporadic Doric elements, such as the use of νιν for μιν.

THE EDITORIAL SITUATION

The study of the Derveni papyrus has regrettably been hindered by the lack of an authoritative edition. As the editorial situation is quite complex, and has not been solved as yet, it seems appropriate to summarise it briefly here. Shortly after the discovery S. G. Kapsomenos received the rights of the edition from C. I. Makaronas, Curator of Antiquities in Central Macedonia. After the first brief announcements, Kapsomenos published in 1964 a more elaborate general description of the find and the content of the text, to which he also added the provisional transcription of parts of coll. 18, 19, 21, 22, 23, 24 and 26 (according to the current numbering; Kapsomenos' numbering was lower by four) in order to give an idea of the character of the text.[24]

Although the exceptional importance of the text for several areas of classical studies had immediately been recognised, Kapsomenos' very incomplete transcription was the only officially available text for a long time. When Kapsomenos died in 1978, K. Tsantsanoglou, who had been Kapsomenos' assistant and collaborator since the first phases of the work, took over the task of the edition with G. M. Parássoglou. In the meantime, unpublished transcripts of larger parts of the text were apparently circulating among scholars working on related subjects.

Events, however, took an unexpected turn in 1982, when a transcript, with a very short German introduction but without signature appeared in *Zeitschrift für Papyrologie und Epigraphik* 47, on separate page numbering following page 300.[25] This transcript contained a number of supplements and emendations, without in most cases indicating their originator; as far as one can reconstruct it, some of them apparently came from the Greek editors, whereas some were suggested by other scholars. In reply to this unauthorised edition, the Greek editors and Sir Eric Turner published an objection in *Gnomon* 54, stating that the text published 'is in part provisional and in part wrong, owing to changes made without the editors' knowledge

[22] For the former view, see most recently Funghi (1997) 36; for the latter, see West (1983) 77, and Janko (1997) 62.

[23] For a full survey, see Janko (1997) 62–3.

[24] Kapsomenos (1964).

[25] [Anon.] (1982).

and consisting of alterations, supplements and joints that are impossible.'[26] They declared at the same time their plan to submit to the press their own complete edition by mid 1984. Unfortunately, this plan has not been realised as yet. And in the absence of an official edition, interested scholars kept using the avowedly inaccurate 'pirate' edition.

The most important step forward since 1982 has been the publication of *Studies on the Derveni Papyrus* in 1997. The volume contains an English translation – the first published translation of the available text – prepared by André Laks and Glenn W. Most. This translation was based on the publicly available text, that is the anonymous *ZPE* transcript, but was checked by K. Tsantsanoglou against the papyrus. Moreover, in footnotes to the translation Tsantsanoglou suggests some significant supplements and modifications based on new readings of the papyrus. As expected, most of the new readings come at the bottom parts of the individual columns. Even more important, Tsantsanoglou published in the same volume the text of the first seven columns, accompanied by a thorough word-by-word commentary. This new publication contains chunks of text which had previously either not been available at all or been presented very incorrectly in the *ZPE* transcription. Although Tsantsanoglou underlines the still provisional nature of the text presented, this work, owing to both its quality and the new evidence it provides, means a crucial advance in the study of the papyrus. It not only establishes the definitive column numbering, but also offers new insights into the nature of the text.

As mentioned already in the preface to my text and translation, the typescript of this book was already finished when Richard Janko's new edition of the text was published in *ZPE* (Janko (2002)). Janko's edition is the first critical edition of the papyrus with full critical apparatus, translation and word index. The edition not only records all the publicly available readings and emendations, but also contains a number of new suggestions by Walter Burkert and Richard Janko. In all respects, it is an exemplary piece of scholarship, and an enormous step forward promoting the study of the papyrus.

The text and translation I present in this book is similar in its objectives and aims to that of Janko. Let it be stated here as well that it makes no claim to being a full edition, but is provided merely as an aid for the interpretation of the Derveni text until the definitive edition of the text by Prof. Tsantsanoglou becomes available.

[26] Turner, Tsantsanoglou and Parássoglou (1982) 855.

ATTEMPTS AT THE IDENTIFICATION OF THE AUTHOR

At different stages of the editorial process, different scholars have advanced different suggestions for the identification of the author of the papyrus. Thus far, five names have been given serious consideration.[27]

When Kapsomenos presented the papyrus for the American Society of Papyrologists, he brought up the name of Epigenes, at the same time also voicing some reservations about this hypothesis. The next hypothesis in chronological order is by Charles H. Kahn, who claims that the most probable candidate for the authorship is Euthyphro, known to us exclusively from Plato's *Euthyphro* and *Cratylus*. Kahn had already mentioned this possibility in passing in a footnote in 1973, and he has recently explored the hypothesis in detail.[28] Walter Burkert, for his part, has argued in a short paper that Stesimbrotus of Thasos could also be considered as author; if so, the text of the papyrus might be identical with Stesimbrotus' Περὶ τελετῶν.[29] According to Maria Serena Funghi, this hypothesis had already been advanced *per litt.* by K. Tsantsanoglou.[30]

Most recently, Richard Janko has introduced an interesting alternative interpretation of the papyrus. On the basis of a different evaluation of the author's attitude towards traditional religiosity and the Orphic mysteries, Janko has made two suggestions concerning the identity of the author: first, in passing, Diogenes of Apollonia, and then, more forcefully, Diagoras of Melos. If this latter is the case, Janko argues, the papyrus can offer us fragments from Diagoras' *Apopyrgizontes Logoi*.[31]

None of these suggestions has received general acceptance.[32] Obviously, any identification can be based only on a thorough analysis of the text; I therefore leave detailed discussion of the question until my last chapter.[33] But, since I do not want to raise any false expectations, let me state already at this point that even though some of those mentioned above seem better candidates than others, none of them, in my view, can muster sufficient persuasive arguments on their side; and for my part, I am unable to come up with a more convincing proposal. I shall therefore be using the term

[27] We can add some further suggestions. Lebedev suggested Prodicus of Ceos in an unpublished paper. This option is given brief mention in Janko (1997) 75. Burkert *apud* Ricciardelli Apicella (1980) 129 has considered also Metrodorus.

[28] Kahn (1973) 156 n. 6. and Kahn (1997). See also Kahn (1986).

[29] Burkert (1986).

[30] Funghi (1997) 36.

[31] Janko (1997) and (2001), see also Janko (1998).

[32] In Janko's case there has not as yet been much time for discussion. But, see, most recently, Bernabé (2002) 7 with n. 29.

[33] I shall, however, consider Janko's hypotheses in the Appendix.

'the Derveni author' throughout to refer to the author of the papyrus (not to the author of the poem commented on in it).

THE PAPYRUS SCROLL AND ITS SEPULCHRAL CONTEXT

The location of the papyrus in the remains of the pyre of tomb A poses an obvious problem. Why was it there? The answer to this question is not independent of the interpretation of the text of the papyrus. It may affect our reading of the text and, conversely, one's assessment of the text is likely to influence one's answer to the question concerning the ritual or other function of the scroll. Thus, we should not expect any definite answer at this point; I shall suggest an answer in chapter nine, based on my interpretation of the text.[34] What follows in this chapter is a survey of different possibilities with their respective necessary or likely implications.[35]

The more sceptical answer, raised first in passing by F. W. Walbank and now considered more seriously by Richard Janko,[36] is that the papyrus was mere scrap, used quite simply to light the pyre. Indeed, we have some evidence about the Roman custom of putting papyrus on the funeral pyre in order to set it alight. Nevertheless this evidence is quite meagre. The two most relevant passages both come from Martial. In *Epigr.* 10.97.1 he writes *dum leuis arsura struitur Libitina papyro* ('While the light-heaped pyre was being laid with papyrus for the flame' trans. Ker). Standard editions of Martial connect this line with *Epigr.* 8.44.14 where we read *fartus papyro dum tibi torus crescit* ('while, stuffed with papyrus, your pyre is growing high' trans. Ker). If the two passages attest the same custom, then the latter text implies that one has to think not of a single roll kept whole, but rather a layer of torn papyrus put underneath the wood to help combustion.[37] If so, these texts cannot provide a real parallel for the suggested use of the Derveni papyrus.

The text of *Anth. Pal.* 9.174.2–6 (fifth century AD) is sometimes also adduced in this context. The author is speaking here about the abysmally low wages of teachers, and develops a nice simile: 'Here the nurse brings, perforce, the fee once a month, tying up the wretched pittance in byblus and paper, and puts the contemptible little paper, like a pinch of incense,

[34] The impatient reader can already turn to page 347 below.

[35] The Strasbourg Empedocles papyrus presents the same problem about the relationship between text and archaeological context. For a recent discussion, see Osborne (2000) 332–4.

[36] Walbank *apud* Kapsomenos (1964–5) 22. Janko (1997) 62 and (2001) 1 n. 1. Janko's interest in this possibility is clearly related to his attempt at giving a non-religious interpretation of the text.

[37] Cf. Lewis (1974) 96.

by the master's seat, as if by a tomb.' (ἔνθα τροφὸς κατὰ μῆνα φέρει μισθὸν μετ' ἀνάγκης, | βύβλῳ καὶ χάρτῃ δησαμένη πενίην | ὡς δὲ κάπνισμα τιθεῖ παρὰ τὸν θρόνον, ὡς παρὰ τύμβον, | τὸν μικρὸν χάρτην, τὸν παραριπτόμενον – trans. Paton). This text, however, does not refer to rolls but to the well-attested use of cheap papyrus (*charta emporetica*), as well as discarded and torn writing papyrus, for wrapping incense.[38] Therefore this text, also very remote from the Derveni tombs in time, cannot be used as a parallel either.

There is, to the best of my knowledge, not a shred of evidence from the classical or Hellenistic period for a comparable Greek or Macedonian practice of lighting a pyre with a roll of papyrus.

It is also important to note that the carbonised roll was not on its own in the remains of the pyre. As mentioned above, some other personal belongings of the person buried in tomb A were left among the ashes on top of the covering slabs. For example, a pair of greaves and some spearheads were found there. We can contrast this with the fact that the greaves and arms of the person buried in the neighbouring tomb B were neatly placed inside that tomb together with other valuable objects. And even if one were to try to explain the location of the greaves by the assumption that the person was cremated in his greaves, and that then they were simply left in the remains of the pyre, the presence of other pretty objects, such as the coloured alabastra, found also among the ashes of the pyre on top of the covering slabs, would need explanation. All this indicates that not all the offerings and valuable objects were put inside tomb A, but some of them were left in the ashes on the covering slabs. The presence of such objects in the remains of the pyre, I would maintain, considerably strengthens the hypothesis that the roll was not treated as waste paper.

The alternative answer to the question concerning the location of the roll, and the one espoused by most interpreters, is that it was designed to be burnt with the corpse on the pyre, and thus that it had some function in the funerary ritual. In this event, its placement can be compared with other Greek texts found in tombs. These, however, apparently fall into different categories. Some of them have no obvious eschatological implication. For example, the papyrus roll from Hawara found in a tomb (P. Bodl. Ms. Gr. Cl. a. I(P), mid second century AD) contains two verses from *Iliad* 1 and the whole of *Iliad* 2. One can compare this find with the testimony of Photius *Bibl.* 190.151a, according to which Cercidas of Arcadia (fourth century BC) wanted to be buried with a copy of the first two books of the *Iliad*. A roll

[38] Cf. e.g. Athenaeus 9.374b.

from Saqqarah, also found in a tomb and dated to the fourth century BC – hence very close to the Derveni papyrus in time – preserves the *Persians* of Timotheus of Miletus (P. Berol. 9875). Although different hypotheses have been advanced, the function of these rolls in graves remains unclear.[39]

On the other hand, some texts in tombs have evident eschatological bearing, as we can see most conspicuously in the case of the gold leaves. More and more of these are coming to light, and help to reshape our view of ancient eschatological beliefs and mystery religions. Furthermore, owing to some momentous new discoveries, such as the gold leaves from Hipponion and Thessaly, these objects and texts are now customarily treated again as 'Orphic' and/or 'Bacchic'.[40] These tiny inscribed gold plates clearly have a place in the tomb, for, most probably, they were meant to be guides for the deceased in their journey to the afterlife. The texts carved on them, despite their brevity and enigmatic wording, evince a distinct core of eschatological beliefs with a conception of the nature and fate of the soul as well as an elaborate topography of the underworld.

Admittedly, a few verses inscribed on gold leaves are not the same thing as the longer, argumentative prose treatise on the Derveni papyrus – even though I shall try to show later that there might be significant connections between the Derveni text and the gold leaves.[41] There are some indications, however, that papyri could be used for comparable purposes. Scholars have surmised on the basis of the archaeological data that the papyrus roll found in the right hand of an inhumated person at Callatis (present day Romania) could contain a Dionysiac/Orphic text possibly with eschatological content;[42] but this is sheer speculation, because the roll was unfortunately destroyed before anyone could read it.

A further indication may come from Euripides' *Hippolytus*. Theseus in his diatribe against Hippolytus depicts him as a hypocritical Orphic. He first alludes to the vegetarian diet customarily associated with Orphism, and then continues with the following words: 'Having Orpheus as your lord, you revel honouring the smoke of many writings' (Ὀρφέα τ' ἄνακτ' ἔχων | βάκχευε πολλῶν γραμμάτων τιμῶν καπνούϲ *Hipp.* 953–4). The expression πολλῶν γραμμάτων . . . καπνούϲ is usually understood as referring to the trifling or worthless nature of the Orphic texts. The

[39] Turner (1980a) 76–8 raises the possibility that the Hawara case shows an imitation of the Egyptian custom of putting a copy of the *Book of the Dead* in the tomb. See, however, the case of Cercidas mentioned above.

[40] For a survey, see Pugliese Carratelli (1993), to which add also Frel (1994). Parker (1995) 496–8 presents good arguments for the case that the gold leaves can be treated as Orphic.

[41] See ch. 9 below. [42] See Pippidi (1967) 210.

word καπνός could indeed have this secondary meaning;[43] yet, I suggest, Theseus' turn of phrase might be taken just as well as a play on words, also alluding to Orphic books burnt on pyres – such as the one at Derveni.

We can also add the evidence offered by pictorial representations, the most notable of which is the funeral vase dated 330–320 BC. The vase painting depicts Orpheus with the heroised deceased who holds a papyrus roll in his hand.[44] The scholars who published the vase suggest that this may be taken as a reference to the Orphic custom of equipping the dead with texts. All these pieces of evidence, both textual and pictorial, refer to the prominent place texts and books had in Orphic circles – a phenomenon that has been analysed by, among others, Marcel Detienne.[45] But all this is admittedly very far from compelling in the case of the Derveni papyrus. We have to acknowledge, on the other hand, that the lack of comparable evidence might just as well be due to the fact that papyri are very rarely preserved in Greek tombs. And this is even more likely to be the case if they were burnt, or meant to be burnt, on funeral pyres. In all events, as we shall shortly see, the first columns of the Derveni text do have eschatological implications, and could indeed be referring to rituals connected with the funeral.[46]

To conclude, acknowledging all these qualifications and uncertainties, I still favour the hypothesis that the Derveni roll did have a function in the ritual. This is made likely by the presence of other valuable objects in the remains of the pyre, by the specific Orphic concern for eschatology, by the well-documented Orphic custom of equipping the dead with texts, and possibly also, as I shall argue, by the special role allotted to fire in the Derveni text.

WAS THE PERSON BURIED IN THE TOMB AN 'ORPHIC'?

The hypothesis that the papyrus had a function in the ritual, and, further, that it had this specific role on account of its content, is likely to imply,

[43] See LSJ 1. So e.g. Barrett (1964) ad loc.; Halleran (1995) ad loc.; Linforth (1941) 52–3.

[44] Cf. Schmidt (1975); see also Burkert (1999a) 72–6, esp. 75.

[45] Detienne (1989).

[46] Yet again, the eschatological relevance of a text cannot in itself guarantee its ritual function in a funeral context. Martin and Primavesi (1999) 36–8 argue for instance that the papyrus containing Empedocles' *Physica*, with its demonological story, was used as discarded paper, regardless of its content, for the fabrication of the funeral crown. But this view is also open to debate. (For the alternative, see Most (1997) 130.) The case of the Empedocles papyrus is certainly different in that the papyrus was torn into pieces for this secondary use, and apparently only a small part of the roll was used, with no attention paid to equipping the deceased with the entire text.

without however necessitating it, that the person buried in tomb A had some association with Orphism. But when we try to answer the question set above, everything depends on what we mean by the term 'Orphic'. Here we find ourselves faced with the notorious problem of social organisation and identities in mystery cults in general, and in Orphism in particular.[47]

First of all it should be noted that Orphism was not a unitary movement, and therefore we should accept differences; the trouble is that our evidence, especially from the classical and early Hellenistic period, is so meagre that we do not know who could count as Orphic, and on what basis, in different communities and in different parts of Greece. Second, it should also be remembered that people could be associated with Orphism in different forms and at different levels.[48] At one level, we have the itinerant priests propagating initiations in the name of Orpheus – apparently, this was a distinct type known as the *orpheotelestai*. On another level, we have the initiated, that is, the clientele of the *orpheotelestai*. Our information is frustratingly sparse on both groups and the interaction between them, the main source being Adeimantus' unsympathetic representation in Plato's *Republic* 2. The text is well known, but is worth quoting here at length:

> Begging priests and prophets frequent the doors of the rich and persuade them that they possess a god-given power founded on sacrifices and incantations. If the rich person or any of his ancestors has committed an injustice, they can fix it with pleasant rituals. Moreover, if he wishes to injure some enemy, then, at little expense, he'll be able to harm just and unjust alike, for by means of spells and enchantments they can persuade the gods to serve them. And the poets are brought forward as witnesses to all these accounts. [There comes a quote from Hesiod *Works and Days* 287–9, and then Homer *Il.* 9.497–501 about the gods being swayed by prayers and sacrifices.] And they present a noisy throng of books by Musaeus and Orpheus, offspring as they say of Selene and the Muses, in accordance with which they perform their rituals. And they persuade not only individuals but also whole cities that the unjust deeds of the living or the dead can be absolved or purified through ritual sacrifices and pleasant games. These initiations, as they call them, free people from punishment hereafter, while a terrible fate await those who have not performed the rituals. (*Republic* 364b–365e trans. Grube)

In any case, it is natural to think that the profession of an *orpheotelestes* required a strong adherence to the so-called Orphic life, with its dietary

[47] On this subject, see e.g. Burkert (1982) and (1987a) ch. 2; Parker (1995).

[48] This important, but often ignored, point is well put by Bernabé (2002) 7: 'Il y a beaucoup de manières d'être "Orphique".'

and other restrictions. On the other hand, the fact that someone once got initiated by a wandering *orpheotelestes* is unlikely to commit the person to any stronger reform in lifestyle; it is unlikely, for instance, that all those who used the services of an Orphic priest became vegetarian.[49] In fact, this is the picture we get from Adeimantus' portrayal in the *Republic*: people who choose this easy path think that there is no need for a more thorough moral improvement, or indeed, for any other lasting change, because it is enough to pay for certain sacrifices and to undergo some playful initiation ritual, since these acts by themselves grant a better fate in this life and after death. In such cases previous initiation would become important at the moment of death, even for someone who has not lived an 'Orphic life'.

We can also assume that there were stronger forms of allegiance to the lore propagated by the *orpheotelestai*. It is a long debated question whether or not there were Orphic communities, as such. The inscription ΟΡΦΙΚ[on one of the bone plates from Olbia strengthens this option but, characteristically enough, a lacuna occurs just at the point where the document might have given a definitive answer to the question.[50] If there were such Orphic communities, then it is again natural to suppose that the members followed the distinctive rules of the 'Orphic life'.

Let us now examine the indications the archaeological context provides on the burial ritual and on the individual buried in tomb A, in order to see whether they could strengthen, weaken or exclude the possibility of the supposed Orphic connection of the tomb and the burial. If we start from the Orphic side, the only positive piece of evidence on burial custom comes from Herodotus. He reports that the Egyptians do not wear wool in shrines and do not bury their dead in wool, and that this custom is also followed in the Orphic, (Bacchic?) and Pythagorean rites.[51] As far as I can see, the archaeological evidence does not give us any indication as to whether or not the person in tomb A had any wool garments on him when he was put on the pyre.

A factor which one may consider relevant is that the person in tomb A was cremated. This, however, cannot be decisive, as it turns out that

49 So e.g. Bremmer (1999) 82.

50 Olbia bone plate no. 1. For the text, see Rusjaeva (1978); Tinnefeld (1980), and Zhmud' (1992) who defends, partly on palaeographic grounds, the reading ΟΡΦΙΚ[ΟΙ against ΟΡΦΙΚ[ΩΙ or ΟΡΦΙΚ[ΩΝ as suggested by West (1982) 22. If the word is a noun, and not an adjective, the bone plates can count as evidence for some kind of Orphic identity.

51 Hdt. 2.81. The passage is extensively discussed in Linforth (1941) 38–50. The text is vexed because the reference to the Bacchic rites and their identification as Egyptian is missing from the Florentine family of MSS. But this question has not much bearing on our present discussion.

people with very similar eschatological beliefs could be buried either way. We can observe this phenomenon in the case of the tombs providing the gold plates: the person buried in Timpone Grande in Thurii was cremated, whereas those in Timpone Piccolo and Hipponion were inhumed.[52] Also, within this closed group of six tombs in Derveni, showing similarities in many respects, A, B, Γ, E and Z had cremation burials, whereas Δ had inhumation (the case of H is unclear because of looting, but cremation is more probable). It is to be noted, however, that the Derveni group is exceptional in the contemporary Macedonian context in which inhumation is far more common than cremation (53 cremations out of 600 burials in Olynthos, and 16 out of 229 in Akanthos).[53] This fact may point to some special importance attached to cremation – divergence from the prevailing norm often carries a meaning.

It is customary to associate Orphism with some kind of ascetism and 'puritanism'. The relative wealth of the offerings found in tomb A seems to be in conflict with such an attitude. This is indeed one of the reasons why Themelis and Touratsoglou preclude the possibility that 'the occupant of tomb A (and the owner of the papyrus) was a devotee of Orphism.'[54] However, the clientele of the itinerant Orphics could of course be of higher economic and social status as well. Adeimantus' begging priests and soothsayers 'go to rich men's doors', and it is unlikely that an initiation itself would cause these people to give up their riches, or to make funerals more modest if the local sociological setting required otherwise. Indeed, as has been stressed by Jan Bremmer, the material of the gold leaves also indicates a higher economic status.[55] The Thracian cultural context may also be relevant at this point. According to Ivan Marazov, the mysteries in ancient Thrace, among which Orphism had a distinguished place,

[52] The burial of Timpone Grande is specific in the sense that the corpse was first put in the coffin, and then the pyre was built upon it. The evidence for the type of burial in the tombs of Timpone Piccolo is indirect. See Zuntz (1971) 292. For a survey of the type of burial with the different gold leaves, see the useful survey in Graf (1993b) 257–8.

[53] Cf. Themelis and Touratsoglou (1997) 202.

[54] Themelis and Touratsoglou (1997) 205. Their conclusion is that 'It might be assumed that the Orphic papyrus accompanied the dead man on the pyre either because of its general religious-eschatological content, similar to that of the "Dionysiac" krater B1, or because its owner liked to read such texts, or was indeed himself the commentator.' The last of these options seems highly unlikely to me, especially if one considers those mistakes in the papyrus which indicate that it was a copy. Moreover, the intentional use of an eschatological text in an eschatological context is a sign of a stronger commitment than a simple favourite reading. It has to be noted that Themelis and Touratsoglou do not seem to know about Tsantsanoglou (1997), which, as we shall see, has important implications also on this question.

[55] Bremmer (1999) 82.

'were of a markedly social nature – they were accessible only to the representatives of the upper crust.'[56] And, after all, one should also bear in mind Morris' assertion, based on the study of thousands of Greek tombs, that 'There can be no one-to-one correlation between grave goods and eschatology.'[57]

The presence of *oinochoiai* in tomb A may, at least according to some, cause another problem. For Burkert maintains that the Orphic dietary restrictions included the prohibition on drinking wine.[58] This is in fact the other and stronger reason why, with explicit reference to Burkert's claim, Themelis and Touratsoglou exclude the possibility that the person in tomb A was a follower of Orphism.[59] Yet I find the textual basis for Burkert's claim too weak. Burkert adduces only Plato's *Laws* 672b:

> There is a little-known current of story and tradition which says that Dionysos was robbed of his wits by his stepmother Hera, and that he gets his revenge by stimulating us to Bacchic frenzies and all the mad dancing that results; and this was precisely the reason why he made us a present of wine. (trans. Saunders)

One can take this as a reference to the Orphic myth about the rending of Dionysos, although it is far from necessary to do so. Yet, surely, it does not contain anything about a prohibition on drinking wine. Furthermore, our other sources on the Orphic life only speak about vegetarianism, and never mention that Orphics were teetotallers – a curiosity perhaps no less noteworthy than total abstention from meat. On the other hand, there is evidence for the Pythagoreans' abstinence.[60] Yet nothing makes it necessary that the Orphics followed the Pythagoreans in this respect.

Moreover, such a restriction on drinking wine would be difficult to square with the important place allotted to Dionysos in Orphic thought. And even though it is still disputed whether or not the gold leaves can count as Orphic, it is significant that the last line on the Pelinna leaves, which bridge the gap between the A and B leaves of Southern Italy, speaks emphatically about the reward of wine for the initiate.

Judging from the spearheads, the pair of greaves and the remains of a horse's harness found in the remains of the pyre, the person had military

[56] Marazov (1998) 114. On the social and economic situation of the persons buried with the gold leaves, see the important correctives in Graf (1993b) 255–6.

[57] Morris (1992) 106.

[58] Burkert (1985a) 301 writes that 'Orphics eat no meat, no eggs, no beans, and they drink no wine.'

[59] Themelis and Touratsoglou (1997) 205; their reference is to the Greek translation of Burkert (1985a).

[60] Cf. Iamblichus, *Vita Pyth.* 107. Some other sources mention specifically that Pythagoras drank water (instead of wine): see, e.g., D.L. 8.13.

functions. Such a function can indeed be in conflict with the strict Orphic avoidance of bloodshed. It is questionable, however, whether it precludes what I have called a looser form of association with Orphism.

In conclusion, the archaeological evidence is not decisive either way. On the other hand, perhaps the text of the papyrus can say something about its place in the sepulchral context? Let us turn now to the text.

CHAPTER 2

The first columns

The subject matter of the first six columns of the extant text of the papyrus is relatively homogeneous and clearly distinct from the rest. In these columns the author focuses primarily on the interpretation of certain rituals with the help of certain eschatological concepts. This part of the text may prove to be of the utmost importance for the general understanding of the papyrus; it is likely to provide a clue about the overall communicational situation, the *Sitz im Leben* of the text, and about the professional identity of the author. These columns, however, made up the outer layers of the roll and were accordingly more exposed to the destructive effects of fire and time. For this reason, this part of the text is in a deplorable state of conservation, broken up into dozens of tiny scraps. This is why these columns were almost entirely missing from the first transcripts. Hence they were largely unknown to scholars until Kyriakos Tsantsanoglou published in 1997 a preliminary, but apparently more or less complete, transcript of the first seven columns. Tsantsanoglou also presented a detailed and circumspect philological and historical commentary with his transcript.[1] This publication marks a true advance in the study of the papyrus.

Yet, due to the physical state of the papyrus, it is extremely difficult to reconstruct the author's main line of argument in these columns even armed with Tsantsanoglou's transcript and commentary. The most one can do in this situation is try on the one hand to contextualise the rituals described in the papyrus in Greek and other religious practices, and on the other to identify the main figures and concepts and raise some hypotheses concerning the possible relations between them. Apart from Tsantsanoglou, Albert Henrichs and Fritz Graf have offered illuminating studies from the point of view of the broader context of the history of Greek religion,[2] while Walter Burkert has taken into account the relevant Persian material as well.[3]

[1] Tsantsanoglou (1997).
[2] See esp. Henrichs (1984) and (1998); Graf (1980) and (1991).
[3] Burkert (1999a).

In her recent book on the interactions between the living and the dead in Greek culture, Sarah Iles Johnston has made an important contribution to the understanding of the eschatological lore forming the background of the Derveni author's views.[4] For my part, I have not much to add to these excellent comparative studies. I shall summarise and put together their main results, but my own focus here will be on the internal logic of the author's argument, as far as the condition of the text permits, and I shall formulate some questions to which we can return in the final chapters, after an examination of the second part of the text, the interpretation of the Orphic poem.

Let us start with a quick overview of the topics of the individual columns, to give a table of contents, as it were. In column one, the only significant word is 'Erinyes', which is, however, a bold reconstruction. In column two, Erinyes are surely mentioned at least twice. Moreover, they are clearly mentioned in the context of some sacrificial ritual, possibly a funerary rite. The topic of column three is manifestly eschatological; it not only mentions *daimones*, but speaks about unjust (men?) and responsibility. In column four, the author evokes Heraclitus' dictum about the sun not transgressing its measures and being overseen by the Erinyes (previously known as frr. B3 and B94 DK). The most obvious link with the previous columns is the role Heraclitus allots to the Erinyes: they would find out the sun, should it try to overstep its boundaries. Column five contains a diatribe directed against the contemptible religious, moral and epistemological attitudes of the ignorant populace, and treats those sources of information – such as oracles and dreams – that should give warning for the errant. Finally, column six seems to weave the different threads together: it speaks about the sacrificial rites performed by the magi, the parallel rites of the initiates, as well as the eschatological rationale of these ritual activities, involving souls, daimones and Eumenides.

This thematic array is impressively rich, but, as far as I can judge, can easily make up a coherent whole. The most probable scenario, it seems to me, is that the ritual activities dealt with in columns 2 and 6 constitute the immediate topic of this part of the text, and the rest is invoked in order to explain these cult activities. So when we turn now to the more detailed analysis, I shall start with these two columns: besides dealing with the sacrifices, it is in these columns that the most important actors, the magi and the initiates, are mentioned.

[4] Johnston (1999).

THE RITUAL

The exact nature and religious context of the rites described are not entirely clear. Among the scanty words of column 2, we find the reasonably reconstructed expression 'funeral libations in droplets' (χ]ϙαὶ cταγόcι .[col. 2.5). This expression makes it likely that the two occurrences of cognates of τιμάω (]τιμῶcι.[col. 2.4 and τιμὰc φ[έ]ρη[ι] col. 2.6) should be taken in a ritual sense, 'to honour with gifts/sacrifices'. It is also noteworthy that the author speaks about a specific form of libation, *choai*. As Fritz Graf and Albert Henrichs have shown, this kind of offering is connected to the sphere of eschatology:[5] *choai* were most often made to the deceased at the grave. This would certainly fit well with the archaeological context of the roll, and raises the possibility that the text refers to such funerary rituals as were conducted also at the tomb at Derveni.

The topic of sacrifices comes back again in column 6:

> . . . prayers and sacrifices appease the souls, and the enchanting song of the magi is able to remove the daimones when they impede. Impeding daimones are avenging souls. This is why the magi perform the sacrifice, as if they were paying a penalty. On the offerings they pour water and milk, from which they make the libations, too. They sacrifice innumerable and many-knobbed cakes, because the souls, too, are innumerable. Initiates make the preliminary sacrifice to the Eumenides, in the same way as the magi. For the Eumenides are souls. On account of these, he who is going to sacrifice to the gods, first birdlike . . . and the . . . (they) are . . . as many as . . .[6]

The author tells us here that the magi 'pour water and milk, from which they make the libations (*choai*), too'.[7] On this basis, one can assume that the *choai* in col. 2 are of this type, too – i.e. wineless libations of a mixture of water and milk. As the verb ἐπιcπένδω indicates, the libation mentioned in col. 6 is, however, another type of libation, *sponde*. In this latter, more common type, not the whole vessel was emptied, as in the case of *choai*, but only a measured portion of the liquid was poured on a higher altar. As Tsantsanoglou notes, it is somewhat curious that the *choai* mentioned here are made in drops (cταγόcι.[).[8]

The other type of sacrifice mentioned in column 6 is in the form of sacrificial cakes. As Henrichs puts it, Greek pemmatology (the study of sacrificial cakes) 'is a complex and confusing business'.[9] It seems though that cakes with one or more 'knobs', decorative protrusions, were used

[5] Graf (1980) 217–18; Henrichs (1984). [6] Col. 6.1–14. [7] Col. 6.6–7.
[8] Tsantsanoglou (1997) 102. [9] Henrichs (1984) 260.

in the mystic cults of Demeter and Dionysos and various other, mainly chthonic divinities.[10]

There may be a third type of sacrifice in the text. Even though the reading is quite uncertain in both cases, the expression ὀρνίθειον seems to occur in col. 2.7 and, with an even more hypothetical supplement, in col. 6.11. The Greek word can mean 'of a bird', 'belonging to a bird' or, alternatively, 'birdlike'. The difference between the first two and the third meaning is that the first two refer to some part of an actual bird whereas the third refers to something resembling a bird. Tsantsanoglou has suggested that the author is speaking about some kind of poultry offered sacrificially. Tsantsanoglou takes the pronoun ἑκάcτο[ι]c, 'to each' or 'for each', to refer to the recipients of the sacrifice of poultry. He thinks, rightly, that it is unlikely that a bird should be offered 'to each of the gods', and hence suggests that the recipients are the dead.[11] Although not impossible, I do not find this hypothesis compelling. As a matter of fact, ἑκάcτο[ι]c does not necessarily refer to the ṛecipients, and one can envisage other grammatical constructions as well. Moreover, the τι, with the meaning 'some', 'something' or 'any', following ὀρ[ν]ίθ̣ε̣ιόν is remarkable. I wonder if there are any parallels for a description of a sacrificial requirement so loose that *any* kind of birdlike creature is suitable for the assumed ritual. Moreover, the expression makes it possible, or even probable, that the sacrifice is not a bird as such, but rather some part of it, some feathers for example. Or, taking the meaning 'birdlike', it is also possible that the sacrifice resembles a bird – one could think for example of cakes made in the form of birds.

Whatever ritual activity the fragment refers to, the stress, I would maintain, is on the nature or appearance either of the Erinyes and Eumenides or the souls (or both) mentioned in the previous lines: they are birdlike. Such a conception is well documented by textual and pictorial representations, and does not need any special pleading. If so, the τι would certainly be in place; the Erinyes or souls cannot be identified with any specific bird, but are 'birdlike'. Quite notably, at both occurrences of the word 'birdlike' in the papyrus the souls and the Erinyes or Eumenides are mentioned in the neighbouring lines.

I have to admit at the same time that my reluctance towards accepting the sacrifice of 'some kind of bird', as suggested by Tsantsanoglou, is not

[10] Clement *Protr.* 2.22.4, who uses the same word, πολυόμφαλα πόπανα, 'many-knobbed cakes', that we find in the papyrus, and mentions it in connection with the contents of the mystic basket in the cult of Dionysos. See also Tsantsanoglou (1997) 114–15, referring also to possible Iranian parallels.

[11] Tsantsanoglou (1997) 102.

simply textual, for much hinges on this question concerning the religious outlook of the text as a whole. The prescription of animal sacrifice would be quite unwelcome in an Orphic context.

In the first lines of the same column, the Erinyes appear twice (Ἐ̣ριν[υ col. 2.3 and Ἐ̣ρινύω̣[ν col. 2.4). The extant text does not let us see what the relationship between the Erinyes and the sacrifices was, but one reasonable guess is that the Erinyes were specified as the recipients of the offerings.[12] But, notably, the recipients of these sacrifices are not the chthonic or other gods. As Tsantsanoglou rightly suggests, the fact that (at least in the remaining part of the text) the gods of the traditional pantheon are not specified as recipients of sacrifices, and even only mentioned as belonging to the belief system of others (col. 3.7: 'are called assistants of the gods'), is most probably related to the author's theology expressed in the commentary on the Orphic poem in the second part of the text.[13] Let it suffice to say now, without anticipating too much, that in the second part of the papyrus the author collapses the divine beings mentioned in the Orphic poem into the figure of one divine cosmic agent.[14] It is highly significant, then, that the explanation of rituals in the first columns does not refer to the traditional gods, and hence is consistent at this important point with the theology expounded in the second part of the papyrus.

THE MAGI AND THE INIṬIATES

Since the magi made their appearance in column 6 with the new readings of Tsantsanoglou, there has been some discussion on the question of whether the sacrifices described in this column represent Greek or Persian cult practices. Tsantsanoglou thinks that the reference is to Persian magi, hence the cult should also be Persian.[15] Burkert, for his part, accepts that the reference is to Persian magi, but maintains that the author speaks about two parallel but distinct systems, that of the magi, on the one hand, and that of the 'initiates', on the other. Burkert also stresses that the relation between these two systems is analogy.[16] So, who are these magi and who are the initiates? And further, how are they related to each other and to the Derveni author?

Let us start with the magi. The term μάγοc had a double sense in Greek usage. The word could refer either to a specific Persian priestly caste, experts on religious matters, or to people who were supposed to be able to perform

[12] So Tsantsanoglou (1997) 102–3. [13] Tsantsanoglou (1997) 99.
[14] For a detailed analysis of the system of identifications, see chapter 5 below.
[15] Tsantsanoglou (1997) 110–15. [16] Burkert (1999a) 106.

some kind of wizardry or black magic.[17] This latter usage practically always carries a negative connotation. In our text, by contrast, the reference is no doubt positive. Does it necessarily follow, then, that the reference in the papyrus is to Persian magi? At any rate, this is the reason why Tsantsanoglou takes the reference to be to the Persian magi and Persian religious practices.[18]

We receive the first set of information about the Persian magi from Herodotus. The functions Herodotus assigns to the Persian magi correspond remarkably well to the activities the author discusses in the first columns of the papyrus. Herodotus speaks about the role the magi play in sacrifices, divination (interpretation of dreams and other forms of oracular activity), and funerary rituals.[19] The first two of these are mentioned explicitly in the papyrus, and the presence of the third is highly probable, while its relevance is indubitable.

There is a further text that might be recalled at this point. When Diogenes Laertius discusses the lore and activity of the magi in the first book of his work, he reports, following Sotion, that 'they practise divination and predict the future, declaring also that the gods visually appear to them, and that the air is full of *eidola* which enter, down with a stream caused by exhalations, into the eyes of those who have sharp sight' (D.L. 1.7). I have left the term *eidola* untranslated intentionally. Most translators render it as 'shapes' (Hicks) or even '*simulacres*' (R. Goulet), according to its technical use in philosophical texts. If this were the correct sense of the phrase, the magi would advocate a theory of perception reminiscent of Epicurus. I doubt, however, if this is the right rendering of the text. It should be remembered that the primary meaning of the term εἴδωλον is 'phantom' or 'ghost', i.e. the insubstantial soul of the dead. This usage, attested from Homer to Plato, is much more appropriate in the context, and makes good sense: the magi claim to be able to perceive the gods and the souls of the dead, and this capacity makes them able to engage in oracular activities. If so, this description would also point to a strong concern about the souls of the dead, together with divination, that we see in column 6.

Yet a reference to the Persian magi in the papyrus might seem surprising. Why should our author go outside Greece in the interpretation of these cult practices? One possible answer is that the Persian magi, at least in

[17] See especially the perspicacious analysis in Graf (1994) 31–45.

[18] Tsantsanoglou (1997) 110 n. 25: 'But as I was unable to find the term used technically and not pejoratively of Greek religious practitioners, I limited my search to the Iranian area.' He notes, however, that 'I cannot rule out the possibility that Greek *magi* did exist as a serious religious profession before the term came to mean 'magician, impostor, charlatan'.'

[19] Sacrifices 1.132, 140, 7.43, 113, 191; interpretation of dreams 1.107, 119, 128, 7.19; funerary rites 1.140.

some contexts, were treated as outstanding experts on ritual activities; this is the view of Xenophon, for example, who refers to them as 'religious professionals' (οἱ περὶ τοὺϲ θεοὺϲ τεχνῖται).[20] We can suppose that the Derveni author, and the Orphic initiators in general, would likewise assume towards their activities an attitude characteristic of a profession or a craft.[21] If so, as Tsantsanoglou has suggested, the author can bestow further authority on the cult practices of the 'initiates' by comparing them to those of the Persian magi.

But what about the other, non-ethnographic usage of the term μάγοϲ? The first thing to note in this respect is that it belongs to a group of designations and is very often used in conjunction with other members of that group, such as 'begging priest', 'diviner' and 'purifier'. Plato draws a memorable image of the itinerant religious experts of cathartic and initiatic rites in the famous passage of book two of the *Republic*, where he calls them ἀγύρται and μάντειϲ.[22] The activities of these priestly figures come up in the *Laws*, where the Athenian proposes to legislate against their practices. According to Plato, these people claim to be able to manipulate the souls of the living and the dead with the help of sacrifices, prayers and incantations, and to persuade the gods for their own purposes.[23] The terms in which Plato describes what these people are doing are the very same that we find in the papyrus describing the activities of the magi – although, of course, Plato's depiction is negative, whereas that in the papyrus is positive.

We find another clearly negative portrayal of the methods practised by these people in the Hippocratic treatise *On the Sacred Disease*. The author of this text criticises the magi, purifiers, begging priests and quacks (μάγοι τε καὶ καθάρται καὶ ἀγύρται καὶ ἀλαζόνεϲ) who claim a superior knowledge and great piety, and who insist that they are able to cure the ill by means of purifications and incantations. The author tries to show that the purported piety of these people is impious, and that their self-proclaimed knowledge is imposture.

[20] Cf. Xen., *Cyr.* 8.3.11 and Eur. *Orest.* 1495 (μάγων τέχναιϲ). See also Plato, *Alc.* 1. 122a1–3. The claim in the *Alcibiades* that *mageia* is something teachable may very well mean in the Platonic context that it can qualify as expert knowledge.

[21] On this point, see Burkert's seminal essay 'Craft versus sect' (Burkert (1982)), with col. 20 of the Derveni text.

[22] Pl. *Rep.* 364b5.

[23] Pl. *Leg.* 909a8–b7: ὅϲοι δ' ἂν θηριώδειϲ γένωνται πρὸϲ τῷ θεοὺϲ μὴ νομίζειν ἢ ἀμελεῖϲ ἢ παραιτητοὺϲ εἶναι, καταφρονοῦντεϲ δὲ τῶν ἀνθρώπων ψυχαγωγῶϲι μὲν πολλοὺϲ τῶν ζώντων, τοὺϲ δὲ τεθνεῶταϲ φάϲκοντεϲ ψυχαγωγεῖν καὶ θεοὺϲ ὑπιϲχνούμενοι πείθειν, ὡϲ θυϲίαιϲ τε καὶ εὐχαῖϲ καὶ ἐπῳδαῖϲ γοητεύοντεϲ, ἰδιώταϲ τε καὶ ὅλαϲ οἰκίαϲ καὶ πόλειϲ χρημάτων χάριν ἐπιχειρῶϲιν κατ' ἄκραϲ ἐξαιρεῖν, τούτων δὲ ὃϲ ἂν ὀφλὼν εἶναι δόξῃ κτλ.

The text that may be the earliest attested occurrence of the word μάγος in Greek literature can add something even more notable to the above. Fragment B14 DK of Heraclitus, as preserved by Clemens of Alexandria, runs as follows: 'Against whom does Heraclitus of Ephesus prophesy? Night-wanderers, magi, bacchants, maenads, initiates.'[24] It is true that some interpreters doubt the authenticity of this enumeration.[25] But if it is by Heraclitus, then the Derveni papyrus and this fragment can give support to each other by establishing a connection between magi and the participants of Dionysiac and other mystery cults.[26] Moreover, both fragment B14 and the Derveni papyrus mention the magi and the 'initiates' (μύσται) in a single breath. This can also be crucial for us because, contrary to the customary Athenian usage, the Heraclitean reference to the 'initiates' is unlikely to be to those who were initiated in the Eleusinian mysteries. If so, it leaves open the possibility that the 'initiates' of column 6 of the papyrus belong to an Orphic/Dionysiac cult.

From this brief survey there emerges a distinct religious and social function, which can bear several names and is associated with different but closely connected ritual activities: a religious professional who claims a certain expert knowledge in matters of sacrifice, divination, initiation, healing, eschatology and the manipulation of souls, and whose activities are manifestly external to the traditional religious life of the polis, even if he can occasionally convert a whole polis, as Plato maintains. This religious expert can be called ἀγύρτης (begging priest), μάντις (diviner), γόης (sorcerer), μάγος (magus), τελεστής (initiator), or καθαρτής (purifier). It is this complex of religious functions and activities that the first columns of the papyrus speak about. The formulations in these columns, especially the use of first person plural in column 5, also suggest that the Derveni author speaks about the art he himself practises. If this is so, we shall not be far off the mark in thinking that a contemporary would use at least some of the terms listed above to designate our author.

On this basis, it seems to me possible that when the author speaks about magi in column 6 he does not refer to Persian priests and practices

[24] 12 B14 DK = Clem. *Protr.* 19: τίσι δὲ μαντεύεται Ἡράκλειτος ὁ Ἐφέσιος; νυκτιπόλοις, μάγοις, βάκχοις, λήναις, μύσταις.

[25] So e.g. Marcovich (1967) ad loc., but see Kahn (1979) ad loc. and, most recently, Pradeau (2002) ad loc., with reference to Babut (1975).

[26] Cf. Graf (1994) 32–3, who claims that Heraclitus' magi are itinerant priests or diviners, experts of private rites, like the ones mentioned by Plato in *Rep.* 2 (364b) and by the Derveni author in col. 20. It is all the more notable as Graf wrote these lines without knowledge of col. 6 of the papyrus, which was still unavailable at that moment. Graf also calls attention to Sophocles, *OT* 387–8, where these denominations, offices and functions are treated as belonging together.

(as Tsantsanoglou and Burkert take it), but speaks about the group that would include himself. If so, the magi could designate leaders of the initiates, priests of a private religious group, and the column could thus be speaking about the parallel actions of magi and initiates within the same cult. Such an interpretation can find good external support in Heraclitus fr. B14 quoted above. It is rendered internally plausible by the fact that the author speaks in the first person plural already in the previous column where he describes what he and his 'colleagues' do for those seeking their religious services ('for them we go into the oracular shrine to inquire for oracular answers, whether it is right etc.' col. 5.4–5). This clearly shows that he considers himself as representative of a group of religious experts. And members of this group, we have just seen, can also be called μάγος. For my part, I do not think it a real problem that the term is used pejoratively in practically all those cases where it does not refer to Persian magi. We may assume that the term μάγος in Greek started out as a neutral or even positive description and self-description, just as καθαρτής and ἀγύρτης. As a self-description, it was capable also of appropriating the authority of the Persian priests, whereas it later received negative connotations through diverse attacks starting with Heraclitus.

Moreover, the general Greek term for the activity attributed to the magi in the first lines of column 6 (manipulation of souls and daimones) is γοητεία. Now there is a very strong connection between mystery religions, and Orphism in particular, and γοητεία.[27] Indeed, some ancient sources call Orpheus a γόης, whereas Plato in *Republic* 2 explicitly attributes goetic activities to the Orphic initiators. This again can show that the reference in column 6 is not to another group, but to the group to which the Derveni author claims to belong.

The description the author gives of his involvement in oracular activities shows that he could also be called a diviner (μάντις). Now what kind of divination is this? The reference to the oracular shrine (μα]ν̣τεῖον, col. 5.4) is notable. It indicates that the activity of the Derveni author is connected to some cult centre of oracles. His insistence on the interpretation of dreams provides another side of the activities of the diviner. Dream interpretation, as Herodotus reports, was also a crucial aspect of the functions of the Persian magi, but it was, of course, part of the Greek practice as well.

The question of the identity of the 'initiates' has also caused some worries. Given that in literary texts the term μύσται is customarily used with reference to initiates of the Eleusinian cult, Henrichs tentatively suggests

[27] On this relationship, see the useful discussion in Johnston (1999) 105–11.

that the Derveni author speaks about Eleusinian initiates.[28] Yet, as Henrichs himself mentions in a footnote, there are notable exceptions to this usage.[29] First, we have fr. B14 of Heraclitus, which we have already discussed. The context makes it quite probable that Heraclitus is referring here to Bacchic, rather than Eleusinian, initiates. Another remarkable piece of evidence is the Hipponion gold leaf where μύσται and βάκχοι seem to form a single group. Tsantsanoglou calls attention to the Pherai gold leaf where the term μύστης appears again in an Orphic/Bacchic context.[30] And even more new evidence has come to light, which was still unknown to Henrichs and was not mentioned by Tsantsanoglou. Macedonian gold leaves from Pella, found in cist graves dated to the end of the fourth century (thus contemporaneous with the Derveni tombs), also contain the word μύστης. As Matthew W. Dickie argues, the gold leaf of Posidippus of Pella (possibly identical with the epigrammatist, or someone from his family) shows that he was an initiate of a Dionysiac cult.[31] Hence the term μύστης on that leaf does not indicate initiation into the Eleusinian mysteries either. These occurrences clearly show that there is no reason to restrict the application of the term to the Eleusinian initiates in the Derveni papyrus.

THE INTERPRETATION OF RITUAL

Thus far I have been discussing the *realia*, the cult practices and religious and social identities of the actors. This is only one side of the text, however. For the author does not merely describe the relevant cult activities, but lays much stress on explaining them. As we can see most clearly in column 6, he provides an explanatory account of the different sacrifices. The interpretation of ritual action can be described in the general formula: 'Actor *a* performs ritual action *R* because *E*'. As Albert Henrichs has shown, the content of the explanatory account *E* could take two substantially different forms.[32] The traditional and much more widely practised form was to provide an aetiological explanation in which a mythical narrative presents the origin, and thus the 'historical' rationale, of the cult practice. In the alternative form, we receive a rationalising account, which should also explain the efficacy of the action. In such an account, the explanation is given not via the mythical antecedents (or mythical paradigm) of the action, but rather by listing those factors and forces that are at work in the actual current performance. Henrichs calls this the 'symbolic interpretation' of ritual.

[28] Henrichs (1984). [29] Henrichs (1984) 267 n. 48. [30] Tsantsanoglou (1997) 116–17.
[31] See Dickie (1995). See also Rossi (1996). [32] Henrichs (1998) 45.

It is unfortunate that even though the fragments of columns 2, 3 and 6 lead us to assume that many aspects of the ritual received an interpretation by the Derveni author, the one explanatory account of a specific ritual action preserved in full in the extant text – which is at the same time Henrichs' only actual example of 'symbolic interpretation' – is somewhat ambiguous. When the author says that 'They [sc. the magi] sacrifice innumerable and many-knobbed cakes, because the souls, too, are innumerable' (col. 6.7–8), it is not clear what type of explanation we get. In other words, what is the assumed connection between the indefinitely large number of cakes and the indefinitely large number of souls? One possible answer is that this relationship is analogy, and as anthropologists have shown, analogy is the principal explanatory tool in accounting for the alleged efficacy of magic.[33] If so, one may speak about a symbolic interpretation – even if the relationship between analogy and symbolism is a more complicated issue, and lies outside the scope of this study.

I am not entirely sure, however, that analogy and symbolism are at play here. It could be that the magi sacrifice such a large quantity of cakes not because the number of offerings symbolically represents the number of souls, but because they are trying to give each of the impeding, angry souls its share. If so, there is no need to speak about analogy and symbolism; it is straightforward realism. It, of course, remains an open question what the souls, in the author's view, are supposed to do with the cakes – but that remains a problem on either interpretation.

The sum of the first six columns, despite the deplorable state of the fragments, indicates that the author has a quite elaborate doctrine about souls, daimones, Erinyes and Eumenides, which were thought of as no less real factors of the world than the heavenly bodies and the elements discussed in the second part of the text. Indeed, the true challenge for the author, and the ultimate aim of the Derveni text as a whole, as I take it, is to come up with an account in which both the souls, daimones etc., and the elements and heavenly bodies find their place in a unified, consistent explanatory theory. I shall venture to formulate a hypothesis about the possible content of such an overall account comprising cosmology and eschatology in chapter 9.

It is reasonable to assume that the ritual activities described by the author were actual cult practices, and not merely the author's inventions. On the other hand, it is an open question whether the belief system on which the ritual action was based, or its individual constitutive elements, were

[33] On the role of analogy in magic, see e.g. Tambiah (1973).

traditional, local or otherwise. It is true that in column 6 the author seems to attribute the same explanatory account to the participants in general. The magi and the initiates appear to possess a body of knowledge about the nature and behaviour of souls, daimones and Eumenides, and this is why they offer the appropriate sacrifices in the proper order. They are presented as knowing, for example, that 'the daimones are avenging souls', and that this is why there is need for a sacrifice as a sort of 'penalty'. It may very well be the case, however, that the attribution of this knowledge is fictitious, and hence that the interpretation of the rite is based, at least to some extent, on the author's personal ideas. As we shall see in the next chapters, the author shows a certain degree of inventiveness in moulding the influences received from different sources into a novel and reasonably consistent theory in the physical interpretation of the Orphic poem; he could also be employing some degree of originality in developing the eschatological aspect of his belief system.

SOULS, DAIMONES, ERINYES AND EUMENIDES

As the main elements of the author's eschatological theory forming the basis of his explanatory account of ritual are the souls, the daimones, the Erinyes and the Eumenides, his principal task is to characterise them and to establish some functional and other relations between them. This is what we have to try to reconstruct now.

The first column offers practically no help in this respect; it can only show (provided we accept Tsantsanoglou's supplement Ἐρι]ν̣ύ̣ων) that the author has already started to speak about the Erinyes. The second column, although still very fragmented, is more helpful. The discussion of the Erinyes is clearly combined with the discussion of sacrifice. This consideration finds further support in the expression at the bottom part of the column that something is 'fitted (or adaptable) to music'. It may very well refer to the hymns or incantations accompanying the ritual.[34] It is important to note that there was a very strong association between rituals evoking the souls of the dead and music: singing was a crucial part of communicating with the dwellers of the underworld, as well as of the techniques to manipulate them.[35] This last consideration makes it highly probable, once again, that the account of the Erinyes forms part of the interpretation of the ritual.

Souls (ψυ[χαί) are also mentioned in line 5, and we have every reason to assume that these are the souls of the dead. This is reinforced by the fact that

[34] So Tsantsanoglou (1997) 104–5. [35] On this point, see Johnston (1999) 111–18.

in the same line we have the mention of *choai*, and, as we have seen above, *choai* were most often offered to the dead. Tsantsanoglou suggests that the souls are identified here with the Erinyes – even though the Erinyes could also refer to the agents of the dead, without being identified with them. The identification of souls and the Erinyes certainly makes sense, but the text is too fragmented to make any definite claim. Also, we should bear in mind that, as Johnston has recently shown, 'there is no good indication that the Erinyes were considered to *be* the souls of the dead in popular belief'.[36] Moreover, it will be important to see whether the assumed identification of Erinyes and souls in the Derveni text is restricted to some souls, for example to those who died a violent death (βιαιοθάνατοι), as Erwin Rohde once maintained (without, of course, knowledge of the Derveni text).[37] Yet, as Tsantsanoglou reasonably suggests, if the identification is qualified, the restriction is more probably made on ethico-religious grounds, and singles out the souls of the 'righteous' and the initiates. On this reading only the souls of the 'righteous' (the content of this term remains to be specified) would function as Erinyes.

Tsantsanoglou does not mention an important factor that can substantially strengthen his case, namely that there was a strong association between the Erinyes and Bacchic initiates, as is amply documented in tragedy. Indeed, the Erinyes are sometimes depicted as wearing maenadic costumes, and in some texts the Erinyes explicitly call themselves maenads, while yet in other texts they are said to act like bacchants.[38] It is important to note at any rate that if Tsantsanoglou's suggestion is correct, and I think it is,[39] the Erinyes are depicted in our text as principally positive agents.

Column 3 supplies the third key concept: daimones. As is well known, the semantic field of the word δαίμων is distressingly wide in Greek usage. In view of both the general thrust of the first six columns and the author's theology as developed in coll. 7–26, we can safely exclude the possibility that the term refers here to the gods of the traditional pantheon. The phrase

[36] Johnston (1999) 274 (her emphasis). [37] Rohde (1925) 178–80 and 269–71.

[38] Aesch. *Th.* 699: μελάναιγις; Aesch. *Eum.* 500: μαινάδων τῶνδε; Eur. *Or.* 411: βακχεύουσι, 835: βεβάκχευται. For an analysis of the evidence, with further references, see Seaford (1993) and Johnston (1999) 253–6.

[39] See however Johnston (1999) 276 where she writes 'I do not find any reason that we should assume that the commentator (or the poet himself) equated the Erinyes with souls.' But, as far as I can judge, Johnston is not quite prepared, except in the case of the identification of the Eumenides as souls in the last lines of col. 6, to accept that it is highly characteristic of the author to depart from traditional views. But in the case of the Eumenides, where the text is unequivocal, Johnston herself alludes to the fact that the author's basic method consists in establishing identifications, sometimes very far from traditional ideas – as is clear from his exegesis of the Orphic poem. So why could he not do the same in the case of the Erinyes?

δαίμ]ω̣ν̣ γίνετα[ι ἑκά]cτωι, if Tsantsanoglou's reconstruction is correct, would reflect the belief that there is a 'guardian spirit' to each soul.[40]

The expression δ]α̣ίμονες οἱ κάτω[is a further indication of the eschatological thrust of the text, for as we are reminded in another context by Johnston, the term οἱ κάτω commonly refers to the dead.[41] Supplementing .]υ δ̣ε̣χόμ̣[ενοι (or δ̣ε̣χον̣[ται, as suggested by Janko), which I consider clearly preferable to Tsantsanoglou's το]ῦδ̣ε̣ χοῦ̣[, these daimones are supposed to receive something. It is reasonable to assume that what they receive is an offering. The .]υ preceding δ̣ε̣χόμ̣[ενοι could preserve the last letter of a genitive ending, defining the source or (with a partitive genitive) the substance of the offering.[42] But it is also possible that the daimones of the netherworld receive the soul of the recently deceased.

It is plausible that the group of the chthonic daimones is constituted by the souls of (some of) the dead. If so, one may wonder how the category of the Erinyes relates to the category of the daimones. Now, the next legible bit of the text 'and are called assistants (ὑπηρέται) of the gods' (col. 3.7) still refers, I suppose, to the daimones. This expression strengthens the possibility that the category of the Erinyes is coextensive with, or alternatively, is included in the wider category of daimones. The author's assertion 'and are called assistants of the gods' can have within its scope the Heraclitean description of the Erinyes, Dike's guardians, auxiliaries or aids (ἐπίκουροι), which we shall encounter in the next column. Moreover, Johnston calls attention to Porphyry's commentary on the *Iliad*, in which the Erinys is called assistant (ὑπηρέτης) of Hades.[43] It is important to note, however, that because of the passive construction 'are called', it is not certain that the author subscribes to the description 'assistants of the gods'; indeed, in view of his commentary on the Orphic poem we have to assume that he does not accept the plural in 'gods'.

Among the first six columns, column 6 provides the longest chunk of text uninterrupted by lacunae. It is also in this column that we find together most of the eschatological agents. The text mentions souls, daimones and the Eumenides. Here, finally, we have clear identifications. In lines 9–10 the Eumenides are plainly identified as souls. On the other hand, the first lines appear to identify also the angry daimones as souls. Here, however,

[40] So Tsantsanoglou (1997) 105. The concept is well-documented; just to quote one example, see Pl. *Phd.* 107d5–7: λέγεται δὲ οὕτως, ὡς ἄρα τελευτήσαντα ἕκαστον ὁ ἑκάστου δαίμων, ὅσπερ ζῶντα εἰλήξει κτλ.

[41] See Johnston (1999) 74 with references in nn. 114 and 115. Because of this usage, I do not find Tsantsanoglou's supplement οἱ κάτω[θεν necessary, although, of course, it is not impossible either.

[42] Janko suggests that it is a negation (ο]ὐ), which cannot be excluded either.

[43] Johnston (1999) 276–7, with reference to Porphyry ad *Il.* 9.571.

there is some uncertainty. The question depends to some extent on the supplement one accepts for the last word in δαίμονες ἐμπο[δών ὄντες εἰϲι | ψ[υχαὶ . . . .]ροί. I fully agree with Janko that the supplement τιμω]ροί 'avenging' is clearly preferable to ἐχθ]ροί finally opted for by Tsantsanoglou. It is a set expression which gives perfect sense in the context. If so, the 'impeding daimones' are also identified as souls. But, indeed, we don't even need to build everything on this supplement, for the preceding sentence, 'prayers and sacrifices appease the souls, and the enchanting song of the magi is able to remove the daimones when they impede', although less unequivocal, has the same effect.

The outcome, I suggest, is that all daimones are souls. Further, daimones can take on different functions, and hence receive different denominations. The Erinyes constitute one such functional sub-group, possibly the most important one. They may or may not be identical with the sub-group of 'impeding daimones' (δαίμονες ἐμπο[δών) of col. 6. The supplement τιμω]ροί 'avenging' would reinforce the feeling that they are indeed the same group, for Erinyes are generally thought of as avenging.

The last question concerns the identity of the Eumenides. They are explicitly said to be souls, and they no doubt qualify as daimones. The question, then, is whether they are also identical with the Erinyes when these are appeased, or whether they are a separate group. As Henrichs has shown, although the two groups were not originally identical, they could be thought of as showing contrasting aspects of the same beings by the middle of the fifth century.[44] Hence, this is an identification that the author could take from traditional ideas.

Can we say anything further about the possible ritual context of the text on the basis of what we have been able to recover from the author's eschatological beliefs? There are at least two possibilities, and both are germane to the textual and archaeological context of the papyrus. One is that the author is speaking about rites that should secure the safe passage of the soul of the dead to the underworld, and to the most blissful part of it. The mention of impeding daimones is entirely relevant in such a discussion; the avenging, angry daimones could block a safe journey to the underworld. The other possibility is that the author is speaking about initiation. Remarkably, frightful and impeding daimones are also supposed to be present at initiations. As Johnston shows, the vindictive daimones could be present at the initiation in order to frighten the initiand and possibly also to hinder the initiation process, especially if the initiand is not pure because of guilt.

[44] See Henrichs (1994).

According to some sources, they were present at initiations also to show what happens to the uninitiated in the underworld.[45] The presence of the initiates mentioned in col. 6 can allow two interpretations according to the option we take. It can either refer to the initiands as joining the group of initiates, or it can designate the thiasos, the group of initiates, as they participate in the funerary rite of one of their fellow members.

But, of course, these two ritual contexts, initiation and funerary ritual, are closely connected. The initiation prepares the blissful post mortem existence of the soul, whereas at death the individual so to speak cashes in the advantages gained by initiation, and the funerary rite is supposed to guarantee this privileged status by reminding the powers of the underworld that the person is an initiate.

UNDERSTANDING AND THE RIGHT BEHAVIOUR

A further significant aspect of column 3 is its ethical perspective. Although the text is too fragmented to allow any more specific hypotheses, the expressions '(they) are, like unjust men . . . and they are responsible' in lines 7 and 8, indicate clearly that the thrust of the argument is ethical.[46] We can guess that the author is speaking now about the disciplining role of the daimones and/or Erinyes. This suggestion is made even more plausible by the topic of the next column. In the first lines of column 4 the author appears to carry on the discussion of the penalisation of evil behaviour, to continue with the claim that the same rules apply not only in the sphere of humans, but also in the cosmos as a whole: 'Is it not on account of these that the cosmos possesses order?' (col. 4.4). This is the claim that occasions the quote from Heraclitus, in which the Erinyes, described as auxiliaries of Dike (Justice), are said to capture the Sun should it transgress its boundaries. The last legible words of the column ('. . . would make a transgression . . . of justice . . .') indicate that the discussion is still within the sphere of improper behaviour-justice-punishment.[47]

In column 5 the author seems to elaborate further on the same subject by bringing in different signs that should make people recognise what they should and what they should not do. First, there is oracular activity.

[45] Johnston (1999) 130–9. Especially interesting is the figure of Empousa – from the same stem as ἐμποδών 'impeding' used of the daimones in col. 6 – who appears also in Aristophanes' *Frogs* 293 where she tries to frighten and obstruct the passage of the initiates in the underworld.

[46] It is true, however, that the subjects of the sentence are compared with (ὅπωϲπερ), and not identified with, unjust men.

[47] I shall discuss the place of the Heraclitus quote more in detail in chapter 8.

Remarkably, the author avers that he and his colleagues can function as mediators in the communication with the divine, as they can turn to the oracle on behalf of the layperson to inquire whether a given action is sanctioned by the divinity.[48] Other possible indications mentioned here are dreams. Both are traditionally thought of as ways in which the godhead could communicate signs to humans. As we have seen above, interpretation of dreams was also part of the activities both of the Persian magi and the Greek initiators. It is an interesting question, to which the extant text cannot offer any answer, whether the author had any theory on the operation of oracles and dreams. In particular, it would be exciting to see if he had an account that could complement his natural theology, expressed in physical terms, as it is elaborated in the second half of the papyrus. It has to be remarked that a physicalist account is not impossible in principle, as Plato's *Timaeus* and some Hippocratic texts show.[49]

The reason the author offers for the alleged fact that people are unheedful of divine signs is remarkable: it is their failing ethical standards that veil their view. As the author puts it, they are 'overcome by fault and by pleasure as well' (col. 5.8–9). Thus, it is not that they do not recognise the divine indication, and henceforth lack moral guidance, but rather that they are unable to recognise divine warning and warrant because of their erring conduct. Hedonism and wrongful behaviour are then also an obstacle for a correct understanding and appreciation of, and trust in, divine indications. Moral betterment is a precondition both for piety and gain in knowledge about the divine. This assertion, I suggest, can also be taken as an intellectualised interpretation of the need for purification before initiation.

The author explicitly claims in the last preserved lines of col. 5 that lack of faith and lack of knowledge are two sides of the same coin. As one gains knowledge about the divine, that is as one understands the way the divinity governs the world, and hence takes up the correct cognitive attitude towards it, one naturally, by the very same gesture, assumes faith in it as well. Piety and comprehension of the nature and functioning of the god are two aspects of the same state of mind.[50]

This point already explains why the author finds it so important not only to describe all the relevant rites, but also to accompany them with an

[48] For my part, I find Janko's supplement θέμι[ϲ ἀπ]ịϲṭẹịṿ [τὰ unlikely. It would be quite paradoxical to inquire of an oracle whether there is divine sanction not to believe.

[49] Pl. *Tim.* 71a–72c; *De victu* 4; *De hebdom.* ch. 45. Aristotle in *Div. somn.*, on the other hand, maintains that dreams are not sent by god. His main argument is that it is not the wisest people who have such dreams, and, surely, god would send such signs to the wisest and the best.

[50] We encounter the same view in the tenth book of Plato's *Laws*. In the same book the Athenian proposes to legislate against the magi.

elaborate explanatory account. And the same point also elucidates why he finds it crucial to provide a meticulous interpretation of Orpheus' poem, which is treated by him as a principal source of knowledge about the nature and workings of the divine. Thus, understanding and interpretation are the key concepts in the author's religious attitude. This is a topic I leave for the time being, but I shall return to it in the concluding chapter of this book, after having discussed also the second part of the text.

CHAPTER 3

The reconstruction of the poem

From the second half of col. 7 onwards and up till the end of the extant text, the Derveni author quotes and discusses hexametric verses. According to the near-consensus of the literature, most of these verses belong to an Orphic poem. The poem is usually described as a theogony. Moreover, the fact that the width of the individual columns was set to correspond to the length of a verse in hexameter suggests that the discussion of such verses was one of the main purposes of the entire text.[1]

The reconstruction of the poem, however, has proved to be notoriously difficult, and a number of crucial questions are still far from being settled. It seems obvious, on the other hand, that for a balanced evaluation of the exegetical methods as well as the theological and cosmological ideas of the Derveni author we have to have, as far as possible, a clear grasp of the poem commented on. As a matter of fact, the considerable variance between the particular reconstructions of the poem suggested by different scholars could, and indeed does, result in significantly different accounts of the Derveni author's own contribution, and, vice versa, different assessments of the author's exegetical method have instigated different reconstructions of the form and content of the poem.

Without in any way pretending to be able to settle once and for all the vexed questions of the reconstruction, I shall now give an overview of the problems related to it as well as the different possible solutions proposed in the literature. I shall also try to argue for those solutions which I find more probable myself. When I think, however, that the arguments for the solutions I prefer are not strong enough completely to discredit alternative views, I shall not ignore these other views in my analysis of the Derveni author's own contribution. All in all, the main purpose of this survey is to make clear what points we can take for granted, what can be considered probable, and what is hardly more than conjecture. This procedure seems

[1] So, e.g., Obbink (1997) 44 n. 9.

necessary, for some points originally suggested as mere hypotheses have subsequently received the status of the probable, or even worse, have been treated as facts, without any additional arguments adduced.

The basic differences in the reconstructions, as will become clear in the course of my argument, result from the different policies adopted in the assessment of the evidence of the papyrus. Most commentators do not have much confidence in the Derveni author.[2] In their view, the Orphic poem just serves him a prop which he can alter or distort in many ways in order to adapt it to his own fanciful theories. This representation is not, at least prima facie, easy to disprove. The text really does contain some very surprising assertions, and the exegetical methods used by the Derveni author rarely conform to our standards. Now, the trouble lies in the fact that, in view of such a negative picture of the author, scholars grant preference to external evidence in the reconstruction of the Orphic poem, i.e., they try to restore the poem used by the Derveni author on the basis of other pieces of Orphic literature.

It is not difficult to see, however, that such a procedure creates a vicious circle. What we can gain from the papyrus differs, in some salient points, from other poems transmitted under the name of Orpheus. Therefore, one can claim, it is obvious that the Derveni author misconstrued the poem in his tendentious exegesis, so we have to restore it in such a way that it gets closer to other Orphic texts. Then, once we have accommodated the poem used in the papyrus to other Orphic texts, it becomes evident that the Derveni author badly distorted the text he was commenting on. And since the author of the papyrus is so unfaithful to the poem, we have every right to take it out of his hands and reconstruct it on the basis of other sources available to us rather than on the basis of the papyrus itself.[3]

I would espouse a different policy, based on the belief that the Derveni author, almost the only, but certainly the earliest surviving representative of a particular exegetic form, does have the right to the presumption of innocence in face of the jury of modern scholarship, if not as regards the contents of his interpretative remarks, at least concerning the way he treats

[2] One could make a nice collection of the depreciatory remarks, e.g. 'The anomaly may arise from a misinterpretation by the commentator, who is in general the least trustworthy of guides' West (1983) 88; 'our poor commentator' Rusten (1985) 139, etc.

[3] For a recent example for this procedure, see Casadesús (1996). A notable exception is the short but, as I regard it, exemplary treatment of the evidence of the papyrus in KRS 30–3. Although KRS do not adduce arguments, the outcome of my lengthy critical scrutiny turns out to be very close to the reading suggested there.

his source.[4] All the more so since, as we shall see in the next chapter, the available comparative evidence of other Orphic theogonic narratives used by other scholars is neither homogeneous in itself nor close to our text in time.

I shall argue, then, that we should give preference to the internal evidence in the reconstruction of the poem in all those cases where this procedure does not result in untenable consequences. True, this is not a strict criterion either, since the borderline between tenable and untenable can be rather fuzzy. The same result can be 'slightly unusual' for some and 'flagrant absurdity' for others. For my part, I shall leave a relatively large margin for the unusual. Accordingly, I shall begin the reconstruction of the poem by a close reading of the papyrus itself and shall then proceed to the evidence of other Orphic theogonic narratives in the next chapter.

THE LEMMATA

We can recognise the verse quotations on the basis of three considerations which, in most cases, reinforce each other. Most obviously, we have the distinctive syllabic pattern of the hexametric verses. Second, the Derveni author uses a set of idiomatic expressions to introduce a new quotation. Of these, we have the following examples in the surviving text:

. . δ]ε̣δήλω[ται ἐν τῶι]δ̣ε τῶι ἔπ̣[ει]·	8.1
ἐν τῶ[ιδε δη]λοῖ·	8.3
τὰ δ' [ἐπὶ τούτ]ω̣ι λέ̣γει·	11.9
τὸ δ' ἐχόμε̣[νον ἔ]π̣οc ὧδ' ἔχει·	12.1
ἀλλὰ δηλοῖ ὧδε λέγων·	13.3
τὸ δ̣' ἐπὶ τούτωι·	14.5
ἐχόμενον δὲ ἔποc·	15.5
cημαίν]ε̣ι δὲ [τ]όδ̣ε̣·	15.12
[καὶ ἐν τῶι ἐχ]ομένωι | [ἔπει τούτ]ου	16.12–13
cη]μαίνει δ' ἐν τοῖc ἔπεcι το̣[ῖcδε]·	17.11
λέγων ὧδε·	19.9
τὸ δ' ἐχόμενον·	23.10
δηλοῖ δὲ τόδε·	24.2
ἐν δὲ [τ]ῶιδε cημαί[ν]ε[ι]·	25.13
δηλοῖ δὲ̣ καὶ ἐν τοῖcδε̣ τοῖc ἔπε̣cιν | ὅτι κτλ.	26.2–3
δηλοῖ δὲ καὶ ἐν τ[ῶ]ιδε·	26.5

According to the analysis of Lamedica, all of these expressions belong to the technical vocabulary of scholia and other commentaries, or at least appear

[4] No matter how far-fetched, and indeed deliberately comic, the interpretation of Simonides' poem in *Protag.* 339a–347a is, we do not normally think that Plato wilfully misquoted the text itself. The flexibility of the interpretative procedure makes the tendentious misquotation of a poem unnecessary.

in such texts in similar contexts.[5] Anyhow, the elliptical form of many of the expressions in the papyrus itself suggests, if not a strictly technical, at least a well-established usage. It is noteworthy, however, that the expressions are not used mechanically, but on the contrary, almost always have slight variations in their phrasing.

Apart from the syllabic patterns and the introductory expressions, the identification of the lemmata is made easier by the fact that the Derveni author, when he quotes at least an entire verse, always starts the quotation on a new line.[6] We have to note, however, that the lines immediately preceding the lemmata do not, in most cases, contain either an unusually low or an unusually high number of characters. The normal number of letters per line in the entire papyrus varies between 30 and 45 letters. In the lines preceding the lemmata we have two cases where the lines contain slightly fewer characters (24 characters in col. 8.3 and 26 in col. 15.5) and two cases where they contain considerably fewer characters (17 characters in col. 26.3 and 5). It is also worth noting that the latter, as we shall see, are two of the three cases in the extant text where the Derveni author introduces hexametric lines that might not stem from the poem commented on. Furthermore, there are lines which apparently do not precede a quotation, and are still significantly shorter, usually at the end of a sentence.[7] On the other hand, we do not find lines with an uncommonly high number of letters; this could have otherwise suggested that the author preferred squeezing more letters into one line to starting a new line with just one or two words in it.[8] Thus, although the number of letters over or below the average in a line can occasionally strengthen the previous factors, it cannot in any case be conclusive in itself.

Paragraphoi may be used as a means of indicating quotations. Indeed, we find that the scribe marks the full-verse lemmata with a pair of *paragraphoi*, one placed before and the other after the quotation. A complete record of *paragraphoi* has not been published yet,[9] but the available evidence already suggests that *paragraphoi* cannot be treated as decisive in the determination

[5] Lamedica (1990) with parallels and a comparative table on p. 91. [6] Obbink (1997) 44 n. 9.

[7] Col. 10.10 (22 characters); 11.7 (11 characters). Incidentally, this consideration also shows that the fact that line 10 in col. 20 contains slightly fewer characters than the surrounding lines (34 letters as against 38–39 letters) cannot prove that a prose quotation starts in the next line, as was proposed by Rusten (1985) 138–40, especially since a sentence ends at the end of line 10. See also Obbink (1997) 44–5.

[8] The numbers of characters in the lines immediately preceding the lemmata are the following: coll. 8.1: ±33; 8.3: 25; 11.9: ±38; 12.1: 30; 13.3: 39; 14.5: 36; 15.5: 26; 15.12: ±30; 16.12: 36; 17.11: ±41; 19.9: 38; 23.10: 29; 24.2: 42; 25.13: 44; 26.3: 17; 26.5: 17. (± means either that there is a lacuna without a conjecture, or that the conjecture proposed is uncertain.)

[9] The *ZPE* transcription does not indicate them, while Kapsomenos' edition from 1964, which does indicate them, is far from complete. Rusten (1985) also indicates *paragraphoi* for the passages treated. I have added some to this record on the basis of autopsy.

of lemmata: the scribe's practice is not entirely consistent, or at least the rule for the use of *paragraphoi* is not immediately obvious. For instance, although most of the quotations are marked by a pair of *paragraphoi*, I could see no trace of a *paragraphos* after the full-verse lemma in col. 13.1. Again, there is a *paragraphos* after the full-verse lemma in col. 12.2, but none preceding it. Moreover, *paragraphoi* can indicate simply a full stop in the author's own text, as for example the ones after col. 10.10, col. 11.7 and col. 13.6. Funghi (1997) 26 reports that *paragraphoi* indicating punctuation are accompanied by a dash; were this the case the two different uses of *paragraphoi* could be distinguished. Yet I could see such dashes, inside the line, functioning with a *paragraphos* as a full stop, only in col. 23.7, and perhaps in col. 22.11. All in all, it seems wiser to leave out the evidence of *paragraphoi* in the determination of lemmata, at least in the current editorial situation.

On the other hand, the previously mentioned factors can be decisive in some cases. There is, for example, at least one case in which a line is too lacunose to display the syllabic pattern and where it is the introductory formula (ἐν δὲ [τ]ῶιδε cημαί[ν]ε[ι]), reinforced by the relatively high number of characters in the preceding line (44), that makes us believe that there must be a quotation there (col. 25.13–14):

ἐν δὲ [τ]ῶιδε cημαί[ν]ε[ι]·
[αὐτ]ὰ̣ρ̣ [ἐ]π̣εὶ δ[ὴ πάν]τα Διὸ[c φρὴμ μή]cατ̣[ο ἔ]ρ̣γα

This must be the reason why both West and Bernabé, even before the new reading of the line by Tsantsanoglou, treated it as a quotation from the poem.[10]

On the basis of the above criteria, we can recognise the following full-verse lemmata in the surviving part of the papyrus:[11]

***L*1**	[ο]ἳ Διὸc ἐξεγ̣ένοντο [ὑπερμεν]έοc βαcιλῆ̣οc	8.2
***L*2**	Ζεὺc μὲν ἐπεὶ δὴ π̣α̣[τρὸc ἑο]ῦ πάρα θέ[c]φατον ἀρχὴν [ἀ]λκήν τ' ἐγ χείρεccι ἔ[λ]αβ[εγ κ]α̣ὶ̣ δαίμον̣[α] κυδρόν[12]	8.4–5
***L*3**	[ἥ δὲ] ἔχρηcεν ἅπαντα τά οἱ θέ̣[μιc ἦν ἀνύcαc]θ̣αι	11.10
***L*4**	ὣc ἄρ̣ξ[ηι κα]τὰ καλὸν ἕδ̣οc νιφόεντοc Ὀλύμπου	12.2

[10] West (1983) 115 *exempli gratia* reconstruction of the 'Derveni Theogony', v. 45; F1 Bernabé.

[11] For a critical edition of the lemmata, see Bernabé's forthcoming edition of Orphic fragments. I shall indicate in my critical examination of the lemmata where I cannot follow Bernabé. I mark the full-verse lemmata by the upper case ***L*** plus a number, in order of appearance in the papyrus. I apply this numbering only in order to make references easier in the discussion.

[12] The syllabic pattern of χείρεccι ἔ[λ]αβ[εγ is problematic; for different emendations, see the apparatus to the text.

L5	Ζεὺϲ μὲν ἐπεὶ δὴ̣ π̣ατρ̣ὸϲ ἑοῦ πάρα̣ [θ]έ̣ϲφατ’ ἀκούϲα[ϲ]	13.1
L6	αἰδοῖογ κατ̣έπινεν, ὃϲ αἰθέρα ἔχθορε πρῶτοϲ	13.4
L7	Οὐρανὸϲ Εὐφρονίδηϲ, ὃϲ πρώτιϲτο̣ϲ βαϲίλευϲεν	14.6
L8	ἐκ τοῦ δὴ Κρόνοϲ α̣ὖτιϲ, ἔπειτα δ̣ὲ μητίετα Ζεύϲ	15.6
L9	μῆτιγ κα[ὶ μακάρων κατέχ]ω̣μ βαϲιληίδα τιμ̣[ήν]	15.13
L10	πρωτογόνου βαϲιλέωϲ αἰδοίου, τῶι δ’ ἄρα πάντεϲ ἀθάνατ̣οι προϲέφυμ μάκαρεϲ θεοὶ ἠ̣δ̣ὲ θέαιναι καὶ ποταμοὶ καὶ κρῆναι ἐπήρατοι ἄλλα τε πάντα, ἅ̣ϲϲα τότ’ ἦγ γεγαῶτ’, αὐτὸϲ δὲ ἄρα μοῦ̣νοϲ ἔγεντο	16.3–6
L11	[νῦν δ’ ἐϲτὶ]ν βαϲιλεὺ[ϲ] πάν̣τ̣[ων, καί τ’ ἔϲϲετ’ ἔπ]ε̣ιτα	16.14
L12	Ζεὺϲ κεφα[λή, Ζεὺϲ μέϲ]ϲ̣α, Διὸϲ δ’ ἐκ̣ [π]άντα τέτ[υκται]	17.12
L13	Ζεὺ̣ϲ̣ βαϲιλεὺϲ, Ζεὺϲ δ’ ἀρχὸϲ ἁπάντων ἀργικέραυνοϲ	19.10
L14	ἶναϲ δ’ ἐγκα̣[τέλε]ξ’ Ἀχελώιου ἀργυ̣[ρ]οδίνε̣[ω]	23.11
L15	ἣ πολλοῖϲ φαίνει μερόπεϲϲι ἐπ’ ἀπείρονα γαῖαν	24.3
L16	[αὐτ]ὰ̣ρ̣ [ἐ]π̣εὶ δ[ὴ πάν]τα Διὸ[ϲ φρὴμ μή]ϲατ̣[ο ἔ]ρ̣γα	25.14
L17	Ἑρμῆ Μαιάδοϲ υἱὲ διάκτορε δῶτορ ἐά̣ων	26.4
L18	δοιοὶ γὰρ τε πίθοι κατακήαται ἐν Διὸϲ οὐδεῖ δώρων οἷα διδοῦϲι, κακῶν, ἕτεροϲ δέ τ’ ἐάων	26.6–7

THE PROVENANCE OF THE LEMMATA

Our next task is to decide whether all of these verses come from the same poem, or whether the Derveni author uses multiple sources. There are two considerations which we have to take into account in answering this question. First, we have to see if the verses quoted in the papyrus appear elsewhere in ancient literature, and, second, we have to consider in what context and in what way the Derveni author introduces the lemma. By answering the first, we can detect possible quotations from other texts known to us. But even if we find that a hexametric line quoted in the papyrus is part of another canonical hexametric poem, we still have to consider the possibility that, as loans are not uncommon especially in hexametric poems, the same line is also part of the poem commented by the Derveni author. This is why it is always important to see, as far as the condition of the papyrus permits us, the way such a lemma is introduced.

As has been shown by Bernabé and others,[13] most of the phrases of the lemmata have exact or close parallels in the canonical epic poems (including the Homeric *Hymns*), and in the various hexametric poems transmitted

[13] Cf. also West (1983) 82–4; Funghi (1997) 27.

under the name of Orpheus, such as the Orphic *Hymns* and the Orphic *Rhapsodies*.

There are, however, only four lemmata which appear elsewhere in the same, or very closely resembling, form. ***L12*** appears in a hymn to Zeus quoted as Orphic by Porphyry on the authority of Eusebius,[14] and in the hymn to Zeus at the end of the Pseudo-Aristotelian *De Mundo* = *OF* 21a v. 2.[15] Furthermore, ***L13*** is identical with the *De Mundo*'s hymn to Zeus, verse 7. We have a version of it, differing only in two words, in the hymn to Zeus as quoted by Porphyry.[16] Although it does not reproduce the entire verse, the main part of ***L4*** is identical with verse 7 of the Homeric hymn to Heracles.[17] ***L17*** and ***L18***, on the other hand, are very close to *Od.* 8.335 and *Il.* 24.527–8, respectively.[18]

As regards the contexts of the lemmata, we find that, as far as the condition of the text allows us to judge, all the lemmata – with the exception of ***L17*** and ***L18*** to which I shall return – are the direct subject of the Derveni author's exegesis. They further the commentary by introducing a new point, and in some cases they are ushered in by words that clearly indicate that the Derveni author carries on his exegesis by the ensuing quotation from the same poem (e.g. τὸ δ' ἐχόμε̣[νον ἔ]π̣οϲ ὧδ' ἔχει· col. 12.1).

We have, on the other hand, an explicit mention of *another* poem (or poems). In col. 22.11–12 the commentator refers to some *Hymns*,[19] from which he evokes a series of divine names:

> ἔϲτι δὲ καὶ ἐν τοῖϲ Ὕμνοιϲ εἰρ̣[η]μένον·
> ¯Δη̣μήτηρ [Ῥ]έα Γῆ Μή̣τ̣ηρ Ἑϲτία Δηι̣ῶι.

This enumeration of divine names is not necessarily a verbatim quotation. In contrast to the other quotations, the series of names does not appear in a separate line, and the syllabic pattern is not an obvious indication in this case.[20]

[14] *OF* 168 v. 2.

[15] ***L12*** makes Diels' emendation (substituting τέτυκται with τελεῖται), also printed by Kern, unnecessary. We find ***L12*** = *OF* 21a v. 2 with a slight variation in a scholium to Plato's *Laws* 715 e: Ζεὺϲ ἀρχή, Ζεὺϲ μέϲϲα, Διὸϲ δ' ἐκ πάντα τέτυκται. It is also not impossible that the rhythmically fitting ἀρχή, replacing κεφαλή, was governed by the Platonic text in which we read: ὁ μὲν δὴ θεόϲ, ὥϲπερ καὶ ὁ παλαιὸϲ λόγοϲ, ἀρχήν τε καὶ τελευτὴν καὶ μέϲα τῶν ὄντων ἁπάντων ἔχων κτλ.

[16] *OF* 168 v. 7 Ζεὺϲ βαϲιλεύϲ, Ζεὺϲ αὐτὸϲ ἁπάντων ἀρχιγένεθλοϲ.

[17] The line in the Homeric hymn is νῦν δ' ἤδη κατὰ καλὸν ἕδοϲ νιφόεντοϲ Ὀλύμπου.

[18] But *Od.* 8.335 begins with Ἑρμεία Διὸϲ υἱέ, whereas *Il.* 24.528 runs δώρων οἷα δίδωϲι κακῶν, ἕτεροϲ δὲ ἐάων. For the other, less close parallels, see Bernabé (forthcoming).

[19] These *Hymns* cannot be identical with the collection of poems known as the *Orphic Hymns*, composed as late as the fourth century AD.

[20] Cf. Rudhardt (1991) 271: 'Serait-ce approximativement, il cite en outre un vers qu'il tire de certains hymnes, pour étayer son argumentation'. Obbink (1994) argues on the basis of these lines that

Apart from the aforementioned reference to the *Hymns*, and the reference to Heraclitus in col. 4, the Derveni author does not name the source of his lemmata. When he speaks about the words he is commenting on, he most often refers to the poet of the line quoted by simply using the third person singular (e.g., cη]μαίνει δ' ἐν τοῖc ἔπεcι τọ[ῖcδε]· col. 17.11; λέγων ὧδε· col. 19.9 etc.). These features strongly suggest that the author is speaking about the same poet.[21]

Who is the poet then? Apart from those names which obviously refer to gods (characters of the poem itself), there is only one suitable proper name in the papyrus. In col. 18, we read the following:

> Now Orpheus named this breath Moira. But all other men according to the common usage say that Moira spun for them and that those things which the Moira has spun will be, on the one hand speaking correctly, but on the other hand not knowing either what Moira is or what spinning is. For Orpheus called wisdom Moira. This seemed to him to be the most suitable out of the names that all men have given etc.

The combination of two facts – that, first, some of the verses from the poem commented upon can only be found in poetry transmitted under the name of Orpheus,[22] and, second, that the only name in the surviving text to which such a poem can be attributed is the name of Orpheus – leads to the conclusion that the Derveni author attributed the subject of his exegesis to the mythical poet Orpheus.[23]

Philochorus (*FGrHist* 328 F 185), as quoted by Philodemus (*De piet.* 248 I. pp. 63+23 Gomperz), used the Derveni text. Philodemus says that κἀν] | ṭοῖc Ὕμνοιc δ' Ọṛφ[εὺc | π]αρὰ Φιλοχόρωι Γῆν [κ]αὶ Δήμητηρα τὴν || αὐτὴν Ἑστίαι. For Burkert's approval of Obbink's hypothesis, see Burkert (1997) 174 n. 32 and (1999a) 79. Nevertheless, I do not find Obbink's arguments entirely convincing, especially when he concludes in Obbink (1997) 49 n. 16 from the same material that 'This shows that the Derveni text was known and used by authors later in the fourth century, and may now be considered a true literary text: it was consulted by scholars and historians, and had a relatively wide audience, rather than the idle scribblings of an unknown person for merely private purposes. It was a book.' As against this, it can be the case, for instance, that the Derveni author and Philochorus drew on the same source. Moreover, given Philochorus' specific role and interest, even his assumed knowledge of the Derveni text cannot be taken as evidence for a 'relatively wide audience'. It seems to me that there are a number of possibilities between 'a true literary text' and 'the idle scribblings of an unknown person for merely private purposes'.

21 Henry (1986) 150, in her attempt to define the text as a 'commentary on multiple texts', treats, first, the term 'text' in a very wide sense, including dreams, oracles, sacrificial rites and initiatory rites. Second, she gives too much weight to Rusten's hypothesis that col. 20 is a long prose quotation from another text. Finally, she passes too lightly over the eventuality that an author can refer to other texts in support of his or her argument even if he or she is basically concentrating on one text.

22 Cf. the parallels for ***L12*** and ***L13*** above.

23 Picard's claim (1966) 739 to the effect that 'le texte paraît plus sûrement appartenir à tout ce qui groupe, peu à peu dans les séries dites *hésiodiques*, si largement dispersés au cours des siècles' is based on a very limited part of the text, provided by Kapsomenos (1963).

However, even if the above conclusion is accepted, one problematic point remains. As has already been mentioned, ***L17*** and ***L18*** have a somewhat special status. Not only are they the sole verses in the papyrus which appear in the Homeric poems in a very similar form, but they seem to be invoked not in order to pursue the exegesis, but to give additional support for an argument. This is marked by the fact that there are only three expressions introducing lemmata in which we find a καί 'also': the expression introducing the lemma from the *Hymns*, and the expressions introducing ***L17*** and ***L18***. The purpose of these latter two lemmata is to give weight to the Derveni author's somewhat surprising contention that the word ἑᾶc, which one would otherwise take to be a reflexive possessive, is, in fact, an adjective, synonymous with 'good'.[24] So he introduces ***L17*** and ***L18*** by simply saying (col. 26.2) that

> δηλοῖ δẹ̀ καὶ ἐν τοῖcδẹ τοῖc ἔπẹcιν
> ὅτι ἀγαθὴ̣ν cημαίνει.

Indeed, both lemmata contain the word ἑάων, the genitive plural of the epic ἐύc, meaning 'good' or 'brave', and phonetically not far from ἑόc. Thus, the context evinces that these two lemmata are adduced for the simple reason that both contain the word ἑάων. Also, as we shall see, these two lemmata do not easily find their place in the narrative structure of the poem. The most natural assumption would then be that the author quotes these two loci from Homer to prop up his argument.

The trouble is that the author of the papyrus uses here the verb δηλοῖ in the same construction as in those cases where, at least on our hypothesis, the implicit subject of the verb must be Orpheus. We have two options at this juncture. Either we accept that the author treats these two passages – which we know from the *Odyssey* and the *Iliad* respectively – to be of Orpheus as well, or we have to assume that the verb is used here impersonally. L&M have apparently opted for the latter solution and hence translate 'And it is also clear in the following words that it means good'. Yet, apart from the fact that they translate the verb personally in all the other occurrences, we should also note that the reference to the impersonal usage of δηλόω in LSJ is not entirely satisfying as there the verb is supplemented with a dative: δηλοῖ τέ μοι κτλ. (Hdt. 9.68). It seems, then, that the Derveni author thinks that the author of these verses is Orpheus, too, even if these lines are not necessarily from the poem commented upon.[25] These passages show also that even if we can identify a lemma in the papyrus with phrases

[24] Col. 26.2–3: ἑᾶc δὲ ὅτι ἀγαθῆc.

[25] So also Obbink (1997) 41 n. 4; Janko (2001) 31 n. 186; and Bernabé (2002) 3 n. 11.

or verses known to us from another poem, this will not guarantee that the Derveni author treats the lemma as coming from that poem. This is, I maintain, why it is always important to take into account the introductory formulae when attempting to define the object of the author's exegesis and the pool he draws on for comparative illustration: relying on the lemmata alone can be misleading.

I have to admit, however, that my arguments from the linguistic formulations of the introductory expressions cannot discredit Edwards' view according to which

> the commentator (. . .) is either treating a variety of authors or is prepared to make long digressions in the vindication of his critical judgements. In either case we cannot premise that every verse which he cites from Orphic writings is of the same provenance.[26]

True, if the author deals with two or more poems, all of which he attributes to Orpheus, he can still simply use the third person singular when referring to their author. He can even treat such otherwise independent texts as forming a whole for the reason that they are all attributable to the same poet: Orpheus. Nevertheless, the recurrent reference to the 'next line', taken together with the fact that, at least in one case, he does say when he refers to another poem (the *Hymns*), seems to guarantee that at least most of the lines are from the same poem. I fully acknowledge that it is possible that only where 'next' occurs does the author remain in the same poem, and even that in the lacunae there were some explicit indications of a shift to a different poem. Yet I would nevertheless maintain that, on the basis of the above arguments, it is more probable that from col. 7 onwards the Derveni author focuses on a single poem known to him under the name of Orpheus, occasionally referring to other poetical works in order to enhance his exegesis. This conclusion is in accord with the majority of the literature.[27]

SHORTER LEMMATA AND PROSE PARAPHRASES

From those cases where we have at least an entire verse quoted and at least a part of the Derveni author's exegetical remarks on it, we can see that his basic method is the following. First he quotes the verse or verses, with some introductory remark, and then he either gives a general paraphrase of the line, or picks up those words, one by one, or phrase by phrase, which he finds crucial for the understanding of the text he has just quoted. In the

[26] Edwards (1991) 205.

[27] Already Kapsomenos (1963) assumes that the papyrus discusses a poem attributed to Orpheus.

latter case he repeats – already as part of his prose commentary – the words he finds worth explicating.[28] It follows, then, that in those cases where the lemma itself is lost, the author's commentary may still contain sub-quotations, words repeated from the lemmata. The problem that remains is how to sort these sub-quotations out.

The most obvious solution would be to expect the author to make it explicit when he is repeating or quoting something from the lemmata by using a verb of saying in the third person singular. But if we take a look at the use of these verbs, we find that they are employed ambiguously. Words or statements introduced by 'he (sc. Orpheus) says' are not necessarily the *ipsissima verba* of the poet, but might just as well be the Derveni author's paraphrase of them, already loaded with his personal and far from obvious interpretation. This should not raise surprise in view of the fact that λέγει can either mean 'Orpheus says this and that' or 'Orpheus means this and that'. When we read, for example, in col. 14 'He says (φηcὶν) that this Kronos was born from the sun to the earth (γῆι)' it is not obvious at first sight whether Kronos' parents were Ge and Helios in the poem (instead of the canonical Ge-Ouranos couple), or whether this is how the Derveni author interprets the words of the poet; all the more so, since Ouranos does appear in other lemmata. It means, then, that we have to examine these cases one by one, and should refrain from attributing to the poem everything that the author claims to be 'said' by Orpheus.

A second possible indication of a sub-quotation is when a distinctively epic, poetic word appears in the prose text of the author. A clear example is πανομφεύουcα in col. 10.9.[29]

Paraphrases can play a similar role to the sub-quotations. Already embedded in the interpretative prose of the Derveni author, they refer back to the full lemma under consideration. Although they can provide precious pieces of information, they almost inevitably involve some interpretation on the Derveni author's part. However, when the author attempts to discount an alternative reading of the poem in order to prepare the ground for his own interpretation, the possibly idiosyncratic interpretative additions seem to

[28] For this method see ***L1*** with commentary from col. 8.6, pursued also in the next column: the repetition of every word of the two verse lemma; ***L4*** in col. 12: two keywords repeated in the surviving text – but the explication apparently pursued in the now lost bottom of the column; ***L6*** in col. 13: we have at least the keyword (αἰδοῖον) repeated in the commentary; ***L8*** in col. 15: we find the repetition of the final phrase of the verse; ***L10*** in col. 16: we have the final, and surely most important final phrase of the four-verse lemma repeated; ***L12*** in col. 17: at least one keyword repeated in the very lacunose bottom part of the surviving text; ***L13*** in col. 19: at least two keywords repeated in the extant text; ***L14*** in col. 23: repetition of the divine name (Achelous) and keywords.

[29] πανομφεύουcα is a hapax: the closest parallel to it is πανομφαίῳ in *Il.* 8.250.

be kept at a minimum. These are the cases where the apparent literal meaning of the verses quoted is described as 'the ignorant's understanding' of the lemma: a naive reading which should be replaced by the Derveni author's own non-evident, allegorical interpretation. We find, for example, a very close paraphrase of ***L2*** in col. 9, described as the interpretation of 'those who do not understand'. For the quotation

And when Zeus took from his father the prophesied rule
And the strength in his hands and the glorious daimon (col. 8.4–5)

the author remarks that

Those who do not understand the meaning of the things said think that Zeus takes the strength and the daimon from his father (col. 9.2–4).

This case shows that we might grasp something of the apparent, literal meaning of the verses referred to from a discarded, allegedly naive interpretation.

On the basis of these considerations, we may try to identify the stretches of text which might have come from the poem, more or less paraphrased by the Derveni author. As at this stage my purpose is to provide as much information as possible, from which we can later select what we accept as genuine, I have collected here everything that might be relevant on the basis of the three considerations just mentioned. I shall quote them with their immediate context, and underline those words which seem to come from the poem. I mark these shorter and more or less conjectural lemmata by the lower case ***l*** in order to distinguish them from the much more probable full-verse lemmata indicated by the upper case ***L***. Furthermore, I have considered here only those parts of the text which provide further information on the poem, and hence have left out those sub-quotations and paraphrases which evidently refer to an existing ***L*** lemma, discussed earlier.

l1 ὡ̣[c δηλοῖ] καὶ ἐν τῶι
[εὐκ]ρινήτω[ι ἔπει· θ]ὑ̣ρ̣ας γὰρ ἐπιθέ[cθαι κελ]εύcας τοῖ̣[c]
[ὠcὶ]ν αὐτ[οὺc οὔτι νομο]θ̣ε̣τ̣εῖμ̣ φη[cιν τοῖc] πολλοῖc . . . 7.8–10

l2 οὕτωc [οὐδὲγ κωλ]ύει πανομφεύουcαγ καὶ πά̣ν̣[τα]
διδά[cκουcαν τὸ αὐ]τ̣ὸ εἶναι. 10.9–10

l3 τροφ[ὸν δὲ λέγων αὐ]τ̣ὴ̣ν̣ α̣ἰ̣ν̣ί̣[ζε]ται ὅτι [ἅ]cca
ὁ ἥλι̣[οc θερμαίνων δι]α̣λύει ταῦτα ἡ νὺξ ψύ[χουcα 10.11–12

l4 [τ]ῆc Νυκτόc. ἐξ ἀ̣[δύτοι]ο δ' αὐτὴν [λέγει] χρῆcαι
γνώμην ποιού[με]νοc ἄδυτ̣ον ε̣ἶ̣ναι τὸ βάθοc
τῆc νυκτόc· (. . .) 11.1–3

χρῆσαι δὲ καὶ ἀρ̣κέσαι ταὐτ̣ὸ̣ [δύ]ναται.
σκέψασθαι δὲ χρὴ ἐφ' ὧι κεῖτα̣[ι τὸ] ἀρκέσαι
κ̣α̣ὶ τὸ χρῆσαι. 11.5–7

l5 τοῦτον οὖν τὸγ Κρόνον
γενέσθαι φησὶν ἐκ τ̣οῦ ἡλίου τῆι γῆι, ὅτι αἰτίαν ἔσχε
διὰ τὸν ἥλιον κρούεσθαι πρὸς ἄλ̣ληλα. 14.2–4

l6 διὰ τοῦτο λέγει· ὃς μέγ' ἔρεξεν. 14.5

l7 δι' ὅ τι δὲ
ἀὴ̣ρ̣ ἐκλήθη δεδήλωται ἐν τοῖς προτέροις. 17.3–4

l8 καὶ ὕστατον ἔφησεν ἔσεσθαι τοῦτον, ἔπειτ'
ὠνομάσθη Ζεὺς κ̣αὶ τοῦτο αὐτῶι διατελεῖ ὄνομα ὄν,
μέχρι εἰς τὸ αὐτὸ ε̣ἶ̣δος τὰ νῦν ἐόντα συνεστάθη
ἐν ὧιπερ πρόσθε̣ν̣ ἐόντα ἠιωρεῖτο. 17.6–9

l9 τοῦτ' οὖν τὸ πνεῦμα Ὀρφεὺς 18.2–3
ὠνόμασεμ Μοῖραν.

l10 Ὀρφεὺς γὰρ
τὴμ φρόνησ[ι]μ Μοῖραν ἐκάλεσεν· 18.6–7

l11 . . . ἓν [ἕκ]αστογ κέκ[λητ]αι ἀπὸ τοῦ
ἐ̣π̣ικρατοῦντος, Ζεὺ̣[ς] πάντα κατὰ τὸν αὐτὸν
λόγον ἐκλήθη· 19.1–3

l12 θόρνηι δὲ λέγ[ων] δ̣ηλοῖ
ὅτι ἐν τῶι ἀέρ̣ι κατὰ μικρ̣ὰ μεμερισμένα ἐκινεῖτο
καὶ ἐθόρνυτο, θορνύμενα δ' ἕκα<σ>τα συνεστάθη
πρὸς ἄλληλα. μέχρι δὲ τούτο̣υ̣ ἐθόρνυτο, μέχρι
ἕκαστον ἦ̣λθεν εἰς τὸ σύνηθες. 21.1–5

l13 Ἀφροδίτη Οὐρανία
καὶ Ζεὺς καὶ ἀφροδισιάζειν κ̣αὶ θόρνυσθαι καὶ Πειθὼ
καὶ Ἁρμονία τῶι αὐτῶι θεῶι ὄνομα κεῖται. ἀνὴρ
γυναικὶ μισγόμενος ἀφροδισιάζειν λέγεται κατὰ
φάτιν. τῶγ γὰ[ρ] νῦν ἐόντωμ μιχθέντων ἀλ̣λ[ή]λοις
Ἀφροδίτη ὠν[ο]μάσθη. Πειθὼ δ' ὅτι εἶξεν τὰ ἐ[ό]ντα
ἀλλήλ̣ο̣[ι]σ̣ι̣ν̣. ε[ἴ]κειν δὲ καὶ πε̣ί̣θειν τὸ αὐτό̣ν̣. [Ἁ]ρμονί̈α δὲ
ὅτι πο[λλὰ 21.5–12

l14 Γῆ δὲ καὶ Μήτηρ καὶ Ῥέα καὶ Ἥρη ἡ αὐτή. ἐκλήθη δὲ
Γῆ μὲν νόμωι, Μ̣ήτηρ δ' ὅτι ἐκ ταύτης πάντα γ[ίν]εται.
Γῆ καὶ Γαῖα κατὰ [γ]λ̣ῶσσαν ἑκάστοις. Δημήτηρ̣ [δὲ]
ὠνομάσθη ὥσπερ̣ ἡ Γῆ Μήτηρ, ἐξ ἀμφοτέρων ἓ[ν] ὄνομα 22.7–10

l15 εὔδηλον ὅτι Ὠκεανός ἐστιν ὁ ἀήρ, ἀὴρ δὲ Ζεύς·
οὔκουν ἐμήσατο τὸ̣ν̣ Ζᾶνα ἕτερος Ζεὺς ἀλλ' αὐτὸς
αὑτῶι σθένος μέγα. 23.3–5

h6 οἱ δ' οὐ γινώσκοντες τὸν Ὠκεανὸν ποταμὸν δοκοῦσιν εἶναι ὅτι εὐρὺ ῥέοντα προσέθηκεν. 23.5–7

h7 ἴσα ἐστὶν ἐκ τοῦ [μέ]σου μετρούμενα, ὅσα δ[ὲ μ]ὴ κυκλοειδέα οὐχ οἷόν τε ἰσομελῆ εἶναι. 24.1–2

h8 μη[τρ]ὸς μὲν ὅτι μήτηρ ὁ Νοῦς ἐστιν τῶν ἄλλων ἐᾶς δὲ ὅτι ἀγαθῆς. 26.1–2

h9 ὁ δ' εἴπερ ἤθελεν ἑαυτοῦ μητρὸς ἐμ φιλότητι ἀποδεῖξαι θέλοντα μιχθῆναι τὸν θεόν, ἐξῆν αὐτῶι γράμματα παρακλίναντι μητρὸς ἑοῖο εἰπε[ῖ]ν· 26.9–12

THE ORDER OF THE LEMMATA

Our next task is to say something about the original order of the verses in the poem in relation to the order in which the lemmata appear in the papyrus. Roughly speaking, there are three options, each of which have found supporters:

1. The Derveni author does not organise his text on the basis of the structure of the poem, but quotes and interprets a verse from here, a verse from there, as his own argument dictates.[30]
2. He basically follows the sequence of the poem, but does not give a systematic interpretation of it. He comments only on those passages that he finds particularly significant from the perspective of his own agenda.
3. He not only follows the sequence of the poem, but gives a systematic running commentary of the poem.[31]

Making our choice among these options is important not only because it tells us something about the internal logic of the author's exegetical method, but also because it determines how far we can rely on the papyrus in the reconstruction of the poem. Remarkably, the author makes an explicit statement in col. 13 that, at least at first sight, seems decisive in favour of the third option. For he says that:

> Since in his whole poetry he speaks about facts enigmatically, one has to speak about each word in turn (col. 13.5–6).

[30] This is the view expressed by Tortorelli Ghidini (1989) 32; Ricciardelli Apicella (1993) 32; Casadesús (1996) 78.

[31] So West (1983); KRS etc.

There is another relevant remark that turns up in the middle of col. 7, at the very point where the author starts the exegesis of the poem. He declares here that

> The poem is something strange and riddling for people. But Orpheus did not intend to tell them captious riddles, but momentous things in riddles. Indeed, he is telling a holy discourse from the first right along to the last word (col. 7.4–8).[32]

This remark entails not only that the poem needs explication, but also that it is of such importance that every word of it can reveal something of interest. This basic contention also speaks for a systematic commentary.

Nonetheless, both of these statements might be merely rhetorical, and they have actually been treated as such by numerous scholars – by all those who opt for the first alternative indicated above. This is also Edwards' view, when he claims that 'the assumption that the commentator has annotated all the lines of his poem in their order is untenable'.[33] Edwards has two arguments in support of this critical stance. First, that 'every scholar who wishes to retrieve the poem finds himself obliged to fill lacunae'. This point is far from conclusive. If we take into account that we have lost the bottom half, or even two thirds, of each column, and accept that there is no reason why the bottom parts of the columns should contain significantly fewer lemmata than the upper parts, we conclude that we have about 50% of those verses which were originally quoted in the roll. But Edwards has a second argument, trying to counter this objection. He continues by claiming that

> even if we argue that these might have been supplied in missing portions of the commentary, we cannot reconcile the sequence of the extant lines with the surviving fragments of any Orphic poem.

Although the comparative evidence can be very useful towards understanding the Derveni text, I do not think we should pretend that all the Orphic poems with theogonic content should necessarily be of the same structure as those we know about from late, mostly Neoplatonic testimonies. More important, we should not forget that what we have of these later Orphic poems is not more than a series of fragments and testimonies, much like what we have of the Presocratics. Thus, the fact that we are able to reconstruct, with a fair degree of confidence, the temporal sequence of the theogonic events on the basis of the Neoplatonic summaries does not in any way mean that we know anything for sure about the actual structure of those poems.

[32] Cf. the detailed commentary in Tsantsanoglou (1997) 117–24.

[33] Edwards (1991) 204–5.

Nonetheless, Edwards does raise an important point here concerning the sequence of the narrative elements. The content of the lemmata strongly suggests that the poem commented on in the papyrus was a theogony. Indeed, most scholars treat it as such. Now, if we place the successive lemmata one after the other, then, even if we allow losses due to the lacunae in the papyrus, we obtain a series of verses. But this series presents the theogonical events in an order other than that of the successive generations of gods – something unattested in other theogonies. This is a genuine difficulty to which I shall return in the next section, where I shall examine whether the sequence of the lemmata provides an acceptable narrative structure for the poem.

Because of all these uncertainties, the best we can do, I suggest, is to return to the expressions introducing the lemmata. Here we find the recurrent use of the expression 'in the next (verse)' or, alternatively 'after that' (coll. 11.9; 12.1; 14.5; 15.5; 23.10). These formulations imply, beyond any reasonable doubt, that the verse or verses thus introduced come in the poem immediately after the ones which were discussed in the respective previous sections of the papyrus. On the other hand, none of the introductory remarks says something like 'somewhat later he says', or 'in another part of the poem', or 'earlier he says'. I fully acknowledge that this evidence permits gaps to be left in the commentary since we should not take it for granted that the author continues the exegesis with the consecutive line of the poem where he does not say so. Yet the two explicit declarations, reinforced by the recurrent expressions implying that the Derveni author continues his commentary with the explication of the immediately following verse of the poem, show that there is at least a strong tendency in the commentary towards systematicity. It follows, then, that one would need strong arguments for maintaining the contrary thesis that the order of the lemmata does not correspond to the order of the verses in the poem but is governed merely by the logic of the Derveni author's own argument.

Obviously, this does not mean that the Derveni author effectively explained each and every word, or even that he could not occasionally skip a verse or two. It does mean, however, that his basic intention was to give a comprehensive, methodical explication of the poem. Moreover, it should also be noted that occasional gaps and dense, systematic commentary are not incompatible, for the commentator can leave out words and lines if he feels that the understanding of those parts of the poem already follows from his previous explanations. Therefore, once the above conclusion is accepted, we can assume that, had the papyrus come down to us in its entirety, we could reconstruct the Orphic poem with a fair degree of certainty and relatively little loss.[34] I note however that this remains a point of considerable disagreement.[35]

CRITICAL EXAMINATION OF THE LEMMATA

If we accept, at least as a working hypothesis, the above conclusion about the systematic nature of the author's commentary, our next task is simple. First, we must arrange those lemmata which seem to come from the same poem consecutively, in the same order as they appear in the papyrus. Then we have to introduce the ***l*** lemmata at the proper places. Finally, it seems appropriate to group together the consecutive lemmata coming from a continuous part of the papyrus, while separating those that are separated by a longer lacuna – these latter may or may not be following immediately upon the previous lemma we have. The result is as follows:

l1 [. . .], ***L1-L2*** [. . .], ***l2-l3*** [. . .], ***l4-L3*** [. . .], ***L4*** [. . .], ***L5-L6*** [. . .], ***l5-l6-L7*** [. . .], ***L8-L9*** [. . .], ***L10-L11*** [. . .], ***l7-l8-L12*** [. . .], ***l9-l10*** [. . .], ***l11-L13*** [. . .], ***l12-l13*** [. . .], ***l14*** [. . .], ***l15-l16-L14*** [. . .], ***l17-L15*** [. . .], ***L16*** [. . .], ***l18-l19***

The question is whether this series of lemmata makes any sense at all. That is, can we possibly imagine a poem which speaks about such things in such an order? This is what we have to examine now. First, we should check if the order of these groups makes up a tolerable narrative sequence for the whole poem (or a longer, continuous part of it assuming that the exegesis continued on a second roll). This is necessary to confirm or discard our general claim that the Derveni author follows the narrative sequence of the poem. We must also test the stronger claim that the immediately consecutive lemmata of the papyrus always, or in most cases, represent immediately consecutive verses of the poem. This we can do by examining whether the consecutive quotations not separated by a lacuna in the papyrus can form a continuous part of the poem. When we are dealing with consecutive full-verse lemmata we have to see if grammar and semantics allow us to take these verses together. When, on the other hand, we are dealing with shorter lemmata and paraphrases, we have to consider whether the juxtaposed lemmata show enough thematic coherence. I shall here give also the English translation, italicising the full verse lemmata as well as plausible reconstructions of the verses of the poem. At this stage of our work, it seems

[34] Whether or not the audience of the papyrus had independent access to the poem is an interesting question, but one that I cannot answer.

[35] For: West (1983), KRS, Bernabé (2002), etc.; contra: Rusten (1985), Tortorelli Ghidini (1989), Edwards (1991), Ricciardi Apicella (1993), Casadesús (1996) etc.

justified – even on the interpretative principles adopted here – to refer to other Orphic texts[36] whenever they can clarify the internal evidence. For the sake of convenience, I shall also indicate in parenthesis the numbering of the fragments in Bernabé's collection (fragment number followed by a capital B.).

I1 ὡ̣[ϲ δηλοῖ] καὶ ἐν τῶι | [εὐκ]ρινήτω[ι ἔπει· θ]ύ̣ρ̣αϲ γὰρ ἐπιθέ[ϲθαι κελ]εύϲαϲ τοῖ̣[ϲ] | [ὠϲὶ]ν αὐτ[οὺϲ οὔτι νομο]θ̣ε̣τ̣εῖμ φη[ϲιν τοῖϲ] πολλοῖ̣ϲ̣ 7.8–10

As he also makes clear in the well-chosen verse: for having ordered them to put doors to their ears he says that he is [? not legislating] for the many . . . [?but only for] those pure in hearing

This lemma is not thematically related to anything that follows. Nonetheless, parallel texts show that it was a set formula to begin a poem connected with mysteries. On this basis, West, followed by Bernabé, reconstructs the following verse:

φθέγξομαι οἷϲ θέμιϲ ἐϲτί· θύραϲ δ' ἐπίθεϲθε βέβηλοι (fr. 3 B.)[37]

I shall speak to those for whom it is lawful; but shut the door, you profane.

We are happy to find this lemma, which was thus probably part of the proem, at the head of our list.

L1-L2:

[ο]ἳ Διὸϲ ἐξεγ̣ένοντο [ὑπερμεν]έοϲ βαϲιλῆοϲ 8.2 = fr. 4 B.
Ζεὺϲ μὲν ἐπεὶ δὴ π̣α̣[τρὸϲ ἑο]ῦ πάρα θέ[ϲ]φατον ἀρχὴν
[ἀ]λκήν τ' ἐγ χείρεϲϲι ἔ[λ]αβ[εγ κ]α̣ὶ̣ δαίμον̣[α] κυδρόν[38] 8.4–5 = fr. 5 B.

Who were born from Zeus the mighty king.
And when Zeus took from his father the prophesied rule
And the strength in his hands and the glorious daimon

This group can form a continuous unit, provided we print a full stop at the end of ***L1***. Such a solution is rendered internally plausible by the μέν

[36] By 'Orphic' I mean a text or narrative that someone in antiquity attributed to Orpheus or treated as Orphic. For a justification of this usage, see 151 below.

[37] Cf. the texts adduced by West (1983) 82, most importantly the first line of the so-called *Testament of Orpheus* (*OF* 245) φθέγξομαι οἷϲ θέμιϲ ἐϲτί· θύραϲ δ' ἐπίθεϲθε βέβηλοι and Plato *Symp.* 218b = *OF* 13 πάντεϲ γὰρ κεκοινωνήκατε τῆϲ φιλοϲόφου μανίαϲ τε καὶ βακχείαϲ· διὸ πάντεϲ ἀκούεϲθε . . . οἱ δὲ οἰκέται, καὶ εἴ τιϲ ἄλλοϲ ἐϲτὶ βέβηλόϲ τε καὶ ἄγροικοϲ, πύλαϲ πάνυ μεγάλαϲ τοῖϲ ὠϲὶν ἐπίθεϲθε. True, the text of the papyrus was reconstructed on the basis of these very texts. But the parallels are so convincing that it does not seem untenable both to reconstruct and interpret the lemma on this evidence.

[38] The syllabic pattern of χείρεϲϲι ἔ[λ]αβ[εγ is problematic; for different emendations, see the apparatus to the text.

in the first line of ***L2***. Furthermore, it is possible to conceive that ***L1*** was the last line of the proem, in which the poet tells us what he will sing about.[39] According to this reconstruction – not more than conjectural – the sequence of ***L1-L2*** can show the articulation between the end of the proem and the beginning of the narrative proper.[40] Therefore, it is at least possible to take these lines together and connect them, with some verses now lost in the missing part of col. 7, to ***l1***. We have to note right now that this reconstruction implies that the narrative begins with ***L2***, that is, Zeus' accession to royal power. The last verse presents a grammatical problem: δαίμον̣[α] κυδρόν may be taken as a second direct object of the verb together with [ἀ]λκήν, or, alternatively it may be construed, possibly as a direct object, with another verb now lost in the subsequent verse.[41]

l2-l3 [. . .]+***l4-L3***

οὕτωϲ [οὐδὲγ κωλ]ύει πανομφεύουϲαγ καὶ πά̣ν̣[τα] |
διδά[ϲκουϲαν τὸ αὐ]τ̣ὸ εἶναι. 10.9–10

τροφ[ὸν δὲ λέγων αὐ]τ̣ὴ̣ν̣ α̣ἰ̣νί̣[ζε]ται ὅτι [ἅ]ϲϲὰ | ὁ ἥλι̣[οϲ
θερμαίνων δι]α̣λύει ταῦτα ἡ νὺξ ψύ[χουϲα 10.11–12

[τ]ῆϲ Νυκτόϲ. ἐξ ἀ̣[δύτοι]ο δ' αὐτὴν [λέγει] χρῆϲαὶ | γνώμην
ποιού[με]νοϲ ἄδυτ̣ον ε̣ἶναι τὸ βάθοϲ | τῆϲ νυκτόϲ· (. . .) 11.1–3

χρῆϲαι δὲ καὶ ἀρ̣κέϲαι ταὐτ̣ὸ̣ [δύ]ναται. | ϲκέψαϲθαι δὲ χρὴ ἐφ'
ὧι κεῖτα̣[ι τὸ] ἀρκέϲαὶ | κ̣α̣ὶ τὸ χρῆϲαι. 11.5–7

[ἥ δὲ] ἔχρηϲεν ἅπαντα τά οἱ θε̣[μιϲ ἦν ἀνύϲαϲ]θ̣αι 11.10

Thus nothing prevents 'all-pronouncing' and 'teaching all things' from being the same thing.

By saying that she is 'nurse', he (sc. Orpheus) expresses in riddling form that the sun dissolves by heating those things which the night unites by cooling . . .

He says that she proclaims the oracle out of the innermost shrine (ἐξ ἀ̣[δύτοι]ο) his view being that the depth of the night is 'never setting' (ἄδυτ̣ον). . . .

Now, 'to proclaim oracles' and 'defend' mean the same. But one has to examine to what 'assist' and 'furnish/proclaim oracles' apply (sc. as a name).

[?She] proclaimed an oracle about all that was right to him to . . .

The lemmata from ***l2*** down to ***L3*** seem to form a larger block, linking the lemmata of two consecutive columns with a suitable thematic unity. The subject of these lemmata is almost certainly the primordial goddess Night.

39 The phrasing could have been very similar to what we find in Hesiod's *Theogony* vv. 106 and 111. Cf. West (1983) 84.

40 This is the reading adopted by West.

41 Bernabé (2002) 12 indicates this problem and finally opts for the former solution.

We have other Orphic texts calling her the 'nurse of gods'.[42] She delivers her prophecies from her shrine in a cave in other texts as well,[43] while the hapax πανομφεύουσα can refer to her ability 'to proclaim all' due to her prophetic power.

This theme (i.e., prophecies by Night) fits well with the word θέ[c]φατον in ***L2***. Furthermore, it is tempting to take the ὡc at the beginning of ***L4***, the next lemma in order, as introducing a subordinate clause governed by a verb of prophesying or declaring in the main clause. West, followed by Bernabé, assumes that this verb is the ἔχρηcεν of ***L3***. Accordingly, they print ***L4*** right after ***L3***. Again, this is not more than a conjecture, but it is a plausible one. This construal, together with some highly likely supplements, results in the following sequence of verses (= fr. 6 B.):

πανομφεύουcαγ . . . [θεῶν] τροφ[ὸc . . . Νύξ[44]
. . . χρῆcαι . . . ἐξ ἀ[δύτοι]ο
[ἣ δὲ] ἔχρηcεν ἅπαντα τά οἱ θέ[μις ἦν ἀνύcαc]θαι
ὡc ἄρξ[ηι κα]τὰ καλὸν ἕδοc νιφόεντοc Ὀλύμπου

all-pronouncing . . . nurse [of the gods] . . . Night
. . . to proclaim oracles . . . from the innermost shrine
[?She] proclaimed an oracle about all that was right to him to hear
so that he may rule on the lovely abode of snowcapped Olympus

L5-L6

Ζεὺc μὲν ἐπεὶ δὴ πατρὸc ἑοῦ πάρα [θ]έcφατ' ἀκούcα[c] 13.1 = fr. 7 B.
αἰδοῖογ κατέπινεν, ὃc αἰθέρα ἔχθορε πρῶτοc 13.4 = fr. 8 B.

Zeus when he heard the prophecies from his father
He swallowed the phallus of [. . .], who sprang from the aither first

These two verses are highly problematic. It is very probable that they did not form a continuous part of the poem. Although the main difficulty is internal to ***L6***, we would run into further pitfalls if we tried to arrange them right after each other. ***L6*** has given rise to an extensive discussion in the literature. As the quandaries associated with this verse will turn out to be of further importance, I shall now discuss the relevant questions at some length.

[42] *OF* 106 = Procl. *in Plat. Cratyl.* 404 b (. . .) ἔχει δὲ παράδειγμα τὴν Νύκτα, – θεῶν γὰρ τροφὸc ἀμβροcίη Νὺξ λέγεται (. . .)

[43] *OF* 105 = Herm. *in Plat. Phaedr.* 284 c (. . .) ἔνδον μὲν γὰρ ἐν τῶι ἀδύτωι τῆc Νυκτὸc κάθηται ὁ Φάνηc· ἐν μέcωι δὲ ἡ Νὺξ μαντεύουcα τοῖc θεοῖc (. . .); cf. also *OF* 103 and 104.

[44] Bernabé prints also the epithet ἀμβροcίη before Νύξ. The supplement is certainly quite likely, but I prefer to leave it out, as there is no trace of it in the surviving text of the papyrus.

Let us begin with what is common ground for the interpreters. Everyone takes it for granted that the subject of the verb κα[τ]έπινεν must be Zeus. They base this assumption on the fact that one distinctive feature of the Orphic *Rhapsodies*[45] is that in that narrative, in contradistinction to Hesiod's *Theogony*, it is Zeus, and not Kronos, who swallows something. This is the most crucial and, at the same time, the most startling moment of the story. Zeus has already obtained the royal power from his father, but, in order to make his kingship enduring, he ought to follow Night's oracular advice and swallow the primordial bisexual divinity Phanes.[46] If we add to this that we find in the papyrus not only the act of engulfment but also its result (that the whole universe with all the beings in it get inside Zeus, and hence Zeus becomes 'the only one', for which see ***L10-L12*** below), it is reasonable to accept that the subject of the verb κα[τ]έπινεν is Zeus.

Although everyone agrees that Zeus was the one who swallowed something, there is intense disagreement about what it was that he swallowed. The question could be decided by determining the meaning and reference of the word αἰδοῖον, the direct object of the verb κα[τ]έπινεν. It is beyond dispute that the Derveni author understands the word αἰδοῖον as meaning 'phallus',[47] but there are certain considerations which can lead us to believe that the word had another sense in the poem itself. On this point, two mutually incompatible views have been developed, with strong arguments and authoritative names on both sides. I shall list the arguments adduced by scholars on both sides, adding some further arguments and counter-arguments of my own.

According to one view, the word αἰδοῖον must have been a masculine adjective in the accusative in the poem itself. On this hypothesis, the adjective qualified the word δαίμων which, in turn, referred to the primordial god Phanes, whose name, incidentally, does not occur in the extant text of the papyrus. Thus, on this reading, the Orphic poem partially preserved in the Derveni papyrus was very close to the Orphic *Rhapsodies*. Or, to put it in another way, the *Rhapsodies*, probably composed some time around the beginning of the Christian era, preserved an important motif of the Orphic theogonies composed as early as the sixth or fifth century BC. The arguments for this construction are as follows:

[45] It is not clear if the theogony reported in Athenagoras *Pro Christian.*, in which the same episode is referred to, is identical with the *Rhapsodies*, or with the so-called *Hieronymian Theogony*, or whether it is an independent version in its own right. Cf. Amersfoort (1981) and 144–6 below.

[46] Cf. *OF* 129 = Procl. *in Plat. Cratyl.* 396b; and three passages of Procl. *in Plat. Tim.* in *OF* 167.

[47] Cf. 13.7–9.

A. Internal arguments

(i) The masculine relative pronoun ὅc in ***L6*** cannot refer to the neuter noun αἰδοῖον meaning 'genital organ'. It must have, therefore, originally been an adjective, qualifying a masculine noun in the accusative in the previous verse. Hence the hypothesis: the Derveni author cut an enjambement, either on purpose or due to inadvertence.[48] The problem with this argument is that the ὅc could just as well refer to some other masculine noun in the previous verse. It could, for instance, refer to a masculine noun in the possessive genitive, subordinated to the neutral noun αἰδοῖον. Take for example the construction [. . . Οὐρανοῦ (or something equivalent with it, such as πρωτογόνου βαcιλῆοc) . . .] | αἰδοῖογ κατέπινεν, ὃc αἰθέρα ἔχθορε πρῶτοc where the ὅc refers to Ouranos and not to the αἰδοῖον. This point, then, cannot discredit the Derveni author's reading.[49]

(ii) The author introduces the interpretation of ***L6*** by declaring that

> Since in the whole poem he speaks about facts enigmatically, one has to speak about each word in turn (col. 13.5–6).

According to Casadesús, this announcement should be treated as an acknowledgement on the part of the Derveni author that the riddling nature of Orpheus' text necessitates unusual exegetical methods, which might include cutting enjambements and treating adjectives as nouns. Had the word αἰδοῖον meant 'genitals' already in the poem, there would be no reason for the Derveni author to insist on the need for an allegorical interpretation.[50] I do not think that this point is of much weight either, since the Derveni author is just about to identify the αἰδοῖον with the sun, and thus his methodological remark can just as well refer to his move to interpret the word 'genitals' as designating 'the sun'.

(iii) Although Phanes' name does not occur in the surviving text of the papyrus, the word μῆτιc does appear in ***L9***. Now, in Hesiod's *Theogony* Zeus swallows Metis in order to make his kingship stable,[51] while Metis is explicitly identified with Phanes at least in some fragments and summaries of the *Rhapsodies*.[52] This is a strong argument, it seems to me, yet certainly

[48] West (1983) 85; Rusten (1984) 334; Casadesús (1996) 83.

[49] Janko (2001) 24 apparently ignores this problem and translates 'He ingested the penis that first procreated the ether.'

[50] Casadesús (1996) 81–3. [51] Hes. *Thg.* 886–90.

[52] *OF* ?60, 65, 168 v9 (Μῆτιc πρῶτοc γενέτωρ), 169 v4 (Μῆτιc πρώτη γένετιc), 170, and especially 85: δαίμονα cεμνόν, | Μῆτιν cπέρμα φέροντα θεῶν κλυτόν, ὅν τε Φάνητα | πρωτόγονον μάκαρες κάλεον κατὰ μακρὸν Ὄλυμπον.

not a conclusive one. First of all, the presence of Metis in the poem commented on in the papyrus does not necessarily mean that this divinity was identified with Phanes already in this early version of the Orphic theogony. Moreover, apart from the fact that we cannot know what the role of Metis was in this narrative, and whether or not she was swallowed in the poem commented on in the papyrus, we can still imagine that there were two entities swallowed by Zeus in this version: Metis on the one hand, and a phallus on the other. Or else – and I think this is the most plausible solution – μῆτις in ***L*9**, probably parallel to τιμή and subordinated to κατέχω, is not the name of a goddess but a common noun explaining Zeus' epithet μητίετα in ***L*8**. Hence, accepting the conjecture of West for ***L*9**, the translation of the two verses:

> *From him* (sc. *Ouranos) in turn Kronos, and then wise Zeus*
> *Holding wisdom and royal honour over the blessed gods*

B. External evidence

(i) The verb καταπίνω is used to describe Zeus' swallowing of Phanes in *OF* 58, 129, 167.[53] The usage is not, however, significant since this seems to be the most appropriate word for describing any act of swallowing, irrespective of its object. Even though the object of καταπίνω is certainly Phanes in the Orphic fragments mentioned above, 'Orpheus' could use the same verb to describe the engulfment of a phallus in the poem commented on in the papyrus. Therefore, I do not think that this parallel is compelling either.

(ii) We have some comparative evidence in favour of taking δαίμον̣[α] κυδρόν, as the expression appears at the end of ***L*2**, to refer to Phanes/Metis. Proclus quotes two verses from the *Rhapsodies* in which Metis is called δαίμονα σεμνόν and σεμνόν is a synonym for κυδρόν.[54] But there are other divine figures called *daimon*, with or without adjectives, in other Orphic fragments. In *OF* 47, for example, the sun is called κλυτὲ δαῖμον, and, a usage to which Proclus explicitly calls attention, in *OF* 155 Zeus himself is called ἀριδείκετε δαῖμον. Hence, we should not infer from the fact that

[53] Rusten (1984) 334; (1985) 125 n. 9; Parker (1995) 490–1. Cf. *OF* 58 = Athenag. *Pro Christian.* 20 22, 10 Schw. (. . .) ἢ αὐτὸν τὸν Φάνητα δέξαιτο, θεὸν ὄντα πρωτογόνον – οὗτος γάρ ἐστιν ὁ ἐκ τοῦ ᾠιοῦ προχυθείς – ἢ σῶμα ἢ σχῆμα ἔχειν δράκοντος ἢ καταποθῆναι ὑπὸ τοῦ Διός κτλ. and *OF* 129 = Procl. *in Plat. Crat.* 396b. ὡς δ' Ὀρφεὺς ἐνθέῳ στόματι λέγει, καὶ καταπίνει τὸν πρόγονον αὐτοῦ τὸν Φάνητα καὶ ἐγκολπίζεται πάσας αὐτοῦ τὰς δυνάμεις ὁ Ζεὺς κτλ. Cf also the three passages in *OF* 167.

[54] *OF* 85 = Procl. *in Plat. Cratyl.* 391d. Cf. Rusten (1984) 334; Casadesús (1996) 80–1.

Phanes/Metis is called δαίμονα σεμνόν in the *Rhapsodies* that δαίμον[α] κυδρόν in ***L2*** can only refer to Phanes/Metis.

(iii) There are parallels for the verb ἐκθρῴσκω referring to Phanes' birth in *OF* 60 and 85.[55] This evidence, some commentators claim, makes it likely that the second clause of ***L6*** (ὃς αἰθέρα ἔχθορε πρῶτος) speaks about Phanes, too. Yet again, ἐκθρῴσκω (or simply θρῴσκω) is used in Homeric hymns to designate a god's birth,[56] and in these cases there is evidently no question of the Orphic Phanes. Thus, this verb in itself cannot prove that we have Phanes in the poem commented on in the papyrus.

(iv) Even the other party agree on the point that the word αἰδοῖον in ***L6*** means the same as αἰδοίου in the first line of ***L10***. ***L10*** describes that most extraordinary element of the story when all the gods and beings of the world get somehow attached to (προσέφυμ) the entity primarily swallowed, and thus the whole world eventually lands in Zeus's belly. Although the genitive construction πρωτογόνου βασιλέως αἰδοίου is ambiguous (αἰδοίου can either be in agreement with βασιλέως or βασιλέως can be subordinated to it), it is important to note that πρωτόγονος was Phanes' fixed epithet and was sometimes used in itself as one of Phanes' many names.[57] From this, commentators infer that πρωτογόνου βασιλέως must refer to Phanes. There is no denying that the adjective πρωτόγονος became unambiguously associated with Phanes in later versions of the theogony and in the Orphic hymns. Yet we have to remember that it was not always so. On a gold leaf from Thurii, for example, Ge (or someone closely associated with Ge) is addressed with the same adjective,[58] which suggests that in early Orphism other gods could well be called πρωτόγονος. Thus, even though the evidence of the later Orphic texts is certainly pressing, the use of the adjective still cannot prove beyond doubt the presence of Phanes in the papyrus.

At any rate, it is noteworthy that those who favour this reading have to assume that the Derveni author misconstrued the text of the poem and, accordingly, they have to emend the lemmata. Two alternative reconstructions have been proposed for the relevant verses of the poem. West's

[55] Rusten (1984) 334. Cf. *OF* 60 = Damasc. *De princ.* 123 (I 316, 18 Ruelle) (. . .) ὅτι ἐκ τούτων (sc. from the entity called the egg or the bright chiton or the cloud) ἐκθρώσκει ὁ Φάνης κτλ. and *OF* 85 = Damasc. *De princ.* 111 (I 286, 15 Ruelle) εἰ δὲ ὁ παρ' Ὀρφεῖ πρωτόγονος θεὸς ὁ πάντων σπέρμα φέρων τῶν θεῶν ἀπὸ τοῦ ᾠιοῦ πρῶτος ἐξέθορε καὶ ἀνέδραμε κτλ. and *OF* 56 = Apion *ap.* Clem. *Roman. Homil.* 6.12 (. . .) ᾠιόν, οὗ ῥαγέντος κατά τινας ἀρρενόθηλυ ἐξέθορεν, Φάνης.

[56] Cf. *Hom. hymn to Apollo* 119; *Hom. hymn to Hermes* 20; Call. *Hymn to Delos* 6.255 (probably taken from the Homeric hymn); and perhaps Call. fr. 43.123 Pfeiffer.

[57] As epithet: *OF* 58, 64, 85.3, 173, 167a1 (to Erikepaios, another name of Phanes); as proper name: *OF* 54, 86.1, 87.1

[58] Tablet C v. 1 Zunz = *OF* 47. On the relevance of this document to the papyrus, see 333–7 below.

reconstruction[59] is rather bold. He supposes that the text the Derveni author used was corrupt, so he gives a reconstruction not of the poem the Derveni author commented on, but of its hypothetical archetype. First he emends ***L5*** on the assumption that Kronos could not give oracular advice to Zeus, then he shifts to this place the second line of ***L2*** on the basis of a resemblance between the first verse of ***L2*** and ***L5***, and finally he prints ***L6*** after the second verse of ***L2***. The result is as follows:[60]

Ζεὺϲ μὲν ἔπει<τ' ἄφραϲτα θεᾶϲ> πάρα̣ [θ]έ̣ϲφατ' ἀκούϲα[ϲ]	13.1 (emended)
[ἀ]λκήν τ' ἐγ χείρεϲϲι ἔ[λ]αβ[εγ κ]α̣ὶ̣ δαίμο̣ν̣[α] κυδρόν	8.5
αἰδοῖογ κατ̣έπινεν, ὃϲ αἰθέρα ἔχθορε πρῶτοϲ.	13.4

Rusten, in response to West, suggests that the text the Derveni author used was not corrupt, but that this 'unscrupulous commentator' altered the text himself in order to trim it to his own theories. According to Rusten's argument, the author did not like the idea of Zeus receiving oracles from any source at all. Indeed, in col. 13.2–3, after having quoted ***L5***, he declares

For neither did he hear this time – but it has been made clear that he had heard – nor does Night command (this time).

Thus, Rusten continues, in order to argue away the oracle given to Zeus the author 'preferred (or invented) an alternative reading θέ̣[ϲ]φατον ἀρχήν' replacing [θέ]ϲφατ' ἀκούϲα[ϲ] because, presumably, in this way the adjective θέϲφατοϲ could be taken in a looser sense like 'divine', not directly referring to oracles. Hence Zeus is not any longer supposed to give heed to any prophecies.[61] Although the underlying explanation is different, the ensuing reconstruction proposed by Rusten is quite similar to that of West:

Ζεὺϲ μὲν ἐπεὶ δὴ̣ π̣ατρ̣ὸϲ ἑοῦ πάρα̣ [θ]έ̣ϲφατ' ἀκούϲα[ϲ]	13.1
[ἀ]λκήν τ' ἐγ χείρεϲϲι ἔ[λ]αβ[εγ κ]α̣ὶ̣ δαίμο̣ν̣[α] κυδρόν	8.5
αἰδοῖογ κατ̣έπινεν, ὃϲ αἰθέρα ἔχθορε πρῶτοϲ.	13.4

As Burkert has noted,[62] the solution advocated by West and Rusten amounts to imputing to the author another misconstrual of the Orphic verses. For if we take δαίμο̣ν̣[α] κυδρόν and αἰδοῖογ together, δαίμο̣ν̣[α] must be the direct object of κατ̣έπινεν. Yet the author, in contrast to this, repeatedly takes δαίμο̣ν̣[α], together with [ἀ]λκήν, to be the object of ἔ[λ]αβ[εγ.[63]

[59] West (1983) 85–6.
[60] I am quoting the verses with the new readings. These do not effect the argument either way.
[61] Rusten (1985) 124–7. Calame (1997) 67 n. 3 concurs with the main steps of Rusten's solution, but claims that the oracle originally referred to in the poem was not proclaimed by Kronos but by Night.
[62] Burkert (1987b) 38 n. 57.
[63] Col. 8.7–8: Ζεὺϲ μὲν ἐπε̣ὶ̣ τ[ὴν ἀλ]κὴν | [πα]ρὰ πατρὸϲ ἑοῦ ἔλαβεγ καὶ δαίμονα̣ [κυδρ]ό̣ν; and col. 9.2–4: οἱ δὲ οὐ γινώϲκον[τεϲ] | τὰ λεγό[μεν]α δοκοῦϲι̣ τ̣ὸ̣ν̣ Ζᾶνα παρὰ τοῦ αὑτο[ῦ] | πατρὸϲ [τὴν] ἀλκήν [τε καὶ] τὸν δαίμονα λαμμά[νειν].

To this I would also add that this is the reading the author attributes to the ignorant. But, as mentioned above, he regularly attributes to the ignorant the *literal* meaning of the lemma. It is, then, a further indication that this is what the author thought to be the primary, obvious meaning of the verse, and not something already manipulated by him.

Edwards, for his part, also takes αἰδοῖον to be an adjective, but does not propose a reconstruction of the relevant verses because he denies that any such reconstruction is possible on the evidence of the papyrus. Furthermore, he challenges the previous unsympathetic evaluations of the Derveni author's exegetical methods by drawing attention to the fact that there are other Orphic passages in which the poet himself plays on the derivation of the noun αἰδοῖον from the adjective.[64] His conclusion is that 'something of this kind it is that the commentator detects in the Orphic poem: a comparison which transpires through etymology, but only such a comparison as bequeaths a deeper riddle to the wise'.[65]

Let us turn now to the alternative interpretation of the lemma, according to which the word αἰδοῖον meant 'genitals' or 'phallus' not just for the Derveni author but already in the poem. As may have become clear from my discussion of the former view, I shall eventually opt for this second interpretation. The arguments for this reading are as follows:

A. Internal arguments

(i) Although it cannot be conclusive in itself, we have to acknowledge first of all that the author himself takes αἰδοῖον to be a noun meaning 'genitals'.[66]

(ii) There is absolutely no mention of Phanes in the surviving text of the papyrus,[67] even if an argument *ex silentio* cannot be compelling in the case of so fragmentary a text as the Derveni papyrus.

(iii) We have also seen that the respective hypothetical reconstructions of West and Rusten show that one is forced to manipulate the received text, with more or less boldness, in order to show that the antecedent of αἰδοῖον was the δαίμον̣[α] κυδρόν of ***L2***. In accordance with the general methodological principles adumbrated above, I would prefer to retain the

[64] Edwards (1991) 205–7. Cf. *OF* 127.4 in which Aphrodite is referred to as παρθένος αἰδοίη born from the severed genitals of Ouranos. In another fragment (*OF* 183 = Procl. *in Plat. Cratyl.* 406c) the word is used as a substantive and refers to Ouranos' genitals. It is noteworthy that the word is connected here with the verb ἐκθρῴσκω: τὸν δὲ πόθος πλέον εἶλ', ἀπὸ δ' ἔκθορε πατρὶ μεγίστῳ | αἰδοίων ἀφροῖο γονή, ὑπέδεκτο δὲ πόντος | σπέρμα Διὸς μεγάλου·

[65] Edwards (1991) 207.

[66] Col. 13.7–9: ἐν τοῖς α[ἰδοίο]ις ὁρῶν τὴγ γένεσιν τοὺς ἀνθρώπου[ς] | νομίζον̣[τας ε]ἶν̣αι τούτωι ἐχρήσατο, ἄνευ δὲ τῶν̣ | αἰδοίων [οὐ γίν]εσθαι, αἰδοίωι εἰκάσας τὸν ἥλιο[ν].

[67] KRS 32; and, more forcefully, Bernabé (2002) 12–13.

text of the papyrus and propose emendations only if there is no alternative left. And, as we shall see, the text does give an acceptable reading without such emendations.

(iv) As agreed by all, this problem should be viewed in conjunction with ***L10***: the expression πρωτογόνου βασιλέως αἰδοίου is likely to give us a clue on the meaning of αἰδοῖον. Now in ***L7***, we are explicitly told that Ouranos was the first king: Οὐρανὸς Εὐφρονίδης, ὃς πρώτιστος βασίλευσεν. It is, then, only natural to take the expression πρωτογόνου βασιλέως in ***L10*** as referring to Ouranos.[68] Furthermore, if Ouranos is the first king, there is no place in the genealogy for a first-born Phanes who rules *after* Ouranos. Hence, we can translate πρωτογόνου βασιλέως αἰδοίου as 'the phallus of the first-born king'.[69] I have to note that this consideration – which I regard as very powerful in relation to the problem of Phanes, i.e. that there is no place for Phanes in this genealogy – does not remove the ambiguity of the genitive construction. Nevertheless, nobody, at least to my knowledge, has proposed a reading to the effect that πρωτογόνου βασιλέως refers to Ouranos but that αἰδοίου is an adjective qualifying the sky-god.

There have however been some attempts to argue away the force of the expression ὃς πρώτιστος βασίλευσεν, 'who first of all ruled', claiming that primacy is only relative in this case. For this purpose, Rusten has adduced two passages from the Orphic *Rhapsodies OF* 111 and 139, that use πρῶτος and πρώτιστος in a qualified sense.[70] In *OF* 111 it is said of Ouranos 'he who was king among the gods after his mother, the Night',[71] even though we know that there were other kings before him. In *OF* 139, on the other hand, we read that 'Kronos was the first to rule over the humans who inhabit the earth',[72] despite the fact that there were other divine rulers before Kronos. But, as Bernabé has rightly argued, in both of these passages the circumstances qualifying primacy are made explicit: Ouranos rules first *after Night*, whereas Kronos rules first *over humans*. There is no sign of such a qualification in the Derveni papyrus.

Brisson, for his part, cites Proclus' explanation of the phrase πρώτιστος γάμος, 'first marriage', referring to Gaia and Ouranos, as a parallel.[73] There had been other unions before the marriage of Gaia and Ouranos (e.g., that of Phanes and Night), but this one, Proclus argues, is the first that can be called marriage because of the specific relationship between Gaia and Ouranos. This is also how, according to Brisson, we should understand the

[68] Bernabé (1989) 169 and (2002) 17–18. [69] KRS 32.
[70] Rusten (1985) 135 n. 31. [71] ὃς πρῶτος βασίλευσε θεῶν μετὰ μητέρα Νύκτα.
[72] πρώτιστος μὲν ἄνασσεν ἐπιχθονίων Κρόνος ἀνδρῶν.
[73] Brisson (2003) with reference to Procl. *In Tim.* 3.176.18–24 = OF 112.

phrase ὃς πρώτιστος βασίλευσεν 'who first of all ruled': there were other divine rulers before Ouranos, notably Phanes, who can be called 'first-born king', but Ouranos is the first king *stricto sensu*, because his rule is the first to show 'the political signification of sovereignty', and this is the first rule that 'implies struggle for power', as it will be contested by Kronos.[74] The argument is no doubt ingenious. Yet I cannot find it persuasive because I have difficulties in imagining that the Orphic poet could make such fine distinctions between the noun βασιλεύς, on the one hand, applicable already to Phanes, and the verb βασιλεύω, on the other, applicable only to Ouranos. Indeed, Proclus in the passage referred to does not say that the relationship between Phanes and Night was also a marriage but in a looser sense, and then the relationship between Gaia and Ouranos is the first marriage *stricto sensu*, but he says that the relationship between Phanes and Night was a union (ἕνωσις), but not a marriage. Moreover, I could not find any evidence that would show that the *Rhapsodies* applied the term γάμος or its cognates already to the relationship of Phanes and Night. If so, I would still maintain that πρωτογόνου βασιλέως and ὃς πρώτιστος βασίλευσεν should refer to the same god Ouranos in the poem commented on in the Derveni papyrus.[75]

B. External evidence

(i) We have evidence for the swallowing of the phallus of the sky-god (the equivalent of Ouranos) in the Hurrian-Hittite sources. First we have to note that the Kumarbi story, as preserved in the poem customarily called the *Kingship in Heaven*, is by common agreement one of the major sources for the Greek succession myths.[76] In this narrative, Kumarbi, the cunning-god (who in many respects is the equivalent of Kronos) severs the phallus (or with the euphemism of the text, the 'thigh' or 'knee') of the sky-god Anu, swallows it, and then becomes pregnant with three gods, among them the weather-god, the equivalent of Zeus. Kronos castrates

[74] This is also the line of reasoning taken by Parker (1995) 491, without however referring to the alleged parallel of divine marriages. He claims that in the phrase 'first born reverend king', the title king 'must here be used loosely, as a mere title of respect'.

[75] There is a specific reason why Zeus can be said to be 'born first' (πρῶτος γένετο), probably also in the poem commented on in the papyrus. For more on this problem, see 172–4 below.

[76] Cornford (1952) had already demonstrated the strong structural resemblances between Greek and Oriental succession-myths in a remarkable paper written before the publication of the Kumarbi narrative. There is an ample literature on the parallelisms between the Hittite narrative and Hesiod from the pioneering study of H. G. Güterbock, *Kumarbi. Mythen vom churritischen Kronos*, Zürich-New York, 1946 to the recent contributions of Burkert (1987b) and Bernabé (1989). For further references, see Burkert (1987b) 21–2. On the origins of the Kumarbi story, see also Komoróczy (1979) 550–2.

Ouranos also in Hesiod, but he does not swallow the severed genitals, which are thrown instead into the sea and give birth to Aphrodite. The myth of the swallowed phallus, on the other hand, is substituted with Kronos swallowing his children – an episode which eventually also results in giving birth to a 'weather-god' (i.e., Zeus) by 'vomiting'. It is possible, however, that the early version of the Orphic theogony preserved the motif of becoming pregnant due to the swallowing of the severed phallus of the sky-god Ouranos. Furthermore, it is also possible that the authors or compilers of later versions found the story unacceptable and replaced the αἰδοῖον of the first-born Ouranos with the first-born Phanes, 'the αἰδοῖος daimon'.[77]

(ii) The late Neoplatonist Olympiodorus refers to an Orphic traditon in which, in contrast to the divine succession of the *Rhapsodies*, there were only four kings: Ouranos, Kronos, Zeus and Dionysos.[78] Even if Olympiodorus does not mention the swallowing of the phallus of Ouranos, his evidence still supports this interpretation for it shows that there was indeed a version of the Orphic narrative in which Protogonos/Phanes does not appear in the succession of divine kings.[79] I have to admit, however, that it is not easy to see how Olympiodorus could have had access to this early version of the Orphic theogony.[80]

(iii) Walter Burkert has recently drawn attention to a passage in the proem of Diogenes Laertius.[81] Diogenes here argues against those who claim that Orpheus was one of the inventors of philosophy, and says the following:

> But for my part, I do not know if we should call 'philosopher' someone who declared such things about the gods, as I do not know either what we should call someone who does not hesitate to impose upon the gods the entire human condition, and even that shameless act that is sometimes committed by some people with the organ of speech (D.L. proem. 5).[82]

As there is no other trace of fellatio in the Orphic fragments, it is tempting to assume that Diogenes is alluding to the act that is described in the poem commented on in the papyrus as αἰδοῖογ κατ̣έπινεν.

77 So Burkert (1980) and (1987b); KRS 32–3; Bernabé (1989).

78 Olymp. *In Phaed.* 1.3: παρὰ τῷ Ὀρφεῖ τέccαρεc βαcιλεῖαι παραδίδονται. πρώτη μὲν ἡ τοῦ Οὐρανοῦ, ἣν ὁ Κρόνοc διεδέξατο ἐκτεμνὼν τὰ αἰδοῖα τοῦ πατρόc· μετὰ δὲ τὸν Κρόνον ὁ Ζεὺc ἐβαcίλευcεν καταταρταρώcαc τὸν πατέρα. εἶτα τὸν Δία διεδέξατο ὁ Διόνυcοc κτλ.

79 E.g. Burkert (1999b) 101 n. 67 supposes that Olympiodorus is preserving an old tradition, corresponding to the Derveni poem.

80 This objection has been raised by Brisson (2003).

81 Burkert (1999a) 81–2.

82 ἐγὼ δέ, εἰ τὸν περὶ θεῶν ἐξαγορεύcαντα τοιαῦτα χρὴ φιλόcοφον καλεῖν οὐκ οἶδα, <οὐδὲ> τίνα δεῖ προcαγορεύειν τὸν πᾶν τὸ ἀνθρώπειον πάθοc ἀφειδοῦντα τοῖc θεοῖc προcτρῖψαι, καὶ τὰ cπανίωc ὑπό τινων ἀνθρώπων αἰcχρουργούμενα τῷ τῆc φωνῆc ὀργάνῳ.

(iv) I would also add a further argument which, it seems to me, can considerably reinforce the previous ones, but which, to the best of my knowledge, has never been voiced in the context of this debate. One of the objectives of allegorical interpretations known to us is to argue away the sexual monstrosities of the mythical narratives.[83] This also can be observed in the Derveni author who is ready to advance a marvellous linguistic *tour de force* in col. 26 in order to save Zeus from committing incest with his mother. It would, then, be rather surprising to see him *introducing* a no less outrageous element at another point of his commentary. If we suppose that Zeus, in the poem, engulfed the 'reverend daimon', we have to admit that it was the commentator who introduced the phallic theme into the text. On this reading he would claim that Zeus did not swallow another divine being but a phallus – just to end up saying that this phallus was the sun. The phallic element would, as a matter of fact, be a detour in identifying Phanes with the sun. If, on the other hand, Zeus had swallowed a phallus in the poem itself, the Derveni author's interpretative manoeuvre is in accordance with both the general attitude of allegorical commentaries and his own efforts in col. 26: he tries to get rid of a sexual oddity in the mythical story by saying that the phallus of Ouranos meant nothing other than the sun. It seems possible to me that the version which eventually became more common, in which Zeus swallows Protogonos, testifies to another attempt to attenuate the archaic story which was felt increasingly outrageous.

I would conclude that both views have powerful arguments on their side, but that neither of them can settle the question definitively. In my view, the occurrence of the adjective πρωτογόνοc is the strongest argument in favour of the first view, whereas the argument from the conjunction of Οὐρανὸc Εὐφρονίδηc, ὃc πρώτιcτọc βαcίλευcεν with πρωτογόνου βαcιλέωc αἰδοίου, together with Diogenes Laertius' allusion to a fellatio attributed to the gods, is almost compelling in favour of the other option. It seems to me that the sum total of the arguments grants preference to the view according to which Zeus swallowed the severed phallus of Ouranos in the poem commented on by the Derveni author.

It is convenient to note right now that the interpretation I have opted for entails a further complication for the mythological narrative of the poem. As we have just seen, Kumarbi both castrated Anu and swallowed the

[83] Cf., e.g., Buffière (1956) 106–15 on the allegorical interpretations of the episode in the *Iliad* of the love of Zeus and Hera on Mount Ida.

genitals. In Hesiod, Kronos castrated Ouranos, as an act of dethronement, and swallowed his own children in an attempt to assure his power. Zeus, at a later point of the story, swallowed Metis, also in order to secure his newly attained kingship. Now if we suppose that, in the poem used by the Derveni author, Kronos was the one who cut off the phallus of Ouranos but it was Zeus who swallowed it, we have to face the following problem. The castration of Ouranos by Kronos was no doubt meant to be an act of dethronement also in this narrative.[84] We have to suppose, on the other hand, that Zeus swallowed the same phallus at the beginning of *his* reign. But what happened to the phallus all through the kingship of Kronos? Where did Zeus find it when it was needed again?[85] I leave this question open for the moment but I shall return to it when speaking about the peculiar relationship between Kronos and Zeus in the Orphic theogonies.

Finally, we have to note that it is unlikely that ***L5*** and ***L6*** were immediately consecutive in the poem, even though they appear in the same column of the papyrus with no lacuna separating them from each other. For if they were consecutive verses, then the phrase ὃϲ αἰθέρα ἔχθορε πρῶτοϲ in ***L6*** would refer to Zeus. Although this option cannot be completely excluded, I cannot see what mythical event the phrase could describe. But let us now return to the subsequent group of lemmata.

l5-l6-L7

τοῦτον οὖν τὸγ Κρόνον | γενέϲθαι φηϲὶν ἐκ ̣τοῦ ἡλίου τῆι γῆι, ὅτι αἰτίαν ἔϲχε | διὰ τὸν ἥλιον κρούεϲθαι πρὸϲ ἄλ̣ληλα. 14.2–4

διὰ τοῦτο λέγει· ὃϲ μέγ' ἔρεξεν. τὸ δ̣' ἐπὶ τούτωι· 14.5

Οὐρανὸϲ Εὐφρονίδηϲ, ὃϲ πρώτιϲτ̣οϲ βαϲίλευϲεν 14.6 = fr. 10.2 B.

He says that this Kronos was born from the sun to the earth because he was the cause through the sun that they were struck against each other.

For this reason he says: '(He) who did a great deed.' And after this:

Ouranos son of Night, who first of all ruled

This sequence concentrates on a previous stage of the narrative, focusing on Kronos. ***l5*** refers back to Kronos' genealogy, while ***l6*** probably alludes to the castration of Ouranos. There is one prima facie disturbing feature of these lemmata however. ***L7*** comes right after ***l6*** in the papyrus and is introduced by the expression 'in the next (verse)'. If this passage, one could argue, really

[84] This is also reinforced by the Derveni author's exegesis of ***l6*** and ***L7*** in col. 14.5–9.

[85] This problem is highlighted by Brisson (2003). Bernabé (2002) 21 speculates that the severed phallus of Ouranos 'resta dans l'espace', and that this circumstance could also help the Derveni author in assimilating it with the sun.

refers to the castration of Ouranos by Kronos then the subject of ὃϲ μέγ' ἔρεξεν must be Kronos while the object of it should be Ouranos. Therefore, Οὐρανὸϲ Εὐφρονίδηϲ should be in the accusative. This must be the reason why West both emends the text and prints Οὐρανὸν Εὐφρονίδην in his reconstruction of the poem. Admittedly, this is not a drastic correction, but such cases could nevertheless undermine one's faith in the reliability of the papyrus' evidence. Without trying to force the issue too far, there is a way to render even this emendation unnecessary. It is just conceivable that the direct object of ἔρεξεν, if it had one in addition to μέγα at all,[86] was stated in the previous verse, now lost, while ***L7*** already starts a new syntactic unit of which Ouranos is the subject.[87]

Moreover, the appearance of Ouranos in this context makes it probable that Kronos is Ouranos' son also in the narrative of the poem. This conclusion, in turn, casts some doubt on ***l5***, suggesting that it is not a real lemma but the Derveni author's allegorical interpretation. Although that cannot be excluded, it does not seem very probable that the identification of Ouranos with Helios was already present in the poem.[88] Accordingly, φηϲίν here is not to be taken as introducing a verbatim quotation, but as accompanying an interpretative paraphrase.

The somewhat awkward formulation of ***l5*** is probably also explainable by this reading. First, why is the emphasis on the fact that we are speaking about the genealogy of *this* Kronos (τοῦτον οὖν τὸγ Κρόνον)? Is there another one? No, but the author wants to emphasise that he is not speaking about the Kronos of the ignorant, but about the Kronos as he himself interprets this divine name in the poem. Kronos should be understood as the Mind that clashes things against each other, and *this* Kronos – i.e., Kronos understood in this way – is born from the sun, since it is the heat of the sun which causes the things to collide with each other.[89]

Second, γενέϲθαι (. . .) ἐκ τοῦ ἡλίου τῆι γῆι is, though not an impossible,[90] still an unusual way of expressing a parental relationship. Now if we read this phrase together with the Derveni author's interpretation of ***L6*** identifying the genitals with the sun, it becomes plausible that the

[86] Cf. the parallel passage in Hesiod's *Theogony* 209–10: φάϲκε δὲ τιταίνονταϲ ἀταϲθαλίῃ μέγα ῥέξαι | ἔργον κτλ.

[87] The asyndeton is not disturbing in this construction.

[88] Even if this identification seems odd to us, I still prefer to call it the Derveni author's interpretation and not, as West (1983) 88 does, his 'misinterpretation'. I do not think we are entitled to make such negative judgements until we have understood the internal rules of an allegorical interpretation. Otherwise, West suggests that there might have been an ambiguous relative pronoun in the poem, something like 'He [Protogonos] generated Ge and Ouranos; and to him [meaning Ouranos] Ge bore Kronos, who did a great deed to Ouranos'.

[89] On the physical theory underlying this explanation, see 254 below.

[90] Cf. LSJ ἐκ *s.v.* III. 2.

construction ἐκ + genitive and the dative does not speak about parental relationships in the literal sense, but rather transforms the mythical genealogy into a physical theory. As we shall see in the reconstruction of the cosmogony of the Derveni author, the collision of lumps of matter (derived from the name of Kronos) is immediately caused by the sun, but has crucial impact on the cosmogonic events going on on the earth. The ἐκ + genitive construction, then, expresses a causal relationship, whereas the dative may be a dative of advantage. Besides, if one accepts our previous conclusion that the word αἰδοῖον in the poem referred to Ouranos' phallus, which was then interpreted by the Derveni author as the sun, then this expression can also signify that Kronos' birth is dependent on the sun, i.e. on the phallus of Ouranos.

L8-L9

ἐκ τοῦ δὴ Κρόνος α̣ὖτις, ἔπειτα δ̣ὲ μητίετα Ζεύς — 15.6 = fr. 10.3 B.
μῆτιγ κα[ὶ μακάρων κατέχ]ω̣μ βασιληίδα τιμ̣[ήν] — 15.13 = fr. 11 B.

From him in turn Kronos, and then wise Zeus
Holding wisdom and royal honour over the blessed gods

The combination of these two full-verse lemmata can give a smooth reading, although much depends on how we fill in the lacuna in the middle part of ***L9***. Even if we do not commit ourselves to West's reconstruction of the verse, it seems convenient to take]ω̣μ as a masculine participle ending, subordinated to Zeus, the last word in ***L8***.

Furthermore, it seems reasonable to follow West in printing ***L8*** and ***L9*** right after ***L7***.[91] The discussion of the change of kingship from Ouranos to Kronos, started in the second half of the existent text of col. 14, appears to continue in the first lines of col. 15. Then ***L8*** is introduced by the expression 'and the next verse'. On this reading, the continuous sequence of ***L7***, ***L8*** and ***L9*** summarizes the royal succession from Ouranos to Zeus. This reconstruction is, by the way, not affected by the question of whether we should emend Οὐρανὸς Εὐφρονίδης into the accusative.

L10-L11

πρωτογόνου βασιλέως αἰδοίου, τῶι δ' ἄρα πάντες
ἀθάνατ̣οι προσέφυμ μάκαρες θεοὶ ἠ̣δὲ θέαιναι
καὶ ποταμοὶ καὶ κρῆναι ἐπήρατοι ἄλλα τε πάντα,
ἅ̣σσα τότ' ἦγ γεγαῶτ', αὐτὸς δὲ ἄρα μοῦνος ἔγεντο — 16.3–6 = fr. 12 B.
[νῦν δ' ἐστὶ]ν βασιλεὺ[ς] πάν̣τ̣[ων, καί τ' ἔσσετ' ἔπ]ε̣ιτα — 16.14

[91] See also Rusten (1985) 137 and Bernabé (2002) 23.

[with?] the phallus of the first-born king, onto which all
The immortals grew (or: *clung fast to) blessed gods and goddesses*
And rivers and lovely springs and everything else
That had been born then; and he himself became solitary.

Now he is king of all and will always be

The author explicitly says that our ***L11*** comes immediately after ***L10***. These consecutive verses must refer to that moment of the story when, due to the engulfment of the generative principle (the genitals of Ouranos or Protogonos according to the alternative interpretation), everything becomes interiorised in Zeus. This astonishing episode, as we have already seen, is well documented in other Orphic texts.[92] ***L11*** refers to the fact that by interiorising the whole universe, Zeus makes his rule stable.

l7-l8-L12 [. . .], ***l9-l10*** [. . .], ***l11-L13***

δι' ὅ τι δὲ | ἀὴρ ἐκλήθη δεδήλωται ἐν τοῖς προτέροις.	17.3–4
καὶ ὕστατον ἔφησεν ἔσεσθαι τοῦτον, ἔπειτ' | ὠνομάσθη Ζεὺς ḳαὶ τοῦτο αὐτῶι διατελεῖ ὄνομα ὄν, | μέχρι εἰς τὸ αὐτὸ ẹἶδος τὰ νῦν ἐόντα συνεστάθη | ἐν ὧιπερ πρόσθεṿ ἐόντα ἠιωρεῖτο.	17.6–9
Ζεὺς κεφα[λή, Ζεὺς μέσ]ṣα, Διὸς δ' ἐḳ [π]άντα τέτ[υκται]	17.12 = fr. 14.2 B.
τοῦτ' οὖν τὸ πνεῦμα Ὀρφεὺς | ὠνόμασεμ Μοῖραν.	18.2–3
Ὀρφεὺς γὰρ | τὴμ φρόνησ[ι]μ Μοῖραν ἐκάλεσεν·	18.6–7
. . . ἓν [ἕκ]αστογ κέκ[λητ]αι ἀπὸ τοῦ | ἐπ̣ικρατοῦντος, Ζεὺ[ς] πάντα κατὰ τὸν αὐτὸν | λόγον ἐκλήθη·	19.1–3
Ζεὺ̣ς̣ βασιλεύς, Ζεὺς δ' ἀρχὸς ἁπάντων ἀργικέραυνος	19.10 = fr. 14.4 B.

And the reason why it was called air has been made clear above.
And he said that this will be 'the last', because it was named Zeus, and it will continue to be his name until the things that are now have got set together into the same form in which they were floating as they were before.

Zeus the head, Zeus the middle, and from Zeus all things have their being

Now Orpheus named this breath Moira.

For Orpheus called wisdom Moira.

[92] Cf. the passages collected under *OF* 167, and especially Procl. *in Plat. Tim.* 28c where the verses adduced contain verbal echoes of ***L10***: μετὰ γοῦν τὴν κατάποσιν τοῦ Φάνητος αἱ ἰδέαι τῶν πάντων ἐν αὐτῷ πεφήνασιν, ὥς φησιν ὁ θεολόγος· τοὔνεκα σὺν τῷ πάντα Διὸς πάλιν ἐντὸς ἐτύχθη, | αἰθέρος εὐρείης ἠδ' οὐρανοῦ ἀγλαὸν ὕψος, | πόντου τ' ἀτρυγέτου γαίης τ' ἐρικυδέος ἕδρη, | Ὠκεανός τε μέγας καὶ νείατα Τάρταρα γαίης | καὶ ποταμοὶ καὶ πόντος ἀπείριτος ἄλλα τε πάντα | πάντες τ' ἀθάνατοι μάκαρες θεοὶ ἠδὲ θέαιναι, | ὅσσα τ' ἔην γεγαῶτα καὶ ὕστερον ὁππόσ' ἔμελλεν, | ἐνγένετο, Ζηνὸς δ' ἐνὶ γαστέρι σύρρα πεφύκει.

. . . existing things have been called each single name by reason of what dominates (them); all things were called Zeus according to the same principle.

Zeus the king, Zeus who rules all with the bright bolt

These lemmata most probably belong to a mini-hymn to the universal, all-governing king Zeus. The expressions are well attested in other hymnic texts, most notably in the Hymn to Zeus at the end of *De mundo* (*OF* 21a) and the other as reported by Porphyry (*OF* 168). On the basis of these parallels, it seems reasonable to suppose that the author is explaining the following verse in the first half of col. 17:

Ζεὺς πρῶτος γένετο, Ζεὺς ὕστατος [. . .] = fr. 14.1 B.

Zeus who was born first, Zeus the last [. . .]

There is one point of disagreement regarding this part of the text. The question is whether the concept of πνεῦμα, which appears in the first lines of col. 18, came from the poem, or whether it was introduced by the Derveni author. Most commentators accept the suggestion according to which it was already present in the poem, in the form of πνοιή.[93] The basis for this reconstruction is a phrase from the *De mundo* hymn, verse 5:

Ζεὺς πνοιὴ πάντων, Ζεὺς ἀκαμάτου πυρὸς ὁρμή.

West, on the other hand, claims that there is no reason to think that this verse, or the first part of it, was already in the poem commented on by the author. This verse, West argues, is absent from the version of the hymn known by the Neoplatonists and looks like a specifically Stoic interpolation. I think West is right, and we can actually provide an internal argument as well. The concept of πνεῦμα turns up in the papyrus in the following context:

(. . .) and all other things are in the air, it being breath. Now Orpheus named this breath Moira (col. 18.1–3).

The phrasing itself strongly suggests that Orpheus himself used the word Moira for the entity the Derveni author calls πνεῦμα. In addition, it is more natural to think that the author follows his usual practice and gives a physical equivalent for a mythic character than to suppose that the identification of Moira and πνεῦμα had already been present in the poem. Nevertheless, the concept of πνεῦμα is an important one, and we shall return to it in the discussion of the Derveni author's own contribution.[94]

[93] This suggestion is by Merkelbach according to Boyancé (1974) 96. Those who accept this reconstruction are Boyancé (1974); Ricciardelli Apicella (1980) 118; Funghi (1980) 80; and Bernabé (2002) 26 with n. 120.

[94] See 200–2 below.

The long series of lemmata from ***l*11** down to ***L*16** apparently speaks about the ensuing events of the story: the births of all kinds of divine beings from Zeus. First we have a number of goddesses (Aphrodite Ourania, Peitho, Harmonia, Ge, Meter, Rhea, Demeter ***l*12-*l*14**), then some divine beings personifying parts of the physical world (Okeanos, the river-god Achelous, and finally the Moon ***l*15-*L*15**). It is noteworthy, however, that Zeus does not give birth to all of these beings by simply regurgitating them in the manner of Hesiod's Kronos.[95] Let us first consider the birth of Aphrodite, which may turn out somewhat problematic.

***l*12-*l*13**

θόρνηι δὲ λέγ[ων] δ̣ηλοῖ | ὅτι ἐν τῶι ἀέρ̣ι κατὰ μικρὰ μεμερισμένα ἐκινεῖτο | καὶ ἐθόρνυτο, θορνύμενα δ' ἕκα<ϲ>τα ϲυνεϲτάθη | πρὸϲ ἄλληλα. μέχρι δὲ τούτου̣ ἐθόρνυτο, μέχρι | ἕκαϲτον ἦ̣λθεν εἰϲ τὸ ϲύνηθεϲ. 21.1–5

Ἀφροδίτη Οὐρανία | καὶ Ζεὺϲ καὶ ἀφροδιϲιάζειν κ̣αὶ θόρνυϲθαι καὶ Πειθὼ | καὶ Ἁρμονία τῶι αὐτῶι θεῶι ὄνομα κεῖται. ἀνὴρ | γυναικὶ μιϲγόμενοϲ ἀφροδιϲιάζειν λέγεται κατὰ | φάτιν. τῶγ γὰ[ρ] νῦν ἐόντωμ μιχθέντων ἀλλ̣[ή]λοιϲ | Ἀφροδίτη ὠν[ο]μάϲθη. Πειθώ δ' ὅτι εἶξεν τὰ ἐ[ό]ντα | ἀλλήλ̣ο̣[ι]ϲ̣ι̣ν̣. ε[ἴ]κειν δὲ καὶ πε̣ίθειν τὸ αὐτόν̣. [Ἁ]ρμονΐα δὲ | ὅτι πο[λλὰ . . . . . ἥ]ρμοϲε τῶν ἐ̣όντων ἑκ̣άϲτω[ι] 21.5–12

By saying 'by mating', he (sc. Orpheus) makes clear that divided up into small (pieces) they were moving and mating in the air, and as they were mating, they all got put together with each other. They were mating until each came to its like. Aphrodite Ourania, and Zeus, and to aphrodise, and to mate, and Peitho, and Harmonia are given as name to the same god. A man mingling with a woman is said by common usage to aphrodise. For (this reason) as the things that are now got mixed with one another, (this god) has got the name Aphrodite. (It has got the name) Peitho, because the beings yielded to one another; and to yield and to persuade is the same. (It has got the name) Harmonia, because many of the beings got fitted to one another.

The denomination Aphrodite Ourania seems to indicate that her first birth was due to Ouranos, much as in Hesiod.[96] In the *Theogony*, Aphrodite was born from the foam formed in the sea around the severed phallus of Ouranos. This detail, if present in the poem used by the Derveni author, makes it even more difficult to imagine the adventures of Ouranos' phallus between its amputation by Kronos and its engulfment by Zeus. It seems, on

95 Cf. West (1983) 90.

96 It has to be noted, however, that it was probably the cult title Ourania which suggested that Aphrodite has to be Ouranos' daughter. Cf. Wilamowitz, *Kleine Schriften* vol. v (2), 174 n. 1; *Glaube der Hellenen* vol. i 95 cited by West (1966) 212.

the other hand, that Aphrodite had two births in the *Rhapsodies*. First she was born as in Hesiod's *Theogony*,[97] but then she had a second birth from the semen of Zeus who was longing for but could not attain Dione. Zeus eventually ejaculates in the sea and (the second) Aphrodite is born from the foam.[98] West thinks that Aphrodite was born from Zeus' ejaculation also in the poem commented on in the papyrus.[99] Moreover, Bernabé calls attention to the Hittite myth in which Kumarbi spits out some of the semen that was contained in the swallowed phallus of Anu, and this act gives birth to a god.[100] Thus it is conceivable that Aphrodite had (a second) birth by Zeus' own semen or the semen spat out by Zeus also in the poem commented on by the Derveni author. The hapax θόρνηι in ***l*12** can refer to either of these. The interpreters suppose that the word θόρνη, a cognate of θρώσκω, means something like 'seed' or 'ejaculation' – but this is no more than conjecture.[101] Peitho and Harmonia, the other goddesses mentioned in the lemma, most probably attended the birth of Aphrodite, as her birth was attended by different divine beings both in the *Theogony* and in the *Rhapsodies*.[102]

***l*15-*l*16-*L*14**

εὔδηλον ὅτι Ὠκεανός ἐςτιν ὁ ἀήρ, ἀὴρ δὲ Ζεύς· | οὔκουν ἐμήςατο τὸν Ζᾶνα ἕτερος Ζεὺς ἀλλ' αὐτὸς | αὑτῶι cθένος μέγα. 23.3–5

οἱ δ' οὐ γινώςκοντες τọ̀ν | Ὠκεανὸν ποταμὸν δοκοῦςιν εἶναι ὅτι εὐρὺ ῥέοντα | προςέθηκεν. 23.5–7

ἶνας δ' ἐγκα̣[τέλε]ξ̣' Ἀχελώιου ἀργυ̣[ρ]οδίνε̣[ω] 23.11 =fr.16.3 B.

it is clear that Okeanos is the air and air is Zeus. It is not the case that another Zeus contrived this Zeus, but that the same one (contrived) for himself great strength.

97 *OF* 127=Procl. *in Plat. Cratyl.* 406c: παράγει οὖν αὐτὴν ὁ Οὐρανὸς ἐκ τοῦ ἀφροῦ τῶν γονίμων ἑαυτοῦ μορίων ῥιφέντων εἰς τὴν θάλαςςαν, ὥς φηςιν Ὀρφεύς· μήδεα δ' ἐς πέλαγος πέςεν ὑψόθεν, ἀμφὶ δὲ τοῖςι | λευκὸς ἐπιπλώουςιν ἑλίςςετο πάντοθεν ἀφρός· | ἐν δὲ περιπλομέναις ὥραις Ἐνιαυτὸς ἔτικτεν | παρθένον αἰδοίην κτλ.

98 Actually, Proclus speaks not about a second birth but about a second Aphrodite in *OF* 183=*in Plat. Cratyl.* 406 c: τὴν δὲ δευτέραν Ἀφροδίτην παράγει μὲν ὁ Ζεὺς ἐκ τῶν ἑαυτοῦ γεννητικῶν δυναμένων, ςυμπαράγει δ' αὐτῷ καὶ ἡ Διώνη· πρόειςι δ' ἡ θεὸς ἐκ τοῦ ἀφροῦ κατὰ τὸν αὐτὸν τῇ πρεςβυτέραι τρόπον· λέγει δ' οὕτως καὶ περὶ ταύτης ὁ αὐτὸς θεολόγος· τὸν δὲ πόθος πλέον εἷλ', ἀπὸ δ' ἔκθορε πατρὶ μεγίςτῳ | αἰδοίων ἀφροῖο γονή, ὑπέδεκτο δὲ πόντος | ςπέρμα Διὸς μεγάλου· περιτελλομένου δ' ἐνιαυτοῦ | ὥραις καλλιφύτοις τέκ' ἐγερςιγέλωτ' Ἀφροδίτην | ἀφρογενῆ.

99 West (1983) 91–2. He suggests in another context ((1983) 216–7) that the birth of Aphrodite from Ouranos' phallus was introduced only in the Hieronymus-Hellanicus version.

100 Bernabé (1987) 148–9.

101 West (1983) 91–2; Bernabé (2002) 28 n. 132 with further bibliography.

102 Hes. *Thg.* 201 (Eros and Himeros); *OF* 127 (Zelos and Apathe).

But those who do not understand think that Okeanos is a river because he (sc. Orpheus) added the epithet 'broadly flowing'.

he placed in it the sinews of the silver-eddying Achelous

This series speaks about the birth, or rather re-birth, of the river gods. What makes these verses particularly interesting is the verb ἐμήϲατο describing the way Zeus gave birth to them. This verb implies not a physical act of engendering but the intellectual act of conceiving.[103] Incidentally, I do not see any harm in following Merkelbach and reconstructing the verse echoed in ***l15*** and ***l16*** as follows:

μήϲατο δ' Ὠκεανοῖο μέγα ϲθένοϲ εὐρὺ ῥέοντοϲ = fr.16.3 B.
and he contrived the strength of the broadly-flowing Okeanos

For the birth of the river-god Achelous mentioned in ***L14***, the parallels are duly listed by Bernabé ad loc.

l17-L15

ἴϲα ἐϲτὶν ἐκ τοῦ [μέ]ϲου μετρούμενα, ὅϲα δ[ὲ μ]ὴ | 24.1–2
κυκλοειδέα οὐχ οἷόν τε ἰϲομελῆ εἶναι.

ἣ πολλοῖϲ φαίνει μερόπεϲϲι ἐπ' ἀπείρονα γαῖαν 24.3 = fr. 17.2 B.

. . . are equal measured from the centre, but those which are not round-shaped cannot be of equal limbs.

which shines for many articulate-speaking humans on the boundless earth

The main interest of these lemmata, which most probably speak about the birth of the Moon, lies in the adjective ἰϲομελῆ. Interestingly, its closest parallel is in Parmenides B 8.44 DK: μεϲϲόθεν ἰϲοπαλὲϲ πάντῃ. Moreover, it seems to be the only explicit reference to the sphericity of the moon in the classical period.[104]

L16

[αὐτ]ὰ̣ρ̣ [ἐ]π̣εὶ δ[ὴ πάν]τα Διὸ[ϲ φρὴμ μή]ϲατ̣[ο ἔ]ρ̣γα 25.14 = fr. 18.1 B.

But once [? the heart] of Zeus devised all deeds

Even though highly conjectural, Tsantsanoglou's new reading for ***L16*** offers welcome support for the claim that the Derveni author follows in the main the structure of the poem and does not impose the internal logic of his own argument on the order in which he discusses the lemmata. For ***L16***, as reconstructed by Tsantsanoglou, could provide the link between the

[103] For a similar use of the verb, describing this time the birth of Selene, see *OF* 91 and the daimon creating Eros in Parmenides' *doxa* (28 B 13 DK). Cf. ***L16*** = col. 25.14 and 180–1 below.

[104] West (1983) 92–3 and 247–8 below.

description of the second creation effected by Zeus and the diverse ensuing episodes.

***l*18-*l*19**

> μη[τρ]ὸc μὲν ὅτι μήτηρ ὁ Νο̣ῦc ἔcτιν τῶν ἄλλων̣ | ἐᾶc δὲ ὅτι ἀγαθῆc. 26.1–2
>
> ὁ δ' εἴπερ ἤθελεν ἑαυτοῦ μητρὸc | ἐμ φιλότητι ἀποδε̣ῖ̣ξαι θέλοντα μιχθῆ̣ναι τὸν | θεόν, ἐ̣ξῆν αὐτῶι γράμματα παρακλίναντι | μητρὸc ἑοῖο εἰπε[ῖ]ν· 26.9–12
>
> '[of?] mother' because the Mind is the mother of the others; and 'of his own (ἐᾶc)' because she is good.
>
> But if he had wished to show the god wishing to mingle in love of his own (ἑαυτοῦ) mother, it would have been possible for him by altering some letters and say of his own (ἑοῖο) mother.

The last episode of the story attested in the papyrus speaks about the desire of Zeus for Rhea. This sexual desire indicates that the phase of solitary procreation is over, and as the world is filled again with fellow gods and goddesses, the more usual sexual model is back. But the act is no less outrageous since Rhea is no doubt Zeus's mother in this account as well. As in the *Rhapsodies*, Rhea might have been identified with Demeter already in the poem and not only in the Derveni author's interpretation of it.[105] This would give further weight to Burkert's suggestion that the rape of Rhea resulted in the birth of Kore or Persephone.[106] But this is already an episode for which there is no evidence in the papyrus.

CONCLUSION: THE POSSIBLE NARRATIVE STRUCTURE OF THE POEM

The order of the lemmata in the papyrus discloses the following narrative structure for the poem. First, we have a proem in which the poet, speaking in the first person, warns the profane to 'put doors to their ears' and then describes the subject-matter of his poem (***l*1-*L*1**). After the proem, he starts with Zeus' accession to royal power, with special emphasis on the fact that all this happened according to a divine decree or prophecy (***L*2**). Then, the poet speaks about Night, the source of the prophecy (***l*2-*l*4** and possibly ***L*3**). In the next lemma we hear something about another divine prophecy, probably issued by Zeus' father (***L*5**). ***L*6** speaks about an act of swallowing; an act which in other theogonies is associated with Zeus, eager to secure his

105 Cf. col. 22.7–16. 106 Cf. KRS 32; West (1983) 93–4.

newly obtained power. It is not clear, however, what exactly is swallowed by Zeus in this narrative. It can either be a first-born divinity called in other Orphic poems by the names of Phanes, Protogonos and Metis, or, more probably, the severed phallus of Ouranos. After the verse or verses describing the act of swallowing, we have some verses referring cursorily to the earlier events and kings: Kronos, Kronos dethroning Ouranos the first king by a 'mighty deed' (***l5***, ***l6***, ?L7). ***L7***, ***L8*** and ***L9*** summarise the line of succession: first Ouranos, then Kronos, after him Zeus. ***L10*** and ***L11*** bring us back to the results of the engulfment: all beings get inside Zeus. Now Zeus is the only one, and the poet praises this 'king of all' in hymnic phrases (***l7***, ***L12***, ***l8***, ***l9***, ***h10***, ***h11***, ***L13***). ***h12***, ***h13***, ***h14***, ***h15***, ***h16***, ***L14***, ***h17*** and ***L15*** then relate the second creation of different kinds of gods by Zeus. The last episode already speaks about a further stage of the story: Zeus' incestuous relationship with his mother.

This structure, although it does display some oddities, is not an impossible one. One might see a difficulty in the succession of the lemmata where a passage describing the previous stages of the royal succession (***l5***-***L9***) is squeezed between verses referring to the act of swallowing (***L6***), on the one hand, and other verses relating the surprising consequences of this act (***L10*** and following), on the other. We can, however, suppose that this is an example of the ring-composition[107] characteristic of ancient narratives: the poet interrupts the chronological sequence of events at the point where he considers that the description of an antecedent event or events is important for the understanding of the present stage of the story.[108]

Another unusual feature of this structure, not unrelated to the previous problem, is that it does not present the story in a chronological order, but starts *in medias res* at the moment when Zeus obtains royal power. If the above reconstruction is valid at least in its main lines and thus the poem had such a structure, this composition indeed marks an important difference from other known theogonic narratives. But this problem already leads us to the next chapter where I shall examine the specific features of the theogonic narrative of the poem commented on by the Derveni author.

[107] So, e.g., Calame (1997) 70.

[108] This interpretation, by the way, is a further argument for thinking that Zeus swallowed the phallus of Ouranos. The poet inserts the episode of Kronos dethroning Ouranos in order to explain the history of phallus now swallowed by Zeus.

CHAPTER 4

The interpretation of the poem

A bare reconstruction of the poem commented on by the Derveni author is still not sufficient for our purposes. In trying to evaluate the Derveni author's exegesis of the Orphic poem, we have to measure his interpretation against our own interpretation of the same text – or at least of what we have of it. This we can only do in a meaningful way if we have a relatively clear idea for ourselves about the theological and cosmological import of the poem. This, I suggest, requires us to go beyond the first level of the meaning, that is, beyond the mere literal meaning of the words and sentences, or the reconstruction of the succession of events the poem describes.

It should be obvious even from a first reading of the papyrus that the Derveni author is not making a bare summary of the plot of the poem, and his aim is not simply to paraphrase the literal meaning of the words and sentences. He states explicitly that, due to the peculiar, enigmatic nature of Orpheus' language, we cannot stop at a literal interpretation, as such an interpretation would not reveal the genuine meaning of the poet.[1] Nevertheless, he sees clearly that his interpretation is not the only possible one. He constantly opposes his own reading to another interpretation which he ascribes to 'the ignorant'. Now, the paraphrases of different passages of the poem he puts into the mouth of 'the ignorant' correspond exactly to what we would take as the literal meaning of the text. It is then reasonable to suppose that the author ascribed to the ignorant that interpretation which he himself thought to be the obvious, primary, literal meaning of the text: this is how far a superficial reading of the poem goes.

[1] Col. 13.5–6. It means that, according to the classification used in Long (1992), the author treated the poem as allegorical in the *strong* sense: its author 'Orpheus' composed it with the intention that its interpreters go beyond its surface meaning. It is worth noting that with this claim the author places himself at the head of a long tradition (or more exactly, what we have of this tradition): many allegorical interpretations from later times start with the assertion that the author of the text commented upon expresses himself enigmatically. This is what legitimates an interpretation which goes beyond the level of literal meaning. Cf. Brisson (1996) 84–7.

Thus, the distinction he introduces between the reading of the ignorant and a more thorough genuine understanding of the text can be taken as the distinction between a literal interpretation and an interpretation which goes further. By discarding the reading of the ignorant, he makes clear that he does not want to stop at the level of the literal meaning. He was well aware of the primary meaning of the text, but wanted to replace it with a 'deeper', more ingenious and unexpected reading.[2] His intention was to give an interpretation and not a paraphrase.

With these factors in mind, I do not think it is legitimate on our part to compare the Derveni author's avowedly non-literal interpretation of the poem with a primary, literal understanding of the same text, just to contend at the end, not without a touch of triumph and contempt, that it is clear from his interpretative remarks that the poor Derveni author was unable to understand the text he was commenting on. This procedure is hardly more legitimate than objecting to a modern literary critic that his or her interpretation does not stick to the literal meaning of the text interpreted. Obviously, the success or failure of such an interpretative enterprise lies elsewhere. In our case, it means that we cannot dismiss the Derveni author purely on the basis that he is making an 'allegorical' interpretation, and consequently necessarily speaks nonsense. Instead, we have to make our own non-literal interpretation – our own version of 'allegoresis', as it were – and measure the Derveni author's results against ours.

But how can we make our own 'deeper' interpretation of the poem? Clearly, this will lead us onto slippery ground. Even if we move with special caution, our results can only be tentative; and this is so for a number of reasons. To begin with, we are dealing with a badly fragmented text, the reconstruction of which, as we have already seen, is far from definitive. Second, and more important, we are dealing with a text which speaks in mythological terms. The language of myths, as was recognised already in antiquity, invites an interpretation which goes beyond the level of literal meaning (weak allegory in the sense of Long (1992)), but there neither is nor probably has ever been a fixed code for deciphering the underlying sense. The interpretation of a mythological text is always a risky business, since it attempts to explain away the inherent ambiguities of metaphorical expression and to make explicit what is intrinsically implicit.[3]

[2] See also Most (1997) 125.

[3] The hermeneutics of a mythical text is an immensely complex matter which cannot be treated here in detail. In a way, however, my attempt to understand the exegesis of the Derveni author is just a case study on this problem. On the limits of eliciting theoretical considerations from a mythical text, see e.g. Lloyd (1987) ch. 1. See, on the other hand, the analyses in Burkert (1999b). For a useful discussion of the different approaches to myth in antiquity, see Brisson (1996).

In view of these hermeneutical problems, I suggest the following procedure. Given that the poem speaks about the different generations of gods and divine kings, and above all, about the way Zeus establishes his kingdom, the obvious way seems to be to compare it to the most widely circulated Greek text describing the same story: Hesiod's *Theogony*. This method rests on the premiss that if a Greek composed a theogony or added modifications to an already existing version, his effort was and was taken to be a reaction to the accepted Hesiodic account, with the aim of giving a different answer to the same implicit question or set of questions.[4] Thus if someone gave a different account of the generations of the gods and of the events that led to the establishment of a divine kingship, this was, at least in some respects, in order to convey some kind of meaning which can be expressed in terms of a departure from the standard Hesiodic version.[5]

As a matter of fact, this approach does not require more than the application to the genre of theogony of the generally accepted assumption according to which the web of Greek myths, until a certain moment in intellectual history, meant a living language: in the absence of a unique, sacred, ultimate and codified version individuals could propose their own versions by rearranging traditional mythems and, occasionally, by adding new ones in order to express complex messages.[6] Such creative and meaningful use of the traditional mythological material is no doubt best testified in the genre of tragedy. Admittedly, the intended meaning or the content of the disagreement is probably considerably more difficult to make explicit in the case of rival theogonies than in the genre of tragedy. Yet, as shown by Burkert and others, such analyses can lead to conclusions of prime interest. Thus by analysing and comparing the Derveni poem to the Hesiodic *Theogony*, I hope to elicit some of the specific tenets of the poem, some of which may prove to be relevant to the evaluation of the Derveni author's reading of the text.

With these qualifications in mind I shall try to give an interpretation of the poem focusing on the following points. Are there any recognisable cosmological ideas expressed in the poem? Are there any specific theological and eschatological views in it? How does it treat the different divine beings,

4 So e.g. Guthrie (1952b) 83–4. On parallel (creation) myths as alternative answers for the question of origin in Mesopotamia, see e.g. the introductory essay in Bottéro and Kramer (1989).

5 This is not countered by the fact that Plato and Aristotle repeatedly refer to a genealogical point in Homer i.e., that Okeanos and Tethys were the first gods. These mentions are always related to a specific point about the primeval role of water or, through Heraclitus' river fragment, to the flux of nature.

6 Cf. Bremmer (1987), esp. 3–4.

and, above all, the supreme deity, Zeus? I shall try to consider these questions in two steps. First, I shall say something about the specific way in which the Derveni poem presents its content. Then in a second phase, I shall speak about the content itself, underlining the distinctive features of the story implied in the Derveni poem.

THE SPECIFICITIES OF THE PRESENTATION

One should note first of all that the Derveni poem is not as easily classifiable as is generally maintained. Although it certainly contains elements of a theogonic nature, it cannot necessarily be defined as a theogony. West, for example, at the beginning of his Prolegomena to Hesiod's *Theogony* defines a text as 'theogonic' if it speaks about 'the origin of the world and the gods, and the events which led to the establishment of the present order'.[7]

It is clear that the Derveni poem – at least as reconstructed above – only partially conforms to this definition.[8] Although it at least mentions the events which led to the establishment of Zeus' rule, it is not primarily interested in the origin of the world and in the birth of the gods preceding Zeus. Apparently, the poet dealt with the earlier developments and generations only in so far as they were significant for the understanding of the deeds of Zeus (or perhaps for the understanding of the ensuing events described in the now lost part of the poem).

This consideration appears to be reinforced by the way the poet defines the subject matter of his poem at the end of the proem. For judging on the basis of ***L1*** ([ο]ἳ Διὸc ἐξεγ̣ένοντο [ὑπερμεν]έοc βαcιλῆ̣οc col. 8.2), the proem seems to specify that the poem will concentrate on the figure, deeds and descendants of Zeus. West's *exempli gratia* reconstruction for the lost lines of the proem is apparently based on this assumption.[9] It is true that we only have the last line of the proem, and have no textual evidence

[7] West (1966) 1. The same stands with regard to West's alternative definition in *The Orphic Poems*: 'by theogony I mean a poem of which the major part consists in an account of the gods from the beginning of the world to the present date' (West (1983) 68). It is noteworthy that these general definitions focus exclusively on thematic features and disregard the formal and functional differences.

[8] On somewhat different grounds, Janko (1986) 156 says that 'It is a legitimate question whether we should call it a theogony at all', to finally claim that it is better to call it a *hieros logos*.

[9] West (1983) 114.

[Ζηνὸc παμμεδέοντοc ἄνακτοc θέcκελα ἔργα,]
[ὅcca μελαίνηc Νυκτὸc ὑπ' ἐννεcίηιcι τέλεccεν,]
[ἠδὲ καὶ ὁπλοτέρων μακάρων γένοc αἰὲν ἐόντων,]
ο]ἳ Διὸc ἐξεγένοντο [ὑπερμεν]έοc βαcιλῆοc.

for the reconstruction of the preceding lines. In principle, it is perfectly conceivable that our ***L1*** figured at the end of a longer list: '[I shall sing of Ouranos . . . Kronos, . . . Zeus, and all the gods] who were born from Zeus the mighty king.'[10] But even this last, extant line of the proem in itself provides an interesting comparison with Hesiod's *Theogony*. For in the proem of the *Theogony*, Hesiod asks the Muses to sing about 'the sacred race of the deathless who live forever, those who were born from Earth and starry Heaven' (vv. 105–6). Thus he specifies the primordial couple, Gaia and Ouranos, as the ancestors of all the gods. ***L1***, by contrast, names Zeus as the ancestor of the divine beings we will hear about. This may either imply that, on this occasion, the poet is not interested in the previous generations, or hint at the fact, right at the beginning, that by the peculiar act of giving new birth to the gods, Zeus can be seen as the ultimate ancestor of the other divine beings.[11] In either case, the last line of the proem clearly suggests that Zeus will be even more important in this poem than he is in Hesiod's *Theogony*.

An additional but related feature is that, at least if the main lines of the reconstruction proposed above are accepted, the Derveni poem does not recount the events in a chronological order. The narrative begins with Zeus gaining his royal power, and the poet only briefly mentions the previous stages of the genealogy at a later point. This feature marks out an important difference not only from Hesiod's *Theogony* but also from other theogonic narratives which belong to a larger family of texts, including the theogonies composed in Babylonia, and the two cosmogonic accounts in Genesis. The basic pattern of these narratives sets out the events and generations in chronological order: 'In the beginning there was . . . then came . . . and then'.[12] Hesiod, after a long proem, also starts off with the question 'who was born first of the gods?' to reply immediately that Chaos was (vv. 115–16). In Hesiod, the chronological principle holds even if in some cases one branch of the family tree is followed to the end before the poet turns back to the previous generation of another branch.[13] It seems likely that the later Orphic poems and prose accounts with a theogonic content also observed a chronological order, although this point cannot be definitely proved on the basis of the available evidence. Anyhow, this apparent peculiarity of the Derveni poem suggests again that we are dealing with a text that does not readily fit into the genre of 'theogony'.

[10] See also the alternative proposed by Calame (1997) 67 n. 3: 'Sing Muses the first entities and those which were born from Zeus, the all-powerful sovereign.'

[11] On the interpretation of this episode, see section 172–4 below.

[12] Cf. Burkert (1999b) 91–2.

[13] West (1966) 37–8.

This poem, then, was not primarily interested in describing 'the origin of the world and the gods' by giving a catalogue of the successive divine generations and by describing the events that led to the establishment of the present world-order in an essentially chronological order, but set out instead to tell us 'the story of Zeus'.[14] This conclusion will turn out to be crucial for the assessment of the Derveni author's theology.

Apparently, a possible way to circumvent this difficulty is to call the Derveni poem a 'hymn' instead of calling it a 'theogony'.[15] This is certainly an option, but not unproblematic either. At any rate, in view of the fact that 'hymn' is a notoriously elusive category, one has to make clear in what sense one is using the term. One has to note, first of all, that the semantic field of the term 'hymn' as used in English and other modern languages does not exactly correspond to the semantic field of the word ὕμνος as it was used in ancient Greek.

Indeed, a plausible supplement in col. 7 suggests that the Derveni author himself may have referred to the poem by the word ὕμνος.[16] Yet, even if he did so, it does not imply much; passages such as *Od.* 8.429 and Aeschylus, *Agam.* 709 show that the word could refer to any sung poem. True, the term ὕμνος more often had the more specified meaning: 'a song directed to the gods'.[17] On the other hand, the author, as noted above, once refers to some *other* poem, or collection of poems, presumably also attributed to Orpheus, by calling them *Hymns.*[18] This reference suggests – without of course proving it – that the poem he was commenting on was not usually thought of as belonging in these (? Orphic) *Hymns.*

It is even more doubtful whether we should apply to the Derveni poem the term 'hymn' according our own usage, as defined by experts on Greek religion. In this narrower sense, 'hymn is a sung prayer', and as such necessarily contains an invocation to the god[19] in which the god is addressed either in the vocative or in the accusative; the first in such formulae as, e.g., 'Hermes, listen to me! . . .' or 'Zeus . . . I hail you!', the latter in the invocation of the type 'I call Dionysos . . .' or 'I pray to Demeter . . .'.[20]

[14] Obviously, the now lost part of the poem could modify this conclusion. It is possible that, if we knew the end of the poem, we should rather say 'the story of Zeus and Dionysos'.

[15] E.g. Most (1997) consistently uses the term 'hymn' to describe the poem. See also Bernabé (2002) 4–6.

[16] Cf. col. 7.2: . . ὕ]μνοϲ [ὑγ]ιῆ καὶ θεμ[ι]τὰ λέγο[ντα· As Tsantsanoglou (1997) ad loc. notes, .cε]μνόν is also possible, but less likely.

[17] See the semantic analysis of the word in Bremer (1981) 193–4 and W. D. Furley (1995) 31–2.

[18] Col. 22.11: ἔϲτι δὲ καὶ ἐν τοῖϲ Ὕμνοιϲ εἰρ[η]μένον κτλ.

[19] So, e.g., Bremer (1981) 194: 'The invocatio is an indispensable element.'

[20] Rudhardt (1991) 264.

As far as I can see, neither of these patterns finds its place in the Derveni poem, and no reconstruction known to me tries to insert an invocation into it.

In a looser sense a mere exaltation of a certain god is also traditionally called a hymn; for example, the hymnic description of Zeus at the end of the Pseudo-Aristotelian *De mundo* is nevertheless usually called a 'hymn to Zeus', even though – at least in the part quoted in the *De mundo* – it does not contain an invocation, and speaks about the god in the nominative. Yet the Derveni poem does not readily fit even in this looser category. For if ***L1*** is indeed the end of that section in which 'Orpheus' determines the subject matter, then, accordingly, the poem does not focus on one god, but on a whole race of them. Moreover, if the poem culminated in the story of the birth, death, and rebirth of Dionysos, as many interpreters assume, then it is not clear whether this part can also be taken as part of the praise of Zeus. Therefore, it is also not unproblematic to call the Derveni poem a 'hymn'; in effect, it turns out to be somewhere mid-way between a short epic (cf. also the use of the *in medias res* technique) and a hymn.

Janko's suggestion that the poem was a *hieros logos* certainly remains an attractive possibility.[21] However, the scarcity of comparative evidence makes it difficult to give substance to this term.[22] All in all, the generic definition of it seems to remain an open problem and thus I shall continue to call it a 'poem'.

THE RECONSTRUCTION OF THE THEOGONIC PLOT

The fact that the Derveni poem cannot simply be called a theogony does not mean that it does not rely on a theogonic narrative, or, at least, that a story of a theogonic nature is not implicit in it.[23] For one cannot relate the story of Zeus without speaking about his accession to the highest power,

[21] Janko (1986) 158. This hypothesis can in fact be reinforced by the fact that, at least on the interpretation I propose on p. 327–8 below, the author of the papyrus refers to Orpheus' text as a *logos* in col. 25.11.

[22] See, however, Baumgarten (1998).

[23] Cf. also West (1983) 85 'The poet knows and presupposes a complete account, differing from Hesiod's, of the earlier part of the divine history, but his interest is concentrated on Zeus and the younger gods.' Although I agree with West on this point I cannot follow him in doubling the theogonic narrative presupposed in the poem commented on by the Derveni author. West claims that there was a theogonic narrative used by the author of the Derveni poem (West calls this hypothetic version the 'Derveni theogony') but that this narrative itself was an abridged version of a more complete version (West calls it the 'Protogonos theogony'), for which there is absolutely no textual support. For a critique of this theory, see Brisson (1985). Perhaps the best way to put it is to say with Bernabé (2002) 5 that the poem commented on in the papyrus presupposes the existence of other theogonic narratives without being an abridged version of a longer poem.

and, for this, the poet has to allude to the succession myth which, in turn, is naturally intertwined with a theogonic narrative. In the same vein, the story of Zeus, his accession to kingship, and the way he consolidated his power is the very backbone of Hesiod's *Theogony*.[24] Thus, the connection between a hymn to Zeus, the succession myth, and a theogonic narrative must necessarily be very close; the succession myth is also the glorification of the ruling god, and conversely, a hymn to the ruling god has to contain the outlines of the succession myth and a theogonic narrative as well.

Let us now make an attempt to reconstruct at least the main elements of the theogonic plot implicit in the poem by taking the divine generations and the events that led to the establishment of Zeus's kingdom in chronological order. My method will be the same as for the reconstruction of the poem. The starting point will always be the evidence of the papyrus, and primarily the lemmata, already discussed in the previous chapter. I will thus reverse the methodology adopted by some commentators, who base their reconstructions of the theogony of the Derveni poem on the information we have on some other versions of the Orphic theogony. West, for example, in his book on the Orphic theogonies, starts off with the reconstruction of the admittedly late Orphic *Rhapsodies*,[25] to use this material for the reconstruction of the other, earlier versions of the Orphic theogonies. The reason he gives for this procedure is that since our information is the most complete in the case of this version, the other versions can be supplemented on the basis of the *Rhapsodies*. A major objection to this method is that it is founded on the unwarranted assumption that the different versions of the Orphic theogony are basically homogeneous, so that what is lacking from one, we can reconstruct on the basis of the others. Specifically, West assumes that the theogonic plot implicit in the Derveni poem is roughly identical with that of the *Rhapsodies*.[26]

I too shall make use of the comparative evidence, such as testimonies on and quotations of other Orphic theogonies, provided by some later ancient

[24] On the point that Zeus's story is the climax and aim of all (Greek) theogonies, see Schwabl (1978) 1203–13.

[25] West (1983) 70–5.

[26] See also Rusten (1985) 122, who makes this assumption explicit. Rusten claims also that the validity of it has been proved by Merkelbach: 'The events of the Derveni Theogony (although not its individual verses) can be assumed to be largely identical with those in the later poem ascribed to Orpheus (called the *Rhapsodies*) which the Neoplatonists Proclus and Damascius quote and summarize at length. Therefore the fragments of these poems (. . .) have considerable (although not absolute) authority for reconstructing the text of the poem which the commentator had before him (. . .)'. This general assumption, and the methodology based on it, have been convincingly challenged by Brisson (1997). We shall see, however, that on some points, notably on the question of the relevance of Aristophanes' account, Brisson too is prone to give too much weight to the external evidence.

authors. I shall, however, evoke these pieces of evidence only in so far as they can corroborate the internal evidence or suggest solutions to problems which we cannot solve on the basis of the papyrus alone. If the internal and external evidence are contradictory, I shall give overall preference to the internal evidence. Also, if there is no trace of an episode or god in the papyrus, I prefer to leave the question open and not to supply information merely on the basis of the other Orphic theogonies. As I shall have to refer to these other texts transmitted under the name of Orpheus, it seems necessary to give a rough outline of their content, as well as of our sources on them. I shall however consider only those parts of these theogonies which are relevant for our purposes.

THE DIFFERENT VERSIONS OF THE ORPHIC THEOGONY

To begin with, it is not altogether clear how many different versions we have to deal with. The traditional threefold partition, used by most modern scholars, goes back to Damascius' account of the different versions of the Orphic theogonies.

In the *De principiis*, Damascius elaborates his metaphysical system, heavily dependent on Proclus' system, concentrating on the top levels of the ontological hierarchy, down to the third level of the intelligible triad. For Damascius, just as for Syrianus and Proclus, the full exposition of such an ontology has to include not just the theological interpretation of the second half of Plato's *Parmenides* but also an allegorical reading of the *Chaldean Oracles* and the so-called Orphic *Rhapsodies*. Yet, towards the end of the *De principiis*, and after a lengthy discussion of the *Chaldean Oracles*, Damascius provides us with a more wide-ranging survey and allegorical interpretation of the different theogonical traditions.[27] Within this survey, he mentions three different versions of the Orphic theogonies known to him: the *Rhapsodies*, the version transmitted under the names of Hieronymus and/or Hellanicus, and the version known by Eudemus of Rhodes.

The theogony of the Rhapsodies

The most common of these three, Damascius claims, and the one customarily interpreted by the Neoplatonists, was the Orphic *Rhapsodies* (ῥαψῳδίαι Ὀρφικαί, probably the same as the Ἱεροὶ λόγοι ἐν Ῥαψῳδίαι κδ' listed in

[27] On Damascius' account, and especially on his use of the material from Eudemus of Rhodes, see Betegh (2002).

the Suda s.v. 'Orpheus').[28] As a matter of fact, the cursory summary given in the *De principiis* accords well with the testimonies and quotations we find in other Neoplatonic sources, notably in Proclus, and can thus be supplemented by them. This means that the material is quite abundant for this version. It has to be noted, however, that most of our information consists of prose accounts, and it is not always easy to tell whether our sources give a close paraphrase of the original verses, or whether they provide their own allegorical interpretation of them.[29]

The first generations and events of the Orphic *Rhapsodies* are as follows. First, there was Chronos or Time. From Chronos, Aither and Chasma or Chaos were born.[30] Chronos places an egg in Aither.[31] The egg is also called white tunic (ἀργῆτα χιτῶνα) or cloud (νεφελήν). The god Phanes breaks out from the egg. Damascius reports that Phanes is also called Metis and Erikepaios,[32] while according to some other sources, he also has the names of Protogonos,[33] Eros,[34] Zeus and Bromios.[35] Phanes is a marvellous creature. He/she is of both sexes, has golden wings, four eyes, four horns and four heads (that of a ram, a bull, a lion and a serpent).[36] Phanes is the first king.[37] This is the point where Damascius, having obtained a sufficient number of divine beings to fill in the top levels of his ontological hierarchy, breaks his summary of the theogony of the *Rhapsodies*.

We can nonetheless continue the story, relying on some other Neoplatonic testimonies. In the next generation, we have Night, the daughter and lover of Phanes.[38] Phanes passes the sceptre, the sign of royal power, to

[28] Ἐν μὲν τοίνυν ταῖς φερομέναις ταύταις ῥαψῳδίαις ὀρφικαῖς ἡ θεολογία κτλ. *De princ.* 1.316.18 Ruelle.

[29] In the following analysis I rely on the reconstructions of West (1983) and of Brisson (1985a), (1985b) and (1990). However, I always reconsider the evidence they cite, and do not always accept the interpretation they give.

[30] Cf. also the verse quoted in *OF* 66 = Procl. *in Plat. Rempubl.* 2.138.8 Kroll.

[31] *OF* 70 = Damasc. *De princ.* 1.111.17 Ruelle.

[32] On the occurrences of this name of unknown origin and etymology see West (1983) 205–6. It is noteworthy that in other, not specifically Orphic sources, this name is connected with Dionysos.

[33] *OF* 86 = Herm. *in Plat. Phaedr.* 247c; and the Hymn to Protogonos in *OF* 87.

[34] *OF* 74 = Procl. *in Plat. Tim.* 31a. [35] *OF* 170 = Procl. *in Plat. Tim.* 29a.

[36] For these features, see the texts under *OF* 76–81.

[37] *OF* 108 = Syrian. in Arist. *Metaph.* 1091b4. Cf. also *OF* 107 Alexand. Aphrodis. in Arist. *Metaph.* 1091b4 and Procl. *in Plat. Tim.* prooem. e.

[38] *OF* 98 = Procl. *in Plat. Tim.* 31a. There is some indication, however, that there were actually three Nights. In John Malalas' account (*OF* 65 = *Chronogr.* IV 89), we have Night at an earlier stage, together with Aither and Chaos. Furthermore, Proclus in one passage speaks about Nights in the plural (*OF* 98 = *in Plat. Tim.* 31a). Hermias explicitly mentions three Nights, although one wonders if it is not just the result of a Neoplatonic allegorical interpretation operating with triads (so Lobeck (1829) 503; Kern (1888) 6; Holwerda (1894) 311–12). Proclus also mentions in a doxographic passage the idea of three Nights (*OF* 99 = *in Plat. Tim.* 39b). In most places, however, both Proclus and Damascius speak about one Night, the daughter, lover, and successor of Phanes. I think, *pace* West

Night.[39] Phanes, at least according to the Neoplatonic accounts, is allotted demiourgic tasks, such as the creation of the heavenly bodies and of the earth.[40] It is not clear, however, how this activity is related to the point that Night gives birth to Ouranos and to Gaia.[41] In any case, Night voluntarily gives the sceptre to Ouranos, who thus becomes the third king.[42] From the marriage of Ouranos and Gaia, a great number of gods are born, among whom are the next royal couple Kronos and Rhea.[43] Kronos castrates Ouranos[44] and becomes the fourth king. The severed phallus is thrown into the sea, and Aphrodite is born from the foam.[45] Kronos, unlike the kings before and after him, rules wantonly.[46] Zeus, son of Kronos and Rhea, but nurtured by Night, takes the royal power from his father by castrating him.[47] Kronos, as well as Night, nevertheless helps Zeus at the next stage with oracular advice:[48] Zeus has to swallow Phanes. With the engulfment of this primordial god, the whole world is swallowed into the belly of Zeus, so that now he can give a new creation to everything.[49]

Once the universe is created anew under the rule of Zeus, and new gods inhabit the world, the king Zeus copulates with numerous goddesses. The result is the birth of a large number of gods, some of whom are from the traditional Greek pantheon, others apparently specific to Orphism. From these relationships, the most important is the one which Zeus pursues with his mother Rhea. Incidentally, Rhea, as mother of Zeus (Διὸc μήτηρ) has been identified with Demeter.[50] Rhea/Demeter bears Kore, with whom Zeus commits another act of incest. The child born from this latter relationship is Dionysos.[51]

(1983) 70, and Brisson (1985a) 393, (1985b) 39, (1987) 58 and (1990) 2885, that the evidence is not enough to establish that there were 'three Nights' in the *Rhapsodies*. I do not think, for example, that the sentence in *OF* 106 = Procl. *in Plat. Cratyl.* 404b ἔχει δὲ παράδειγμα τὴν Νύκτα, – θεῶν γὰρ τροφὸc ἀμβροcίη Νὺξ λέγεται –, ἀλλ' ἐκείνη μὲν νοητῶc is enough to establish that Night was also Phanes' mother.

39 *OF* 101 = Procl. *in Plat. Cratyl.* 396b.

40 *OF* 98 = Procl. *in Plat. Tim.* 32b.

41 *OF* 109 = Herm. *in Plat. Phaedr.* 247d.

42 See the different texts from different sources collected under *OF* 101, *OF* 107 and *OF* 111.

43 West (1983) 71 reconstructs the Titanic and other generations born from Ouranos and Gaia with reference to Athenagoras' account (*OF* 57 = *Pro Christianis* 18 p. 20.12 Schw.), which he also believes to be a summary of another version of the Orphic theogony.

44 *OF* 154 = Porphyr. *De antro nymphar.* 16 p. 67.21 Nauck; and Procl. *in Plat. Tim.* 35b.

45 *OF* 127 = Procl. *in Plat. Cratyl.* 406c.

46 *OF* 101 = Procl. *in Plat. Cratyl.* 396b.

47 *OF* 137 = Procl. *in Plat. Cratyl.* 396b; *OF* 154 = Porphyr. *De antro nymphar.* 16 p. 67.21 Nauck.

48 *OF* 101 and 129, both from Procl. *in Plat. Cratyl.* 396b; *OF* 155 = Procl. *in Plat. Cratyl.* 391a. Cf. also the testimony of Damascius under *OF* 155.

49 *OF* 82 = Procl. *in Plat. Tim.* 33c; *OF* 85 = Procl. *in Tim.* 29a; *OF* 129 = Procl. *in Plat. Crat.* 396b; *OF* 167 = Procl. *in Plat. Tim.* 29a etc.

50 *OF* 145 = Procl. *in Plat. Cratyl.* 403e.

51 It is worth noting that the evidence on these latter episodes is scanty, and unclear. Brisson (1990) 2885 thinks for example that there is just one relationship implied, but the same Rhea is called

The young Dionysos, sixth and last in the series of kings, is enthroned by Zeus.[52] The jealous Titans kidnap Dionysos, tear the child into seven pieces and eat from his flesh.[53] Athena saves the still beating heart of Dionysos, and from the heart a new Dionysos is created.[54] Zeus punishes the Titans with his thunderbolt, then chains them up and sends them into Tartarus.[55] Mankind, at least according to Olympiodorus, is created from the ashes of the Titans hit by Zeus's thunderbolt.[56] The story may have continued,[57] but the rest is not directly relevant for our purposes.

The Hieronymus-Hellanicus version

After having set out the first generations of the *Rhapsodies*, as well as two different allegorical interpretations of them, Damascius turns to another version of the Orphic theogony known to him. This version was 'reported under the names of Hieronymus and Hellanicus, if they were not the same person'.[58] In this theogony, Damascius relates,

> From the beginning, there was water, and the matter (ὕλη)[59] out of which earth was coagulated. (. . .) A third principle, after the two, was born out of these two (I mean out of water and earth), which was a serpent, which had a head of a bull, that of a lion, and, in the middle, the face of a god, and it had wings on its shoulders. His name was Unageing Time (Χρόνοϲ ἀγήραοϲ), and also Heracles.

Demeter as the mother of Zeus, and Kore as Zeus' daughter. Hence, on this reading, we would have the same triplication of a goddess that Brisson supposes also for Night, with the same type of doubly incestuous relationship that Phanes had with Night.

52 *OF* 107 = Procl. *in Plat. Tim.* prooem. e; *OF* 207 = Procl. *in Plat. Tim.* 42d; *OF* 208 = Procl. *in Plat. Cratyl.* 396b.

53 *OF* 210 = Procl. *in Plat. Tim.* 35a and 35b.

54 Cf. the different texts collected under *OF* 210 and *OF* 214, especially Procl. *in Plat. Tim.* 35b in *OF* 210.

55 Damasc. *in Plat. Phaed.* 1.7. But how could the Titans be chained and be sent to Tartarus once they were reduced to ashes? We might have here a conflation of different traditions. Cf. Linforth (1941) 328–9.

56 On the details of this anthropogony, on its relationship to Orphism, and especially on the theory of Olympiodorus (*OF* 220 = *in Plat. Phaed.* 1.3–6), according to which there is an element of the divine present in the human body, because the Titans had eaten from Dionysos before they were hit by the thunderbolt, and thus their ashes contain something of the god, see Brisson (1992). Brisson casts doubts on the authenticity of the episode in which humankind was born from the ashes of the Titans by pointing out that, on the one hand, this episode appears only in Olympiodorus and, on the other hand, this account shows traces of an alchemical procedure. For Burkert's vindication of authenticity, see Burkert (1999b) 101 n. 67.

57 Cf. West (1983) 75.

58 We do not really know any more about the persons mentioned. On the very vague hypotheses concerning their identity, see West (1983) 176–8.

59 Zoëga proposed ἰλύϲ, which, in a sense, would be much more appropriate in this context. The reading is rejected by Diels. Kern adopts ὕλη, while West translates ἰλύϲ.

United to him, there was Ananke, being of the same nature (*or*: being the same as Nature), and double-bodied[60] Adrastea, having her arms stretched out in the whole cosmos, and touching its limits. I think that it stands for the third principle, according to the essence, with the difference that she was made of both sexes indicating the generative principle of the whole. And I suppose that the theology in the *Rhapsodies* left out the first two principles together with the one preceding the two and left in silence, and started off from the third after the two, because this was the first effable and appropriate for human ears. For this is the highly honoured unageing Time in it (sc. in the theogony of the *Rhapsodies*), the father of Aither and Chaos. At any rate, also according to this (sc. the theogony of Hieronymus and Hellanicus), this Time, the serpent, has a triple offspring – moist[61] Aither, it says, and unlimited Chaos, and the third after these, misty Erebos (. . .) and among these, it says, Time generated an egg; also this tradition declares that it was born from Time and among these, because it is from these that the third intelligible triad comes forth. But what is this triad? The egg; the dyad of the two natures in it, the male and female, and the plurality of the various seeds in between; and thirdly, after these, the incorporeal god, having golden wings on his shoulders, which has the head of a bull grown to his flanks, and a gigantic serpent, showing the form of all kinds of animals. And also this theology praises the First-born (Protogonos) in a hymn and calls him Zeus the minister of the whole and of the entire cosmos, wherefore he is also called All (Pan). This much is offered by this theology on the intelligible principles. (1.317.15 Ruelle)

The last sentence implies that the story continued, but with the being called Protogonos/Zeus/Pan Damascius has reached the third member of the third triad (that of the intelligible-intellective), and he does not want to pursue the allegorical interpretation into the next triad (that of the intellective).

The outline given by Damascius largely agrees with the plot described by the second-century Christian apologist Athenagoras.[62] A notable difference is that in Athenagoras' account we do not have two principles at the beginning: water is instead the absolute beginning,[63] while mud is formed

[60] Reading διcώματον (instead of ἀcώματον) with Gruppe, followed by Brisson. In Athenagoras' account (see below) Phanes is called διcώματοc. It seems that the expression means 'having both sexes' in both of these passages.

[61] Or 'intellectual', if we read νοερόν instead of νοτερόν.

[62] Athenagoras, *Pro Christian.* 18, p. 20 Schwartz. Most probably Athenagoras and Damascius used the same source. Cf. West (1983) 179–80. KRS 24–6 treats the evidence of Athenagoras as being a variant of the Hieronymus-Hellanicus version.

[63] The beginning of Athenagoras' account might actually suggest that the first principle was not water *simpliciter*, but the river-god Okeanos: 'The gods, as they (the Greeks) say, did not exist from the beginning, but each of them was born just as we are born. And this is agreed by them all, Homer saying: "Okeanos the genesis of the gods, and mother Tethys" (*Il.* 14.201), and Orpheus – who was the original inventor of the gods' names and recounted their births and said what they have all done, and who enjoys some credit among them as a true theologian, and is generally followed, above all about the gods – also making their genesis from water: "Okeanos, who is the genesis of them all" (Ὠκεανόc, ὅcπερ γένεcιc πάντεccι τέτυκται *Il.* 14.246)' (trans. West). Cf. Jaeger (1947) 220 n. 57. The question is discussed at length by West (1983) 184–90. After a survey of the water-gods

from water. This, I suppose, can be explained by the fact that Damascius needed a dyad (composed of the 'one-all' and 'the all-one') before Time (the 'unified') to complete his first triad. Athenagoras' testimony, in turn, can be supplemented by a long scholium to Gregory of Nazianzus, apparently directly dependent on it.[64]

Athenagoras gives some clues on the later developments of the story as well. According to his evidence, the egg split into two, the upper part becoming the sky or Ouranos, while the earth or Ge was formed from the bottom part. The first-born and serpent-shaped double-bodied Phanes leaps forth from the egg. From some verses, quoted in another passage, we learn *en passant* that Phanes gives birth to Echidna. From the marriage of Ouranos and Ge, the Moirai (Klotho, Lachesis and Atropos), the Hundred-handers (Kottos, Gyes, Briareos) and the Cyclopes (Brontes, Steropes, Arges) were born. The children were sent to Tartarus because Ouranos learnt that he would be overthrown by one of his sons. The angry Ge gave birth to the Titans. There is a lacuna at this point in Athenagoras' account, but further on we learn that Kronos takes power by castrating Ouranos. Kronos then swallows his sons, while Zeus in turn sends Kronos to Tartarus and makes war on the Titans. From a list of all sorts of details judged by Athenagoras to be horrendous, we can learn that Zeus swallowed Phanes in this version, too. Further on, Zeus, in the form of a snake, makes love to his mother Rhea. He also mates, again in the form of a snake, with Persephone, who is also called Athela and Kore, and Dionysos is born from the relationship.

Although the similarities prevail, we should note a number of differences between this version and that of the *Rhapsodies*.[65] (i) There is water, or water and mud, at the beginning, Time being secondary to these elements.[66]

in the different Oriental theo-cosmogonies, as well as their relationship to the respective time-gods, West concludes, I think unconvincingly, that the water-gods – presumably Okeanos and Tethys – were present at the earliest stage not only in the Hieronymus-Hellanicus version, but already in the earliest form of the Orphic theogony, which West calls 'the Protogonos theogony'. If it were the case, it is not clear why the source of Athenagoras and Damascius would have given an allegorical interpretation only of Okeanos and Tethys. For had the names of these deities been present in their common source, I see no reason why both Athenagoras and Damascius would have left them out.

64 Schol. in Gregor. Nazianz. *Or*. 31. c. 6.

65 Brisson concludes at the end of a detailed analysis of the sources that the Hieronymus-Hellanicus account is not a different version, but a Stoicising allegorical interpretation of the *Rhapsodies* (Brisson (1990) 2897–914, esp. 2911–12).

66 West (1983) 183 assumes that the introduction of water before Time is the result of the incorporation of Zeno's allegorical interpretation of Hesiod: 'Zeno says that Hesiod's "Chaos" is water, from the settlement of which mud comes into being, and when that solidifies the earth is established' (Schol. A. R. 1.496–8b = *SVF* 1.29.17). Provided that, as we shall see, there is a constant interplay between the different Orphic theogonies and the allegorical interpretation of the earlier theogonies, this seems to be a plausible explanation. The reference to Zeno is, however, superfluous if one follows West in supposing that Okeanos and Tethys were the primeval deities; already Plato and Aristotle interpreted the water-gods as implying primitive elementary water.

(ii) Iconographically, Phanes gets very close to Chronos. (iii) Besides Aither and Chaos, Erebos appears among the children of Time. (iv) Night does not appear at all.[67] Correspondingly, Ouranos and Gaia are not Night's offspring, but are formed from the two halves of the egg. The absence of Night also affects the royal succession. If this theogony, or the interpreters of it, wanted to have the traditional six kings, they must have considered Chronos or some other early divinity a king. (v) There might also be a problem with the Titans. West remarks that, on the one hand, Zeus fought a war against the Titans before becoming king, and probably victoriously sent them into Tartarus, and that, on the other hand, there is no mention of the Titans eating young Dionysos (a detail Athenagoras, hunting for monstrosities, would probably have mentioned).[68] It would imply that the end of the story, and the account of the birth of man, were different in this version. This argument *ex silentio* is not absolutely compelling, however.

The version known to Eudemus

Having summarised the genealogy of the Orphic *Rhapsodies* and that of the Orphic theogony according to Hieronymus and Hellanicus, Damascius turns to the testimony of Eudemus:

> The theology described in the Peripatetic Eudemus as being that of Orpheus is silent about the entire realm of the intelligible for it is completely inexpressible and unknowable by the method of exposition and narration: it made its start from Night, from whom also Homer begins, although he (sc. Homer) did not make his genealogy continuous. For we should not believe Eudemus when he says that he (sc. Homer) begins from Okeanos and Tethys. For he too manifestly knows that Night is the greatest divinity, so that even Zeus feels awe before her:
>
> 'for he feared to do what would be hateful for swift Night' (*Il.* 14.261) But let also Homer himself begin from Night.[69]

If we ignore Damascius' allegorical, Neoplatonic interpretation of the data provided by Eudemus, we are left with the meagre but significant piece of information that the Orphic theogony known to Eudemus began with Night. The second half of the passage, dealing with Homer, shows that Eudemus himself subscribed to the view, present in Plato and Aristotle, and probably stemming from Hippias, according to which the Homeric

[67] West (1983) 208 includes Night in his summary of the reconstructed plot of the Hieronymus-Hellanicus theogony on the analogy of the *Rhapsodies*. I cannot find Night in the relevant testimonies.
[68] West (1983) 181.
[69] Damascius, *De princ.* 3.162 Combès-Westerink = 1.319 Ruelle.

genealogy started with Okeanos and Tethys.[70] This is an important indication that in his report on the Orphic theogony, Eudemus was not influenced by the alternative tradition according to which Night was the most ancient deity in Homer as well.[71]

There have been some attempts to guess the rest of the genealogy. Alderink (1981) 37, for example, assumes without explication that Ouranos and Gaia were the offspring of Night in the theogony referred to by Eudemus. West, on the other hand, claims that we find the following generations of the same genealogy, ascribed to 'descendants of gods' in Plato's *Timaeus* 40e:[72]

> From Ge and Ouranos the children born were Okeanos and Tethys; from these, Phorkys and Kronos and Rhea and all of the brood; from Kronos and Rhea, Zeus and Hera and all their brothers and sisters we hear tell of; and again from these more children.

West does not consider it a difficulty that Eudemus tells us only that, in the version he knows, Night was the first deity, while there is absolutely no mention of Night in Timaeus' account. Plato had to leave out Night, West argues, for in Timaeus' account all gods are engendered by the Demiurge and we have just learnt, on the other hand, that night is something merely produced by the earth's shadow (40c) and is a unit of time.

To be sure, Timaeus ascribes this account to the offspring of gods, i.e. poets – note the plural – considered to be 'divine'. Furthermore, these divine poets are neither Hesiod nor Homer, as we have already seen that in other Platonic dialogues the Homeric genealogy started with Okeanos and Tethys, and not with Ouranos and Ge. The two most probable, but not the only, remaining candidates would thus be Orpheus and Musaeus. Furthermore, there are indications that Plato was familiar with some version of the Orphic theogony,[73] while he never refers to Musaeus in such a context. These considerations certainly suggest that if we really want to ascribe this account to a particular mythical poet, the most probable candidate is Orpheus.[74]

I think, however, that it is not necessary to identify the account in the *Timaeus* with any particular, existing theogonical account. Its characters and events seem to be left intentionally unspecific. Furthermore, it could

[70] Plato *Crat.* 402a–c; *Tht.* 152d–e; Arist. *Metaph.* 983b20–984a5. On the dependence of these passages on Hippias' *Synagoge*, see Snell (1944) and Mansfeld (1986).

[71] Cf. KRS 19. [72] West (1983) 117–19.

[73] Cf. e.g. the famous passage in *Phlb.* 66c about the 'sixth generation'.

[74] KRS 16 claim that the phrase 'off-spring of gods' clearly shows that Plato is describing here an Orphic view.

actually weaken Plato's case if Timaeus' account could have been unambiguously identified with any of the known theogonies. I assume, therefore, that it is not more than an *exempli gratia* theogony, which nevertheless resembles, in varying degrees, some of the theogonies then known. Moreover, I do not see why it is necessary, or even probable, that we should connect Eudemus' testimony with this passage. As West himself argues, in all probability there were numerous theogonic accounts attributed to Orpheus already at an early stage; even if we accept that the genealogic scheme in the *Timaeus* was based on a theogony attributed to Orpheus, why is it so certain that Eudemus referred to the same account?

Brisson, in turn, seems to give too much weight to the comic ornitho-theogony in Aristophanes' *Birds*. Brisson actually takes Aristophanes to be our primary source on the 'early version' of the Orphic theogony, only corroborated by the evidence of Eudemus on Night.[75] The theogony, by which Aristophanes' birds try to testify their high descent, runs like this:

> Chaos and Night and dark Erebos were first and broad Tartarus; but neither Ge nor Air nor Ouranos existed. In the infinite bosoms of Erebos, dark-winged Night begets a wind-egg, from which, with the seasons coming around, the much-desired Eros comes to light, his back shining with golden wings, like the swift whirling of winds. This one (sc. Eros), mingling with winged Chaos in a secluded recess[76] over broad Tartarus hatched out our race, and first led it into light. There was no race of immortals before Eros mixed everything together; but when things mingled with others, Ouranos was born, and Okeanos and Ge and the imperishable race of all the blessed gods. In this way, we are by far the oldest of all the blessed ones. (Ar. *Birds* 693–703 = *OF* 1)

Clearly, this account shows important similarities with the later Orphic theogonies we have just reviewed. Most notably, the birth of the shining, winged Eros out of the egg prefigures the birth of the winged Phanes, as described both in the *Rhapsodies* and in the Hieronymus-Hellanicus version. Furthermore, we have Night in the first generation, a feature which could suggest that this account refers to the same Orphic theogony about which Eudemus informs us.

However, the first point does not necessarily imply that Aristophanes is referring specifically to an Orphic cosmogony. The evidence for the egg-motif in Orphic theogonies is late;[77] and, on the other hand, Eudemus, as reported by Damascius, tells us that there was a cosmic egg both in the

[75] Brisson (1985b) 38; (1990) 2876–7. [76] On the reading μύχιος, see Dunbar (1997) ad loc.
[77] Dunbar (1997) ad v. 697 suggests that the influence could go the other way around as well: Aristophanes' language about Eros could influence later Orphic theogonies.

theogony attributed to Epimenides, and in the theogony of the Sidonians.[78] This means that, for a fourth-century author, the concept of a cosmic egg was not necessarily connected to Orphism. To the second point, we can object that in the theogony of the *Birds*, Night is not an absolute beginning, but just one of four such deities who regularly appear in the first generations of different theogonies. Moreover, as Kirk, Raven and Schofield observe, most of the *Birds* cosmogony is derived from Hesiod's *Theogony*.[79] At any rate, it is more natural to assume that Aristophanes created a comic pastiche of different theogonies[80] than to think that he faithfully reproduced a version of the Orphic theogony. And even if one sticks to the idea that Aristophanes' main source was an Orphic theogony, one still cannot determine how much comes from that theogony and how far Aristophanes altered it. Consequently, I do not think we are entitled to use the *Birds* to reconstruct the rest of the theogony referred to by Eudemus, or to use it as a source for an independent version of the Orphic theogony.

Some other versions?

In Apollonius Rhodius' *Argonautica* (1.496–511 = *OF* 29) Orpheus sings a theo-cosmogony to the Argonauts. In the first stage of this account the earth, the sky and the sea were all mixed into one form, until they got separated by strife. In the next stage the celestial bodies (stars, moon and sun) appear, then the mountains, the rivers and the animals were born. The first to rule on snowcapped Olympus are Ophion and Eurynome. Kronos and Rhea yield the kingship by force, and the first rulers are thrown into Okeanos. Kronos and Rhea rule over the Titans. Zeus is still in the Dictean cave, not yet armed with the thunderbolt and lightning, which he will receive from the Cyclopes.

West claims that 'we have no reason to expect a précis of an actual Orphic theogony known to Apollonius'.[81] His main argument consists in observing that the Ophion-Eurynome episode is taken from Pherecydes. I cannot see why it is more likely that it was Apollonius himself who conflated an existing Orphic theogony (which West identifies as his hypothetical Eudemian theogony) with Pherecydes' story than that there was such an Orphic theogony. I do not think we have enough evidence to decide between these two options, and that hence we can rule out the possibility of such a theogony.

[78] We also find an egg in the Phoenician theogony Damascius reports on the authority of Mochus.
[79] KRS 28. So also Dunbar (1997) 437–9.
[80] So Guthrie (1955) 104; Schwabl (1978) 1473; West (1983) 117.
[81] West (1983) 127.

The theogony sung by Orpheus in Apollonius has sometimes been related to a fragment from Euripides' *Melanippe the Wise*:

This account is not my own; I had it from my mother. Heaven and Earth were once in a single form; but when they were separated from each other into two, they bore and delivered into the light all things: trees, winged creatures, beasts reared by the briny sea – and the human race.[82]

It is noteworthy that one part of the fragment appears with three other Orphic texts on an alabaster bowl dated to the third century AD.[83]

Furthermore, we find a genealogy explicitly ascribed to Orpheus in (?pseudo-) Alexander of Aphrodisias' commentary on Aristotle's *Metaphysics* N 1091b4 (= *CAG* 1.821.5 = *OF* 107). Kern places this text among the testimonies on the *Rhapsodies*, whereas West, while excluding the possibility that it was an independent version, considers it to be a testimony on the Hieronymus-Hellanicus version. The account itself is rather puzzling. We have two lists, one for the genealogy, and another for the succession of kings. The former has Chaos, Okeanos, Night, Ouranos, Zeus, in order of birth, the latter has the royal succession Erikepaios, Night, Ouranos. Now neither of these two lists agrees with any of the other versions we have seen. Neither of them has Chronos, so important in the *Rhapsodies* and in the Hieronymus-Hellanicus version, and neither of them begins with Night, as the version known to Eudemus had it. They do not accord with the allegedly Orphic accounts of the *Birds* or that of the *Timaeus* either.

We also find a theo-cosmogony in the *Homilies* of pseudo-Clemens of Rome, transmitted in Greek, and in the Latin translation by Rufinus of its slightly modified version known as *Recognitions*. Although in different contexts, both versions contain a description of the first generations of gods and cosmic beings expressly attributed to Orpheus.[84] The fact that in these texts the presumably primary mythological material is mixed with physical allegoresis makes the evaluation difficult. Furthermore, the version in the *Recognitions* is apparently closer to the Hesiodic *Theogony*. There are, however, a number of indications which suggest that these accounts are to be treated as versions independent of both the *Rhapsodies* and the

[82] Fr. 484 Nauck (trans. Collard, Cropp and Lee).

[83] Delbrueck and Volgraff (1934), quoted by Alderink (1981) 103, n. 16. Collard, Cropp and Lee (1995) ad loc. voice some doubts about the Orphic nature of Euripides' text: 'the combination need not have been specifically 'Orphic' for Eur[ipides], although Hippo's [sc. Melanippe's mother's] cosmogony, astrology, and healing practices together do have an exotic character'.

[84] Cf. *OF* 55 and 56.

Hieronymus-Hellanicus version.[85] The story starts with Chaos; then comes the egg; the bottom part of the egg submerges and becomes Pluton, and Kronos – not a separate god but identified with Chronos – swallows this heavy matter. The middle part, covering the first sediment, becomes Poseidon. The upper part of the egg, being purer and lighter, fiery in nature, goes upward and is called Zeus, and so forth. Even if we try to remove the features resulting from a physical allegoresis, what remains seems to be difficult to harmonise with the other versions we have seen.

As an outcome of this survey it has, I hope, become evident that, *pace* West, one cannot establish a linear descent between the different versions. It seems to be more appropriate to assume that the individual accounts were related to each other by a loose scheme of 'family resemblance'.[86] Indeed, we cannot find a single mythem which would occur invariably in all the accounts and could thus create the core of Orphic theogonies. And again, it is problematic whether we can distinguish at all between 'authentic' Orphic accounts on the one hand, and travesties and forgeries on the other. For, clearly, all Orphic texts are pseudepigrapha. So when, for example, Apollonius Rhodius puts a theogony into the mouth of his character Orpheus, or when pseudo-Clemens makes his character Appion recount a theogony attributed to Orpheus, I do not see why these accounts should be 'less authentic' than, say, the *Rhapsodies*. The only criterion for treating a text or a narrative as 'Orphic' is that someone at some moment attributed it to Orpheus.[87]

So, as I take it, there was a very flexible general scheme and a largely open set of mythems, which could allow almost any variation and admit any kind of influence from different corners, always according to the specific purposes, philosophical filiations, and imagination of the respective authors. This flexibility and receptiveness to philosophical and other influences is, I believe, the main feature of these theogonies. In contradistinction to Hesiod, there was no codified version of the Orphic theogony. In this respect, one should not be misled by the fact that most of the texts collected

[85] Kern, followed by Brisson, lists these texts under the heading 'the theogony of Hieronymus-Hellanicus'. West treats the texts as related to the *Rhapsodies*. Amersfoort (1981) argues that we should treat these two texts as evidence for an independent version, or versions.

[86] Alderink (1981) 20–3 tries to define Orphism as such by the means of the concept of family resemblance.

[87] A possible objection to this assertion would be that most people in antiquity thought of some of these texts that they were 'really' by Orpheus, whereas few would have said the same of, say, the version in Apollonius' text. It still seems impossible to draw a divide between 'authentic' and 'inauthentic' Orphic narratives along these lines. Another possibility would be to associate authenticity with a certain kind of religious feeling or role in religious group-organisation or cult practice. However, especially in the lack of relevant information, these cannot be hard criteria either.

in Kern's *Orphicorum Fragmenta* probably come from the *Rhapsodies*. This is only due to the contingency that the majority of the fragments and testimonies were transmitted in the works of fifth- and sixth-century Neoplatonists and that these thinkers – highly interested in Orpheus, and treating him as the main authoritative figure of a sacred pagan tradition, a possible rival to Christianity – mostly used one text, the *Rhapsodies*.[88]

So, from the sixth century BC to the end of the pagan era, in the absence of a codified version, and under the name of a near-divine figure of high authority, one could forge theogonic narratives always according to the specific theological or philosophical context. These different theogonies could effectively re-mythologise the outcome of the allegorical interpretations of previous theogonies, or could integrate philosophical or theological tenets in general by taking over or modifying verses and passages from older versions, and adding new material.[89] This flexibility is the very reason why the theogonies circulated under the name of Orpheus could continue to be operative, adequate, and influential until the time when pagan mythology ultimately faded in the Western world. The power of the Orphic theogonies lay in their ability to remain always relevant, and this was due to their extreme versatility. This flexibility and applicability, to a great extent, is a generic feature of myths; nonetheless, two factors – that the different versions were circulated under the name of a single author, the authoritative and somewhat mysterious Orpheus, and that theogony was always felt to be able to convey philosophical or quasi-philosophical messages – combined to give these narratives a particular status.

This conclusion can also be taken as a justification for the methodology I shall apply. For in accordance with the above contentions, I prefer to accept the existence of a higher number of relatively independent versions rather than try to argue away their differences by trimming the testimonies. It also follows that, without some kind of internal textual evidence, it is not possible to make a conclusive deduction about a feature in one version by reference to the content of the other versions. This obviously also applies to the theogony implicit in the Derveni poem.

88 Even this might be too broad a generalisation for it is possible that, for instance, Hermias or Olympiodorus did not use the same version as Proclus and his disciple Damascius. Cf. Burkert (1999b) 101 n. 67.

89 Brisson (1987) and (1991) are highly instructive in this respect. These studies show that Proclus and Damascius could 'prove' the Orphism of Plato precisely because the *Rhapsodies* had already contained not only the re-mythologisation of the results of Stoic physical allegory, but also basic elements of Middle- and Neo-Platonism. On successive strata of myth and allegorical interpretation within the Orphic tradition, see also Kingsley (1995) chapters 9 and 10.

THE THEOGONY IMPLICIT IN THE DERVENI POEM

In this section I shall try to assign a place in a genealogical scheme to the divine names we find in the papyrus, and, if possible, to indicate the interactions between them. This can be done in most cases on the basis of the internal evidence, occasionally corroborated by parallels in other texts. There are some less prominent deities, however, whose place in the family I cannot define conclusively. It is in the nature of such narratives that definite places are not necessarily allocated to all the divine beings, and the picture may be blurred further by occasional inconsistencies. The following reconstruction of the theogonic plot implied in the Derveni poem naturally involves some overlaps with the previous chapter.

THE FIRST GENERATION

The remaining text of the poem does not state explicitly who the first divinity was. Thus we have to combine two methods. First we have to decide who is the oldest among the divinities mentioned in the extant text, and then we have to compare our results with the external evidence.

Night

On the basis of the remaining text of the papyrus, corroborated by independent testimonies, we can assert with fair certainty that Night was in the first generation of gods. The reasons for this conclusion are as follows.

We saw in the previous chapter that the poem commented on in the papyrus had the divine succession Ouranos-Kronos-Zeus.[90] Moreover, the verse speaking about Ouranos (***L8*** = col. 14.6) states explicitly that he was the first king. In the same verse, Ouranos is called Εὐφρονίδης, 'the son of Night'.[91] It seems we have to stop at this stage since we do not find any indication in the text suggesting that there was a generation before Night.

We have, moreover, independent evidence showing that there were Orphic and other theogonies which put Night at the beginning of the divine genealogies. As noted above, Damascius reports that such an Orphic

[90] The main pieces of evidence are ***L8*** = col. 15.6 and ***L7*** = col. 14.6 as discussed 124 and 122–3 above.

[91] The divine name Οὐρανὸς Εὐφρονίδης appears only in an inscription from Bithynia, dated to the first century AD. On the inscription, see Kaibel (1878) n. 1029, 6. The matronymic Οὐφρονίδης has been extensively discussed by Tortorelli Ghidini (1985), where she argues, not very convincingly, for an independent Orphic divinity Οὐφρόνη.

genealogy was known to Eudemus. We learn from the same source that Epimenides' theogony also started with Night,[92] and Aristotle too alludes to such genealogical constructions.[93] To this we can probably add the testimony from Philodemus to the effect that Chrysippus, in the first book of his *Physics*, said that Night was the first goddess;[94] this was probably part of an allegorical interpretation of a theogonical genealogy, Orphic or otherwise.[95]

On the basis of these pieces of evidence, internal and external, it seems fairly certain that Night was in the first generation of gods in the theogony of the Derveni poem. This conclusion is shared by the majority of scholars.

Aither?

Curiously enough, commentators usually do not even mention that there is some evidence showing that Night was not alone in the first generation.[96] However, it does not seem unlikely to me that we have to include Aither in the first generation. The evidence for this is as follows.

First, we have the following verse from the poem quoted by the Derveni author:

αἰδοῖογ κατ̣έπινεν, ὃc αἰθέρα ἔχθορε πρῶτοc (***L6*** = col. 13.4)

This verse poses two problems relevant to our present discussion: first, the reference of the pronoun ὅc, and, second, the meaning of the verb ἔχθορε+acc. The first, at least from this perspective, is relatively simple. Depending on the interpretation we give to the notorious αἰδοῖον, the relative pronoun ὅc can either refer to Phanes or to Ouranos. Yet, on either reading, the reference is to the 'first-born king'.[97] The second problem is more complicated, but crucial when we try to determine the temporal relationship between the aither and the 'first-born'. To begin with, the verb generally refers to the birth of a god.[98] Yet the construction with the accusative remains unclear. The accusative can, in principle, here mean two things. It can either mean the source or the direction of the movement

[92] Eudemus fr. 150 Wehrli; Philodemus *De pietate* 47a, both in 3 B5 DK. Philodemus probably used for his own doxographical surveys the data collected by Eudemus. On this point cf. Henrichs (1972) 78 n. 28; KRS 19 n. 2.

[93] Arist. *Metaph.* 1071b27; 1072a8; 1091b4.

[94] Philod. *De pietate* 14, (col. 18) = *SVF* 2.636: [καὶ δ]ὴ κἀν τῶι πρώτ[{ιcτ}ω]ι τὴν Νύκ[τ]α̣ θεάν φηcιν [εἶ]ν̣α[ι] πρωτίcτην (Henrichs (1974) 18).

[95] So e.g. Hahm (1977) 239 and Henrichs (1972) 79 n. 36.

[96] Cf. Betegh (1996) 464; see also Burkert (1999b) 8.

[97] Cf. col. 16.3: πρωτογόνου βαcιλέωc αἰδοίου.

[98] For the parallels, see 115 n. 56 above.

described by the verb. Edwards adopts the former, while Rusten the latter view.[99] In this case, I think Edwards' construction is better attested. Yet, on either construal, it is clear that the aither must have existed before the birth of the 'first-born'. Even though both Rusten and Edwards ignore this consequence, it implies that we should count the aither as a primordial being.

Walter Burkert has recently suggested an alternative hypothesis, however, which would reverse this temporal order. Burkert adduced Aeschylus' fr. 15 Radt θρῴcκων κνώδαλα that Hesychius explains as ἐκθορίζων καὶ cπερματίζων. On this basis, Burkert claims, the phrase in the papyrus should be understood as 'produced the aither by ejaculation.'[100] Hence, the 'first-born' comes first and then creates the aither. This interpretation, moreover, has an important bearing on the understanding of the first lines of the next column. The commentator gives here his exegesis of the same line we are discussing now. He says that [. . .] || ἐ]{κ}χθό̣ρηι τὸ{ν} λαμπροτάτον τε [καὶ θε]ρ̣μ̣ό̣[τ]ατον | χωριcθὲν ἀφ' ἑωυ̣τοῦ (col. 14.1–2). Even though the rest is just as problematic (and I shall have to return to these problems when discussing the author's physical interpretation),[101] it quite clearly echoes the same construction with the accusative. I have three problems with Burkert's reading. First, I think it indubitable that, in a theogonic context, the verb refers regularly to the birth of a god.[102] Second, I find it difficult to insert πρῶτοc in this construction. We have to note that this relative clause qualifies the god who originally possessed the phallus[103] swallowed now by Zeus. It seems to me much more natural to describe this god by saying 'who was born first' than by saying 'who created the aither first'. As I shall argue later, what is important here is that Zeus internalises the primordial source of becoming and not what the first creation of this primordial god was. Moreover, the phrase 'who created the aither first' would naturally require some sequence, such as 'and then . . .', which has absolutely no place in this part of the narrative.[104]

Thirdly, we have the evidence of other versions of the Orphic theogony to reinforce the hypothesis that aither was there at the very beginning. As

99 Edwards (1991) 207 cites as parallel *Anth. Pal.* 9.371.1 δίκτυον ἐκθρώιcκοντα . . . λαγωόν. Ts. ap. L&M construes the phrase in the same way. Rusten (1985) 125 n. 9 cites as parallel *Hymn. Apoll.* 119 ἐκ δ' ἔθορε πρὸ φόωcδε. L&M accepts Rusten's reading.

100 Burkert (1999a) 82. Both Janko (2001) 24 with n. 125 and Bernabé (2002) 21–2 accept Burkert's reading.

101 See 234 below.

102 See all the parallels cited in 115 nn. 56 and 98 above.

103 Or Phanes on the alternative reading.

104 I think the clumsiness of this dangling πρῶτοc becomes apparent also in Bernabé's translation: 'il dévora le phallus (du Ciel), qui avait *préalablement* éjaculé l'éther' (my italics).

the son of the primordial god Chronos, Aither appears at a very early stage in the Orphic *Argonautica*,[105] in the Hieronymus-Hellanicus theogony,[106] and in the *Rhapsodies*,[107] the latter two as reported by Damascius. We also have a more detailed account of the role of Aither in the *Rhapsodies* by John Malalas.[108] Moreover, in all these accounts Aither is related to Phanes, who is usually called 'the first-born' in these later versions. In some transmitted verses, quoted most probably from the *Rhapsodies*, Aither is explicitly called the father of the first-born Phanes:

πρωτόγονος Φαέθων περιμήκεος Αἰθέρος υἱός[109]

Apart from the evidence from later sources on the different versions of the Orphic theogonies, we can also refer to Hesiod. In the *Theogony*, Aither appears at an early stage. He is son of Night, and, together with Hemere, is part of the first generation of 'bright' beings.[110] Furthermore, Philodemus claims that the author of the *Titanomachy* assigned an absolute priority to Aither.[111] Also, Chrysippus and Posidonius appear to be acquainted with and give the allegorical reading of a theogony in which Aither was in the first generation.[112]

Therefore, it seems highly probable that we have to deal with Aither from the very beginning in the theogonic plot implicit in the poem commented on by the Derveni author. It is true that the phrase ὃς αἰθέρα ἔχθορε πρῶτος would rather suggest a medium or region of the world than a divine being, but this distinction is hardly ever operative in the first generations of any of the different theogonies.[113] If so, Aither/aither can be taken not just as the male[114] partner of Night, but also as the 'bright' counterpart of the 'dark' goddess.

[105] *OF* 37 = Orphic *Argon.* 13. [106] *OF* 54 = Damasc. *De princ.* 1.317.15 Ruelle.
[107] *OF* 60 = Damasc. *De princ.* 1.316.18 Ruelle.
[108] *OF* 65 = John Malalas *Chronogr.* 4.89 p. 74 Dind.
[109] *OF* 73 = Lactant. *Divin. instit.* 1.5.4–6 p. 13.13 Brandt. Cf. *OF* 74 = Procl. *in Plat. Tim.* 31a: ὁ Φάνης "περικαλλέος Αἰθέρος υἱὸς". In a scholium to Proclus' commentary on the *Timaeus*' proem we find that Aither is called the fourth king, after Phanes, Night and Ouranos. The provenance and signification of this isolated view is not clear to me.
[110] Hes. *Thgn.* 124.
[111] Philod. *De piet.* 137.5: ὁ δὲ τὴν Τιτανομαχίαν γράψας ἐξ Αἴθερος φησίν (sc. τὰ πάντα).
[112] D.L. 7.138 Χρύσιππος . . . καὶ Ποσειδώνιος . . . τὸν αἰθέρα . . . ὃν καὶ πρῶτον θέον λέγουσιν.
[113] Also in the prose account of *OF* 54 = Damasc. *De princ.* 1.317.15 Ruelle Aither appears to be more like a medium in which the egg, from which Phanes will break out, is placed. This, however, might already be the interpretation of Damascius.
[114] Although αἰθήρ is feminine in Homer (with the exception of *Od.* 19.540), the word is masculine in Hesiod's *Theogony*. Cf. West ad *Thgn.* 124.

The problem of Chronos

It has been suggested by scholars that the first god in the genealogy implicit in the poem was not Night, but Chronos, the primordial time-god. Let us take a look at the evidence listed for this option.

The internal evidence, if it can be taken as evidence at all, is very meagre. The case was presented in the most elaborate form by Tortorelli Ghidini,[115] and received detailed criticism by Brisson.[116] The comparative evidence is much stronger. For, both in the Hieronymus-Hellanicus version and the *Rhapsodies*, Chronos occupies a distinguished role. Chronos is the absolute beginning in the *Rhapsodies*, while in the Hieronymus-Hellanicus version Chronos is the first god coming from the primordial water (or, alternatively, from Okeanos). Yet this evidence becomes compelling only if we base our reconstruction on the assumption that, as Rusten and West have claimed, the theogony implicit in the Derveni poem is identical in its main lines with the theogony of the *Rhapsodies*. If, on the other hand, we do not accept this assumption, the evidence of Eudemus and Aristotle, showing that there were theogonies with Night as first god, offers equally strong support for a contrary hypothesis.[117]

[115] Tortorelli Ghidini (1991), cf. also (1989). [116] Brisson (1997).

[117] Brisson (1985b) and (1997) examines in detail the figure of Chronos in the Orphic theogonies. His tenet is that the presence of Chronos is the main indication of the difference between the earlier and the later versions. As must have become clear from my own account of the evidence, I agree with this contention. Let me nevertheless briefly voice some objections concerning two of Brisson's additional points. (i) Brisson (1985b) proposes the hypothesis that the appearance of the figure of Chronos was the result of a Mithraic influence on Orphism. Furthermore, he claims that this could not have happened before around the second half of the first century AD, the date of the introduction of Mithraicism into the Roman Empire. Yet, we could object that Eudemus' review of the different theogonical traditions shows that a well-informed Greek could know already in the fourth century BC that Chronos was in the first generation of the 'Magi's theogony'. And the Derveni papyrus shows that, at the same period, at least some Orphics were interested in the Magi's practices. As a matter of fact, this point could have been raised by those who think that there was a primordial divinity Chronos in the Derveni theogony. (ii) Brisson, especially in (1997) 159–65, wants to make his claim concerning the Orphic theogonies stronger by arguing that there was no primordial deity called Chronos at all in the Greek context before the first century AD. One major contrary piece of evidence could be Pherecydes. According to some of our sources (D.L. in 7 A1 and B1 DK and Damascius in 7 A8 DK), Pherecydes had Chronos, together with Zas and Chthonie, in the first generation of gods. Brisson tries to discard this evidence in the following way. The explanation of Probus (in 7 A9 DK) suggests that Kronos was in the text of Pherecydes and that the reading 'Chronos' is just the result of a widespread allegoresis identifying Chronos with Kronos. In my view this still does not explain Chronos in D.L. and in Damascius. I still think that in this context Kronos is the *lectio facilior*; besides D.L. and Damascius do not change every Kronos to Chronos in other accounts. Furthermore, the names Zas and Chthonie, Re, Ogenos, and the rest show that Pherecydes found the traditional divine names dull and replaced them with idiosyncratic forms. Why should he then stick to the usual form only in the case of Kronos? I would like to stress, however, that the two points I have criticised are not basic to Brisson's main

I conclude then that since the contrary external evidence is of equal weight, and there is no sign of a divinity called Chronos in the papyrus itself, the most probable hypothesis is that the theogony implicit in the poem had Night and Aither in the first generation.

THE SECOND GENERATION

The problem of the egg and Phanes

I have already discussed the problem of Phanes at some length in the previous chapter where I have tried to show that it is more likely that there was no Phanes in the Derveni poem. In addition to this, I should now briefly treat a further point on the basis of which some commentators, even those who do not find the comparative evidence of the *Rhapsodies* compelling, try to maintain the Phanes-hypothesis.

In the *Rhapsodies* and in the Hieronymus-Hellanicus version, Phanes is born from a cosmic egg fabricated by Chronos. We have a cosmic egg in the cosmogony of Aristophanes' *Birds*, and from this egg Eros is born. Given that the *Birds* cosmogony, the commentators claim, reflects the early version of the Orphic theogony, we should conclude that the egg and Phanes were already present in the theogony implicit in the Derveni poem.[118] However, if one accepts my previous conclusion that the comic ornitho-theogony does not necessarily mirror a then-existing Orphic theogony – a conclusion based in part on the very fact that the egg motif appears in other, non-specifically Orphic theogonies – then one has to acknowledge that the presence of the egg in the *Birds* theogony does not prove that there was an egg, or a Phanes born from the egg, in the theogony of the Derveni poem.[119]

Ouranos and Gaia

As we have seen above, Ouranos is the son of Night and he is the first in the succession of divine kings (col. 14.6). We do not find any indication as to how he obtained royal power or who the subjects of this first king were.

As the forms Gaia and Ge appear in the Derveni text (col. 22), I do not see any reason to doubt that Gaia was Ouranos' partner in this theogony too. This hypothesis may also be strengthened by the Derveni author's affirmation in col. 14 that 'this Kronos was born to the earth'; yet, as

contention about the absence of Chronos in the theogony implicit in the poem commented on in the papyrus.

118 Cf. Brisson (1985a) 391; (1985b) 38; this is also his conclusion in Brisson (2003).

119 Thus I have modified the view I held in Betegh (1996).

noted above, this remark is probably already part of the author's physical interpretation, and accordingly is not necessarily in conformity with the mythological genealogy of the poem.

Moira?

The name of Moira appears in columns 18 and 19 of the papyrus. It seems almost certain that this figure was already present in the poem, because the author gives an elaborate argument on the point of why Orpheus should have chosen this name for speaking about 'wisdom' and 'breath'.[120] However, the text does not offer much help in determining the genealogical position of Moira; we cannot be sure that the poem itself was explicit on the point. The case is rendered more complicated by the fact that the Moirai (always in the plural) had a double genealogy in Hesiod's *Theogony*. They are presented as the progeny of Night in v. 217; whereas in v. 904 they appear as the daughters of Zeus and Themis. If we want to choose a genealogy for Moira in the poem on the basis of this evidence, the first is definitely preferable, for an interpretative remark of the Derveni author assures us that Moira appeared before Zeus.[121]

This hypothesis finds external support in a remark by Proclus. When he reflects on the double genealogy of the Moirai, he claims that for Orpheus the Moirai were part of the primordial stock of divinities.[122]

THE THIRD GENERATION: THE CHILDREN OF OURANOS AND GAIA

Kronos and Rhea

As has been shown above,[123] Kronos was the son of Ouranos and Gaia in this theogony, in accordance with the canon of Greek mythology. We have also seen that he castrated Ouranos and seized the kingship. Rhea also appears in the Derveni text, and so it seems reasonable to suppose that she was Kronos' sister and later became his wife.

Okeanos?

We have found that ***h15***, ***h16*** and ***L14*** = col. 23.3–11 refer to that part of the poem in which Zeus creates anew Okeanos and the other river-god, the

[120] Cf. col. 18.2–9. [121] Col. 18.9–10.

[122] Proclus *in Plat. Republ.* 2.207.14 = *OF* 126. See also Ricciardelli Apicella (1980) 117–23.

[123] 122–3 above.

'silver-eddying' Achelous. However, this only speaks about the new creation effected by Zeus; and Okeanos' original position in the genealogical scheme is not self-evident. Indeed, he apparently did not have a fixed genealogical position in the different narratives. In some of the early, pre-Platonic or even pre-Homeric theo-cosmogonies, Okeanos and his companion Tethys were at the very beginning of the divine generations.[124] This is what we also find in the genealogy of traditional divinities in the *Timaeus*, considered by some to be Orphic. There is also a quotation in Plato, explicitly ascribed to Orpheus, according to which

> Beautiful-flowing Okeanos first began the marriages,
> who wed Tethys, his sister from the same mother.[125]

As KRS (p. 16) remark, this does not necessarily imply a very early position for Okeanos and Thetys: it can simply suggest that the previous generation(s) did not generate their offspring as 'normal' bisexual divine couples.[126] As we have seen, it is not impossible on the other hand that the Hieronymus-Hellanicus version follows the alternative tradition according to which Okeanos was at the beginning of things. Furthermore, in the account of (?pseudo-)Alexander of Aphrodisias, Okeanos is in the second generation, born from Chaos, and is the father of Night.[127] The Hesiodic *Theogony*, in contrast to these theogonies, puts Ouranos at a fairly late stage in the generation of Titans, in this heterogeneous set of deities born to Ouranos from Gaia.[128] The *Rhapsodies* seems to follow this tradition.[129]

The lack of internal evidence and the equivocal nature of the external evidence make it difficult to decide beyond doubt the place of Okeanos in the theogony of the Derveni poem. I would nevertheless suggest that, in accordance with Hesiod and the *Rhapsodies*, he was in the generation of Kronos and Rhea. The main reason for this is no more than an argument *ex silentio*, i.e. that we do not find any sign of the privileged position or the importance which the alternative tradition ascribes to him.

124 The evidence for such a construction, and the Oriental parallels, are discussed in KRS 10–17.

125 Plato *Crat.* 402b7–c1: Ὠκεανὸϲ πρῶτοϲ καλλίρροοϲ ἦρξε γάμοιο, | ὅϲ ῥα καϲιγνήτην ὁμομήτορα Τηθὺν ὄπυιεν.

126 For a similar argument about the first marriage being that of Ouranos and Gaia, see 118–19 above.

127 Cf. 150 above.

128 West (1966) 36 remarks that the composite nature of the group of the Titans is an indication that it is not based on tradition.

129 Cf. *OF* 114 = Procl. *in Plat. Tim.* 40e. It is not clear why Alderink (1981) 39, in his genealogical diagram of the *Rhapsodies*, represents Okeanos as the child of Kronos and Rhea.

Aphrodite?

I have already discussed at some length the birth, or double birth, of Aphrodite while speaking about ***h*11** = col. 21.5.[130] The problem is whether Aphrodite Ourania had a first birth from Ouranos' severed phallus, as in Hesiod's *Theogony* (vv. 154–210) and also in the *Rhapsodies* (*OF* 127 = Procl. in Plat. *Cratyl.* 406c), and then a second birth by Zeus, or only one birth from the semen of Zeus as West and following him Bernabé claim.[131] I am inclined to prefer the first option.

West bases his suggestion on the allegorical interpretation of the word θόρνηι in col. 21 which also speaks about the birth of Aphrodite. θόρνηι, the meaning and grammatical form of which remains vexed, is supposed to be related to a word meaning 'ejaculation' or 'semen'. Although I am inclined to accept this interpretation of the word θόρνηι, I am not so sure it can be decisive concerning the genealogical place of Aphrodite. Why can θόρνηι not refer to the first birth of the goddess, attested in Hesiod and in the *Rhapsodies*, but at the moment when Zeus is about to give her a second birth after having interiorised everything, including Aphrodite? If so, the poet would speak about Zeus creating Aphrodite anew, just as he will speak about the second creation of Ge, Meter, Rhea and Hera in the next column, and the second creation of Okeanos and Achelous in the column after.

Bernabé also thinks that Aphrodite was born only from Zeus' semen, that is, Aphrodite Ourania was not Ouranos' daughter.[132] Zeus needed a feminine generative principle in order to create the world anew, Bernabé argues, and this is why he created Aphrodite from his semen before giving birth to the other gods. Aphrodite could then fulfil the same function as Eros in Hesiod. But why would Zeus need this feminine generative principle if he continued to create the universe anew not sexually, but by some kind of intellectual planning, as the verb μήσατο, discussed below, suggests? Moreover, Bernabé also follows West in thinking that Harmonia and Peitho, mentioned in the same column, attended the birth of Aphrodite just as Eros and Himeros attended her birth in Hesiod (*Thgn.* 201). This, however, is very difficult to square with the suggestion that Aphrodite was the first to be born after the moment when all the divinities were inside Zeus.

In view of the above arguments I still consider it most probable that Aphrodite Ourania was originally Ouranos' daughter also in this theogonic scheme.

[130] 127–8 above. [131] West (1983) 91–2; Bernabé (2002) 28–9. [132] Bernabé (2002) 28–9.

THE FOURTH GENERATION

The children of Okeanos (and Tethys?)

Achelous?

It is reasonable to assume that the silver eddying Achelous, mentioned in ***L13*** = col. 23.11, is the son of Okeanos, as all the river-gods and spring-nymphs are in Hesiod. Although Achelous does not appear in the fragments of other Orphic theogonies, we find an interesting remark in Serv. *in Verg. Georg.* 1, 8 p. 131.19 Th (= *OF* 344):

'Acheloia' (Vergilius) non praeter rationem dixit: nam, sicut Orpheus docet, generaliter aquam veteres Acheloum vocabant.

The provenance of this piece of information is, however, uncertain.[133]

Metis?

As I mentioned above, editors and commentators assume that the word μῆτιγ in ***L9*** = col. 15.13, a badly fragmented line, refers to the goddess Metis. If there was a divinity Metis in the poem, she might belong to this generation, for Zeus' epithet μητίετα in the previous lemma speaking about the royal succession (ἐκ τοῦ δὴ Κρόνος αὖτις, ἔπειτα δὲ μητίετα Ζεύς ***L8*** = col. 15.6) may suggest that Metis has a similar role both in Hesiod and in the *Rhapsodies*[134] where Zeus has to swallow her in order to make his rule stable. But even though her function is the same, the genealogical position and identity of Metis is different in the two cases. In Hesiod, she is daughter of Okeanos and Tethys (v. 358). Having obtained royal power, Zeus marries and then, on the advice of Ouranos and Gaia, swallows Metis in order to prevent the birth of too bright and too bold a progeny, which could have jeopardised his rule (vv. 886–900). In the *Rhapsodies*, in contrast, Metis is expressly identified with Phanes (*OF* 60; 65 etc.). The act is basically the same: Zeus swallows an early deity to secure his power, although no marriage is involved. The consequences in the *Rhapsodies* are different, however: the world becomes interiorised in Zeus. Conceivably, precisely this similarity in pattern led to the identification of Metis with the first-born Phanes.

[133] Bernabé (2002) 30 with n. 141 and 143 suspends judgement, quite reasonably, on the question of whether or not Achelous was Okeanos' son.

[134] It might also be the case in the version of the pseudo-Clementine *Homilies*, where Metis appears as the powerful and divine *pneuma*, drawn by Zeus to the upper regions. Cf. *OF* 56. It has to be noted on the other hand that some authorities, like Wilamowitz (*Sitz.-Ber. Berl. Akad.*, 1921, 957) and Jacoby (*Hesiodi Carmina*, Berlin, 1930, 37–8), treat the Metis episode as later interpolation.

If we do not accept that there was a Phanes in the theogony of the Derveni poem, we might conclude that this Metis was more likely to be analogous to the Metis, daughter of Okeanos and Tethys, in Hesiod. This, however, is bound to remain conjectural. It should also be noted that Tethys' name does not occur in the papyrus.

However, I have tried to show above that the word μῆτιc in the papyrus is not the name of a divinity but a common noun;[135] in this case the whole argument about Metis' genealogical position is obviously futile.

The children of Kronos and Rhea

I have already discussed the arguments to the effect that, in accordance with standard Greek genealogy, Zeus was born from Kronos and Rhea. On the other hand, we have surprisingly little information on the acts of Zeus. We learn from ***L2*** = col. 8.4–5 that he gains power from his father, and that this coup was foretold in a prophecy. Furthermore, the exchange of power is marked by the fact that, at least on one possible reading of the verse, Zeus grasps in his hand not only the strength, but also a 'glorious *daimon*'. As we have already seen, West, followed by many scholars, takes the daimon to be Phanes. On this hypothesis, Zeus would swallow – and not take in his hand – the first-born Phanes. Bernabé, on the other hand, thinks that the poet indicates the change of power by saying that Zeus took in his hand the 'glorious daimon', that is, the previous ruler Kronos.[136] A third option would be to identify the daimon as Metis. For my part, I have no view as to who this daimon was.

In a next phase, Zeus receives oracular advice from his father (***L5*** = col. 13.1), and probably also from Night (***L3*** = col. 11.10). Then comes the swallowing (***L5*** = col. 13.1 and ***L6*** = col. 13.4), the assimilation of the whole world (***L10*** = col. 16.3–6 and ***L12*** = col. 17.2), and the new creation. Specifically, we hear about the act by which Zeus creates the river-gods Okeanos and Achelous (***h15*** = col. 23.3–5, ***h16*** = col. 23.5–7 and ***L14*** = col. 23.11).

The papyrus gives little clue on possible brothers and sisters of Zeus. Only Hera and Demeter (daughters of Kronos and Rhea in Hesiod's *Theogony* v. 454) appear in col. 22, identified with each other and with further goddesses by the Derveni author.

[135] See 114 above. [136] Bernabé (2002) 12–13.

The fifth generation?

The only reference in the extant text of the papyrus to a fifth generation is in ***h*17** = col. 26.1–2 and ***h*18** = col. 26.9–12), where Zeus is about to have an incestuous affair with his mother. His mother is not named, but it is reasonable to suppose that she was still Rhea. It is reasonable to suppose also that whenever Zeus wanted to have sexual intercourse with any of the goddesses, his desire eventually led to the birth of a further god or goddess. Yet the papyrus breaks off before we can learn who this new god or goddess would be. It is tempting to supplement the theogony of the Derveni poem with the episode – present in both the *Rhapsodies* and the Hieronymus-Hellanicus version – where Rhea, now identified with Demeter, bears Persephone or Kore, who in turn bears Dionysos for Zeus. If it was present in the Derveni poem, it surely was of high importance, but this is already outside the scope of the extant part of the interpretation of the Derveni author. (It is to be noted, however, that the extant part of the proem does not refer to Dionysos.)

Furthermore, the parallel accounts of Zeus's oedipal incest contain the identification of Rhea and Demeter. Now, Rhea, Demeter, Ge and Hera are identified in col. 22.7–11 of the Derveni papyrus. At this point, the question inevitably arises whether the identification of divine names, so characteristic of later Orphic poetry,[137] was already present in the Derveni poem. Unfortunately, this has to remain an open question as the extant lemmata of the Derveni poem do not provide an answer. But clearly, this question has an important bearing on our assessment of the Derveni author's exegesis.

Apart from the divine figures I have just discussed, there are some further goddesses mentioned in the papyrus. Peitho appears, as the name of a goddess, in col. 21.6 where the Derveni author merges Aphrodite Ourania, Zeus, Peitho and the act of sexual intercourse. The extant text of the papyrus does not provide any further clue to Peitho's position in the family. She does not appear in other Orphic fragments either, except in a passing and unspecific remark in *OF* 202 = Procl. *in Plat. Tim.* 33a. For these reasons, if Peitho was a member of the divine brood in the theogony implied in the Derveni poem, our major point of reference could still be Hesiod, where she is the daughter of Okeanos and Tethys.

[137] Cf. Rudhardt (1991) 271–4. Funghi (1979) 22 explains the success and vitality of Orphism in terms of its capacity to embrace and assimilate different traditions by identifying a high number of divine beings with each other.

Yet Peitho is often more directly connected with Aphrodite. Although not in the Hesiodic genealogical account, she and Eros are Aphrodite's closest companions in the cultic representations and ritual practices. The connection was so close that 'Peitho' could even become one of Aphrodite's epicleses.[138] Peitho had a statue in the temple of Aphrodite Pandemos in Athens,[139] and in that of Aphrodite Praxis in Megara,[140] and it is Peitho who places the crown on Aphrodite's head in the representation which Pausanias saw on the base of the throne in the temple of Zeus in Olympia.[141] Peitho was probably also an object of honour in the shrine of Aphrodite at Daphne, on the sacred road from Athens to Eleusis.[142] According to the analysis of Pirenne-Delforge, Peitho, together with Eros, had the fullest character among the escorts of the goddess of love.[143] Moreover, a scholium to Hesiod's *Works and Days* v. 57 tells us that for Sappho, Peitho is Aphrodite's daughter.[144] The theogony of the Derveni poem may or may not correspond to either the Hesiodic account or the genealogical scheme present in Sappho.

We encounter the same uncertainty with regard to the genealogical place of Harmonia. She does not appear at all in the Orphic fragments collected by Kern. On the other hand, she is daughter of Aphrodite in Hesiod's *Theogony* (v. 937), which may explain why she appears in Aphrodite's (and Peitho's) company in col. 21. of the papyrus. We cannot ignore, on the other hand, the possibility that Aphrodite, Harmonia and Peitho were already identified in the poem, and not only in the Derveni author's interpretation. Nevertheless, West's hypothesis according to which Peitho and Harmonia were two deities attending the (first) birth of Aphrodite – as Eros and Himeros did in Hesiod's *Theogony* (v. 201), and Zelos and Apate in the

138 See e.g. *IG* XII, 2.73 from Lesbos.

139 Paus. 1.22.3: Ἀφροδίτην δὲ τὴν Πάνδημον, ἐπεί τε Ἀθηναίους Θησεὺς ἐς μίαν ἤγαγεν ἀπὸ τῶν δήμων πόλιν, αὐτήν τε σέβεσθαι καὶ Πειθὼ κατέστησε· τὰ μὲν δὴ παλαιὰ ἀγάλματα οὐκ ἦν ἐπ' ἐμοῦ, τὰ δὲ ἐπ' ἐμοῦ τεχνιτῶν ἦν οὐ τῶν ἀφανεστάτων.

140 Paus. 1.43.5: μετὰ δὲ τοῦ Διονύσου τὸ ἱερόν ἐστιν Ἀφροδίτης ναός, ἄγαλμα δὲ ἐλέφαντος Ἀφροδίτη πεποιημένον Πρᾶξις ἐπίκλησιν. τοῦτό ἐστιν ἀρχαιότατον ἐν τῷ ναῷ· Πειθὼ δὲ καὶ ἑτέρα θεός, ἣν Παρήγορον ὀνομάζουσιν, ἔργα Πραξιτέλους· Σκόπα δὲ Ἔρως καὶ Ἵμερος καὶ Πόθος, εἰ δὴ διάφορά ἐστι κατὰ ταὐτὸ τοῖς ὀνόμασι καὶ τὰ ἔργα σφίσι.

141 Paus. 5.11.8: ἐπὶ δὲ τοῦ βάθρου <τοῦ> τὸν θρόνον τε ἀνέχοντος καὶ ὅσος ἄλλος κόσμος περὶ τὸν Δία, ἐπὶ τούτου τοῦ βάθρου χρυσᾶ ποιήματα, ἀναβεβηκὼς ἐπὶ ἅρμα Ἥλιος καὶ Ζεύς τέ ἐστι καὶ Ἥρα, <ἔτι δὲ Ἥφαιστος,> παρὰ δὲ αὐτὸν Χάρις· ταύτης δὲ Ἑρμῆς ἔχεται, τοῦ Ἑρμοῦ δὲ Ἑστία· μετὰ δὲ τὴν Ἑστίαν Ἔρως ἐστὶν ἐκ θαλάσσης Ἀφροδίτην ἀνιοῦσαν ὑποδεχόμενος, τὴν δὲ Ἀφροδίτην στεφανοῖ Πειθώ κτλ.

142 One of the inscriptions in the shrine was dedicated not to Aphrodite but to Peitho: Πειθοῖ, Καλλίμα[χος] | τήνδ' ἀνέθηκε Σολεύς. *IG* 2^2.4583. See also Pindar's encomium to Xenophon of Corinthus, as quoted by Athenaeus (Athen. 13.33.4 Kaibel = Pind. fr. 122 Snell-Maehler).

143 Pirenne-Delforge (1991) and (1994) 456.

144 Cf. also the badly fragmented fr. 96 vv. 26–9 of Sappho.

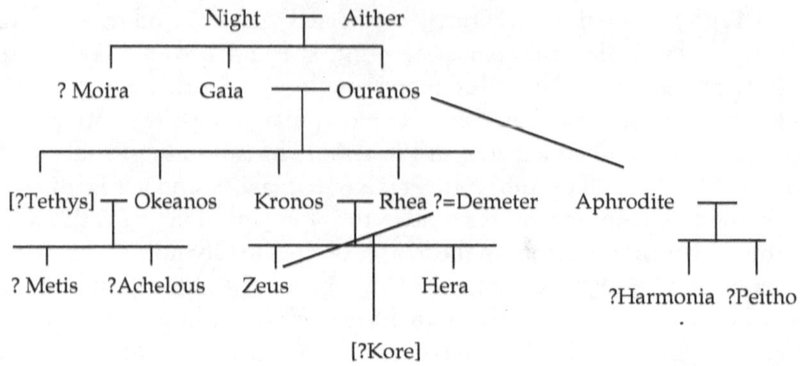

Figure 4.1.

Rhapsodies (*OF* 127 = Procl. *in Plat. Cratyl.* 406c) – is not impossible either.

The results of the above analysis can be tentatively summarised in a diagram (fig. 4.1). But we should not expect that the theogonic narratives that underlie the poem commented on in the papyrus were altogether consistent or explicit about the genealogic position of each and every divinity mentioned in the poem. Therefore, this diagram remains no more than a rough approximation.

SPECIFICITIES OF THE DERVENI POEM'S THEOGONIC PLOT

Although the evidence of the Derveni poem does not allow us to decide how detailed its theogony was, it is clear already from the above account that it shows important similarities with Hesiod's *Theogony*. It is noteworthy, first of all, that all the divine names occurring in the papyrus are present in Hesiod. Thus we do not find in the text of the papyrus any mention of Phanes, eggs, Chronos, Erikepaios, or Adrasteia; it seems, on the other hand, to be fairly certain that the Derveni poem mentioned far fewer gods than Hesiod's *Theogony*. Second, the chronological succession, as far as we can reconstruct it and in at least its main outlines, conforms to Hesiod, in so far as an older god in Hesiod is an older god in the papyrus as well. We do not find such reversals as for example in the Hieronymus-Hellanicus version, where Okeanos comes prior to Chaos and Ouranos. There are, however, some prominent differences between Hesiod's story and the one implicit in the Derveni poem. Let us take these points of divergence in the order of generations.

THE FIRST GENERATION

Chaos in Hesiod is a 'yawning gap', generally imagined as between the would-be earth and the would-be sky – a huge 'not yet'.[145] If such a distinction makes any sense at this stage, Chaos is more a primordial divine medium than a personal anthropomorphic goddess. As regards characteristics, she (for although the word is grammatical neuter Chaos is treated as feminine) suggests a huge gloomy, dark and airy stuff. These features are underlined by her immediate offspring, Night and Erebos.[146] The first entities implying brightness or light come only in the next generation as the children of Night and Erebos. By giving priority to entities implying darkness and lack of visibility, the Hesiodic genealogy seems to lay stress on the indistinctness of the pre-cosmic situation. Furthermore, the first masculine deities appear only in the second generation. If my interpretation of aither is accepted, we have in the Derveni poem's theogony, as opposed to the Hesiodic conception, two deities in the first generation, feminine Night and masculine Aither. This pair instantiates at once a double polarity: light-dark opposition on the one hand, and sexual difference on the other.[147] The primordial pair thus suggests a basic dualistic conception.

Remarkably, Aristoxenus and Diodorus of Eritrea, as reported by Hippolytus, claim that Pythagoras received a doctrine from Zaratas (i.e. Zoroaster) according to which there are two principles, a 'father' and a 'mother'. The 'father' is light, whereas the 'mother' is darkness, and they also represent the basic physical pairs of opposites respectively: hot-cold, dry-moist, light-heavy, swift-slow.[148]

It must be noted that the Presocratic cosmogonies can also be grouped according to this opposition. Most of them posit an original state of indistinctness out of which the world arises through successive steps

145 See Komoróczy (1979) 546–7 for the idea that the 'not yet' – i.e. the abstraction from the presently existing elements of the world in order to arrive at a speculative initial state – is a standard feature of cosmogonic speculation.

146 On the location and characteristics of Chaos, see Stokes (1963) 17–23; West (1966) 193. Stokes suggests that Chaos was below the would-be earth, creating a potential support for her.

147 Funghi (1979) 21, by referring to *OF* 65, attributes some kind of generative and motive energy to Aither which leads to the distinction of the primordial mass. On this interpretation Aither is assimilated to Phanes of the next generation. It should be noted however that the text referred to is (*a*) about the *Rhapsodies*, and (*b*) not a quotation but an allegorical interpretation by John Malalas. So, although the interpretation is tempting, I cannot find sufficient textual support for it.

148 Hippol. *Ref.* 1.2.12 = Aristoxenus fr. 13 Wehrli: Διόδωρος δὲ ὁ Ἐρετριεὺς καὶ Ἀριστόξενος ὁ μουσικός φασι πρὸς Ζαράταν τὸν Χαλδαῖον ἐληλυθέναι Πυθαγόραν. τὸν δὲ ἐκθέσθαι αὐτῷ δύο εἶναι ἀπ' ἀρχῆς τοῖς οὖσιν αἴτια, πατέρα καὶ μητέρα. καὶ πατέρα μὲν φῶς, μητέρα δὲ σκότος, τοῦ δὲ φωτὸς μέρη θερμὸν ξηρὸν κοῦφον ταχύ, τοῦ δὲ σκότους ψυχρὸν ὑγρὸν βαρὺ βραδύ. ἐκ δὲ τούτων πάντα τὸν κόσμον συνεστάναι, ἐκ θηλείας καὶ ἄρρενος.

of differentiation, while some other cosmogonies – such as the one in Parmenides' *doxa* – posit an original pair of opposites.[149] The basic intuition underlying this conception is that to account for the cosmic developments, you need some kind of dynamism, such as the tension between a pair of opposites.

In Hesiod, Chaos does not have a real personality and will not play any role in the later stages of the story. This is not the case with the Orphic Night. Moreover, Night has considerably different degrees of importance in the two stories. In Hesiod, Night is mainly the ancestor of those divine creatures who are not objects of cult, and are more like mythological abstractions (e.g. Eris, Death and Sleep). Apart from that, she does not have any further role in the subsequent stages of the story. In the theogonic plot of the Derveni poem, she continues to be a high authority, possibly the highest, who determines the course of events in more than one way. She gives refuge to and nurtures the child Zeus,[150] and helps the development of the story by proclaiming oracles. In this way she takes the role that Gaia (and to some extent Ouranos) has in Hesiod: Gaia receives the newborn child (v. 479), and suggests at a later stage, together with Ouranos, that Zeus should swallow Metis (vv. 886–900). Note that the Homeric reference in the 'Deceit of Zeus' episode to the high authority of Night – whose will even Zeus has to comply with (*Il.* 14.258–62) – may suggest that the version testified in the Derveni poem has historical precedence.

In Hesiod, Gaia (or the Earth) is the second created entity.[151] In the phrase describing Gaia's birth, Hesiod underlines her capacity to support or give place to the other beings. Then, it will be Gaia who brings forth the starry Heaven (or Ouranos) by herself. We have again a feminine deity first, and the masculine counterpart, characterised by brightness, only in a second step. We have seen that in the theogony of the Derveni poem Ouranos comes immediately in the second generation, being the son of Night. It

[149] Without entering into the details of the vexed question of the purpose and sources of the Parmenidean *doxa*, or trying to formulate a historical hypothesis, let me note that the *doxa* shows at least two notable points of contact with the Derveni poem. First, if my reconstruction of the first generation is valid, the two first deities of the poem, Night and Aither, are closely echoed by Parmenides' two cosmic principles. To this, one should also add the parallel – but otherwise almost unattested – use of μήδομαι for intellectual creation on which see 181 with n. 197 below. See also Burkert (1999a) 79–80 who does not mention the Derveni papyrus in this context, but suggests that Parmenides knew (a version of) the Orphic *Theogony*.

[150] Cf. τροφ[όν as applied to Night. West (1983) 87 suggests that Night also nurtured the Titans for Gaia. This suggestion is based on the assumption that the text had θεῶν τροφόν, but θεῶν is just West's conjecture, on the basis of a parallel with a fragment from the *Rhapsodies* (*OF* 106 = Procl. *in Plat. Cratyl.* 404b).

[151] Hes. *Thgn.* v. 117.

seemed also reasonable to assume that Ge was born as Ouranos' sister. If this is the case, then in this generation too we have a similar preference for antithetical or complementary entities, conceived of as divine couples, born in the same generation.[152]

Thirdly, after Chaos and Erebos, we have Eros in Hesiod.[153] Eros' presence is prerequisite for sexual generation, and thus already presupposed in the first action of this kind: in the birth of Aither and Hemere out of the intercourse of Night and Erebos. As far as we can judge on the basis of the remaining text of the Derveni papyrus, the Hesiodic Eros has no tally in the theogony of the Derveni poem.[154] This hypothesis is reinforced by the fact that, in speaking about the conjunction of physical particles, the Derveni author only refers to Aphrodite, Peitho and Harmonia (col. 21), while one would clearly expect Eros to be mentioned in this context.

THE SUCCESSION MYTH

Thus far we have been speaking about basic structural differences, such as concepts of original homogeneity versus polarity. The interpretation of the convergences and divergences becomes more complicated as we proceed to the generations of more anthropomorphic gods with more complete 'biographies'.

Kronos' basic function in the theogony of the Derveni poem is the same as in the succession myth recorded in Hesiod. He is the son of Ouranos and father of Zeus, an intermediate figure, who seizes power but is bound to lose it again. Furthermore, Kronos seems to obtain power from his father by the same means in both stories: by a crude act of castration. The castration of Ouranos has received substantial attention from experts of mythical thought. It has generally been interpreted as the mythical expression of the ultimate separation of Sky and Earth.[155] The violent separation of heaven and earth seems to be a widespread motif, with parallels in Polynesian, African and Japanese myths, but we can also refer to the 'deadly strife' which separates earth, heavens and sea in the song of Orpheus in Apollonius

[152] We have early evidence of the conjoint birth of Ouranos and Gaia, possibly in an Orphic context, in Euripides' *Melanippe* fragment, mentioned on p. 150 above.

[153] Tartarus in v. 119 is generally considered to be a red herring, not to be treated as one of the primordial divinities. Cf. Stokes (1963) 1–5 and West (1966) ad loc.

[154] But see Bernabé's hypothesis about Aphrodite mentioned on p. 161 above.

[155] The interpretation seems to be first suggested by A. Lang, taken over, among others, by Frazer (*Adonis* 1.238); Nilsson (1925) 73; Cornford (1950) 103–4 and (1952); Maróth (1951–2); West (1966) 211–12; Burkert (1999b) 94.

Rhodius.[156] In Hesiod, the source of the conflict is that Ouranos completely envelops Gaia and does not leave space for their children. This oppressive embrace causes distress to her, so she plots against Ouranos with Kronos.[157]

Again, it is interesting to note that the appearance of the principle of sexuality, and hence of generation, is linked to this original act of separation. This feature is actually doubled in Hesiod. Eros is born immediately after Gaia, while Aphrodite's birth is directly related to the castration of Ouranos. In some of the later Orphic theogonies, Phanes, identified with Eros, is born when the egg is torn apart. Conspicuously, Athenagoras identifies the upper and lower part of the egg with heaven and earth respectively. Thus, the original separation, most usually conceived of as the separation of sky and earth, brings about the possibility and indeed the necessity of conjunction, mixing and generation.[158]

The closest parallel to the story of Kronos can be found in the Hittite text *Kingship in Heaven*, also an epic of succession.[159] In this text Kumarbi, an intermediate god like Kronos, defeats the sky-god Anu, the equivalent of Ouranos, by castration. The difference lies in the fact that in the *Kingship in Heaven* Kumarbi castrates Anu by biting off and swallowing, whereas Kronos uses a knife for the same purpose in Hesiod. Yet the Hittite connection becomes strikingly close when in the so-called *Song of Ullikummi*, a sequel of the *Kingship in Heaven*, we hear about 'the olden copper knife with which they separated heaven and earth'.[160] Anyhow, although we have cutting instead of biting in Hesiod, the motif of swallowing is present. Kumarbi gets pregnant by the swallowed genitals and gives birth to three gods, among whom there is the weather-god, the Hittite parallel of Zeus (the other two gods are Tasmisu and the river Tigris). In Hesiod, Kronos swallows his children (with the exception of Zeus who is brought up in Crete),[161] and vomits them when he is cunningly made to swallow a stone. Incidentally, Kumarbi apparently also swallows a stone, but, as opposed to Kronos, he does it intentionally. For, warned by the sky-god that he, Kumarbi, would be overthrown by one of the gods growing inside him, he

[156] Ap. Rhod. *Argon.* 1.498. The parallels are conveniently collected and analysed in Staudacher (1942) – although the important Hittite material had not become available at the time of writing.

[157] Hes. *Thgn.* vv. 154–82, with West (1966) ad loc.

[158] For an elaborate analysis of the birth and role of Eros and Aphrodite, see Rudhardt (1986).

[159] The following comparative analysis is based on Burkert (1987b) 19–24, with additional material from West (1966) 20 and Bottéro-Kramer (1989). On the double separation of sky and earth in Hesiod (Chaos and the birth of Gaia, on the one hand, the castration of Ouranos, on the other), see e.g. Cornford (1950) 98f; KRS 34–41.

[160] *Ullikummi* III-c, *ANET* 125.

[161] Burkert (1987b) 19 writes 'Kronos swallows his own children, including the weather god Zeus'.

wants to destroy his would-be rivals by swallowing a stone. The plan does not work: the weather-god is born and dethrones Kumarbi.[162]

The similarities and differences between the Babylonian and Greek material become even more interesting when we find the same motifs in yet another arrangement in the theogony of the Derveni poem. As we have already seen, in this version, Kronos castrates Ouranos, and then, most probably, Zeus swallows the severed phallus. As a consequence of this act, Zeus gets pregnant inside his belly with the entire cosmos, to which he gives birth. In this way, we have the swallowing of the phallus, as present in the Hittite story, but absent from the Hesiodic version. The act of engulfment, also in the Orphic version, leads to pregnancy and to the birth of divine creatures.

Furthermore, we have four generations of divine kings in the Hittite story: Alalu, Anu the castrated sky-god, Kumarbi, and the finally victorious Weather-god. In Hesiod, we have three successive rulers, who structurally correspond respectively with the three latter kings of the Hittite story. If we accept that the plot implicit in the Derveni poem had again four kings (with the addition of Dionysos), we have the following modification of the pattern. The first god of the Hittite story is dropped, as in Hesiod. Then we have the sky-god, who is castrated by the next king; while the swallowing of the phallus and the gastric pregnancy belong to the penultimate king, as in both the Hittite and the Hesiodic versions. The constant features of the scheme could in some sense account for the strange fact that Kronos castrates Ouranos but Zeus swallows the phallus: the castration must belong to the king following upon the sky-god (for the castration is the symbolic representation of the separation of sky and earth), whereas the engulfment and gastric pregnancy pertain to the penultimate king, who prepares the ground for the ultimately victorious young ruler.

The above structural analysis, suggesting that Kronos and Zeus respectively executed two halves of an action which originally belonged to one god – to Kumarbi in the Hittite text and to Kronos in Hesiod – can actually help us in understanding a further specific point in the poem. For it seems that Kronos has a much more positive and active role in establishing Zeus's new rule than in other accounts. Firstly, he prepares the new creation by providing the object to be swallowed. True, he did not necessarily castrate Ouranos with a view to helping Zeus. Yet, although the Derveni author tries to explain away this feature of the poem (col. 13.2–3), we learn that Zeus receives advice not only from Night but also from Kronos

162 West (1966) 20.

(***L5*** = col. 13.1). And as the immediately subsequent lemma speaks about the engulfment (***L6*** = col. 13. 4), it is reasonable to suppose that the advice was about this act.[163] Furthermore, the verses describing Zeus's ascent to power (***L2*** = col. 8.4–5) do not suggest anything violent. As a matter of fact, I cannot find anything in the text about a hostility between Zeus and Kronos. It is true that Kronos was not an utterly negative figure in Hesiod either. I would, however, still maintain that in view of the above considerations, Kronos is closer to Zeus in the theogony of the Derveni poem than he is in Hesiod.[164]

ZEUS AND THE NEW CREATION

The previous stages of the succession story were only a preparation for the most important and possibly most radical feature of the theogony of the Derveni poem. I mean the episode when the entire world gets interiorised in Zeus to be born anew in the next moment. This, I shall argue, is not just a flight of fancy but the mythological expression of some important theological intuitions.

The problem of priority

Already Aristotle has pointed out that the mythological stories which operate with successive generations of divine rulers run inevitably into a problem.[165] The diachronic principle operating in these narratives clashes with the systematic principle, or, in other terms, with the respect for hierarchy. For the first in terms of power, the highest in hierarchy, is not identical with the first in terms of time. Or, conversely, the ultimate source or origin becomes unimportant while the sovereignty will belong to a god who, from the point of view of the genealogy, is just a member of a later generation. There are two kinds of distinction and priority here, and some sense of symmetry would dictate that the two should correspond.

[163] In the *Rhapsodies*, or at least in the Neoplatonic interpretation of it, Kronos actively helps Zeus in the creation of the new world-order. Cf. e.g. *OF* 155 = Procl. *in Plat. Cratyl.* 391a: μαντεύει μὲν γὰρ αὐτῷ (sc. Zeus) καὶ ἡ Νύξ, ἀλλ'ὁ πατὴρ προcεχῶc, καὶ πάντα τὰ μέτρα τῆc ὅληc δημιουργίαc αὐτῷ ἐνδίδωcι.

[164] On the puzzling mixture of positive and negative features in the myths and rites connected with the figure of Kronos, with a possible explanation of this ambiguity, see Versnel (1987) and now the most expanded version in Versnel (1993) 90–135.

[165] Arist. *Metaph.* 1091a33–1091b7. For a modern formulation of the point see e.g. Schwabl (1978) 1319 and Vernant (1962) ch. 7 entitled 'Cosmogonies et mythes de souveraineté'.

There are some indications in Hesiod's *Theogony* itself that the problem had already been sensed within the mythological discourse. In the proem of the *Theogony* (vv. 43–9) we hear that the Muses sing:

αἳ δ' ἄμβροτον ὄccαν ἱεῖcαι
θεῶν γένοc αἰδοῖον πρῶτον κλείουcιν ἀοιδῇ
ἐξ ἀρχῆc, οὓc Γαῖα καὶ Οὐρανὸc εὐρὺc ἔτικτεν,
οἵ τ' ἐκ τῶν ἐγένοντο, θεοὶ δωτῆρεc ἐάων·
δεύτερον αὖτε Ζῆνα θεῶν πατέρ' ἠδὲ καὶ ἀνδρῶν,
[ἀρχόμεναί θ' ὑμνεῦcι θεαὶ λήγουcαί τ' ἀοιδῆc,]
ὅccον φέρτατόc ἐcτι θεῶν κάρτει τε μέγιcτοc·

West, following Guyet, condemns v. 48 on the basis that 'it is nonsense to say that the Muses sing of Zeus second, both first and last'.[166] Although I do not want to defend either the authenticity or the clarity of the expression, there is a reason for the line to be there. The author of it (be it Hesiod or a later interpolator), felt that if the Muses follow the chronological order, Zeus will appear at a later stage, yet, on the other hand, it would be more proper to start with Zeus who φέρτατόc ἐcτι θεῶν κάρτει τε μέγιcτοc. As, indeed, the Muses at the beginning of *Works and Days* are asked to praise Zeus first. The awkwardness of the expression 'second, first and last' shows that there was an attempt to accommodate two incongruous approaches, i.e. the chronological and the hierarchical.[167] This asymmetry between the chronological and the hierarchical seems to dissolve with the Presocratic cosmogonies where it is the prime form of matter or the starting point which also somehow governs (κρατεῖ or κυβερναῖ)[168] the cosmos. Thus the tension between the two principles fades as 'starting point' and 'sovereignty' get united.[169]

Arguably, the episode in the theogony of the Derveni poem (reappearing in other Orphic theogonies) which describes how Zeus swallowed into his belly and then gave new birth to the whole world with all the beings in

166 West (1966) ad loc.

167 Two Doric lines quoted by Clemens of Alexandria attributed to Terpander (*Stromateis* 6.11.88) could offer an interesting parallel: Ζεῦ πάντων ἀρχά, πάντων ἀγήτωρ, | Ζεῦ cοὶ πέμπω ταύταν ὕμνων ἀρχάν. Epicharmus B1, generally accepted as original, gives a stunning criticism of the idea that the gods were born at all. The character in the play first objects to the idea that Chaos was born and was not existing for ever, and then seems to react directly to *Thgn.* v. 47: οὐδὲ μὰ Δία δεύτερον | τῶνδέ γ' ὧν ἁμὲc νῦν ὧδε λέγομεc, ἀλλ' ἀεὶ τάδ' ἦc.

168 For the two verbs see e.g. Heraclitus B41; Parmenides B12; Diogenes of Apollonia B5; and Heraclitus B114; Anaxagoras B12 and Diogenes of Apollonia B5.

169 It is irrelevant in this respect whether or not the term ἀρχή actually appeared in the Presocratic texts, and that in Hesiod ἀρχή is used exclusively in the temporal sense.

it, can be conceived of as an attempt to provide a solution for this conflict between temporal and hierarchical priority. For on the basis of this episode, one can say that the whole world derived not from some gloomy first entity, with no religious significance or cult and with hardly any personality, but from the supreme king Zeus himself. The idea of a second birth of the world accepted, Zeus can be treated not just as the supreme sovereign, but also as the source of all things. The episode allows 'Orpheus' to say both in the Derveni poem and in the Hymn to Zeus at the end of the pseudo-Aristotelian *De mundo* that

Zeus is the head, Zeus is middle and from Zeus all things have their being.[170]

In the *De mundo* hymn, the verse immediately preceding the one just quoted is even more explicit. It expressly says that 'Zeus was born first, Zeus is the ultimate, god of the bright bolt'.[171] Now the second half of this verse was certainly there already in the Derveni poem, just preceding the verse quoted above. Furthermore the discussion of the meaning of 'the ultimate' in the papyrus is immediately preceded by the discussion of the meaning of the expression 'being born'.[172] It is thus tempting to follow West and Bernabé in supplementing the lemma on the basis of the *De mundo* hymn. That would mean that 'Zeus was born first' appears also in the Derveni poem.

Thus Zeus becomes not just the supreme sovereign but also the first god and the source of all things. Furthermore, the interiorisation of the then-existing world is a sophisticated way of getting the older gods under his control. It is also interesting to note the means by which this assimilation of the whole world is said to happen. Zeus – on the interpretation defended above – has to swallow the severed phallus of Ouranos. The object swallowed can easily be seen as some kind of origin or principle of all generation. All the more so, since we have already seen that the appearance of other mythical generative principles, such as Eros and Aphrodite, is also linked to the separation of Earth and Sky, Gaia and Ouranos; and the separation is manifested in this case by the cutting of the phallus. Zeus can be the first and source of all becoming because he has interiorised the ultimate source of all generation. Note that the same logic works also if we take it that Zeus swallows not Ouranos' phallus but Phanes, as Phanes is customarily identified with Eros.

[170] ***L12*** = col. 17.12 = *OF* 21a v. 2.

[171] *OF* 21a 1: Ζεὺϲ πρῶτοϲ γένετο, Ζεὺϲ ὕϲτατοϲ ἀργικέραυνοϲ.

[172] Col. 17.4–6.

The One-Many problem

It has often been suggested that one of the keynotes of 'Orphic thought' is a reflection on the One-Many problem. Already Gruppe tried to locate the central teaching of Orphic theogonies in the words by which Diogenes Laertius summarises Musaeus' doctrine:

ἐξ ἑνὸϲ τὰ πάντα γίνεϲθαι, καὶ εἰϲ ταὐτὸν ἀναλύεϲθαι

All the things come from one, and get resolved again to it. (D.L. prooem. 3)

Although scholars tend to agree that the One-Many problem had an important role in the 'Orphic movement', there has been relatively little specific comment on this point.

Guthrie, who gave the most elaborate treatment of the question, localises the One-Many dialectic in Orphic spirituality, i.e. in the allegedly Orphic doctrine that there is a mystic union of the individual soul and the divine outside. However, he rarely gives arguments based on textual evidence for this view.[173] As far as I can see, his only textual support for the 'mystic union' is Aristotle's testimony on the 'Orphic' view on the soul. According to this account 'the soul comes into us from the whole as we breathe, borne by the winds' (*De anima* 410b28).[174] So, in Guthrie's interpretation, air was the link between the individual soul and the divine outside us. Although this evidence will be of some importance for us at a later stage,[175] I do not think it is particularly instructive for the One-Many problem. Guthrie's general conclusion is that Presocratic philosophy, especially its Ionian branch, and Orphic thought are two manifestations of the same problem. The religious-minded Orphics were concerned with the question 'What is the relation of each individual man to the divine, to which we feel we are akin, and how can we best realise and actualise the potential unity which underlines the two?', whereas the philosophers pondered the question 'What is the manifold variety of the world in which we live and the one primary substance out of which, as we are convinced, it must in the first place have arisen?'[176]

Interestingly, Finkelberg, in his polemic directed at Guthrie's presentation of the problem, does not question Guthrie's view on the mysticism of the Orphics, and only mentions in passing the cosmogonic aspect of

[173] He mentions the obviously relevant fragments *OF* 164 and 165 from the *Rhapsodies* in which Zeus asks and receives advice from Night on how to secure his rule: 'How may I have all things one and each one separate?' However, he quotes the fragment not in connection with the One-Many problem, but in speaking about the concept of a creator-god (Guthrie (1952b) 81).

[174] Guthrie (1952a) 92. [175] See 346 below. [176] Guthrie (1955) 316. Cf. also *HGP* I, 132.

the One-Many problem. He tries to show instead that the Presocratic doctrines originated from the same mystic conviction about the unity between the microcosm and the macrocosm. Rationalism, claims Finkelberg, only comes into the picture because rationalism was the particular way in which the Ionians shaped the popular mystical outlook.[177]

Although these considerations might be helpful on a more general level, they cannot immediately further the study of the Derveni poem, and the theogony in it. If one could show that the doctrine of mystic unity between the individual soul and the divine was central for the Orphics, it would be crucial to see whether, and if so how, this tenet was expressed in the theogonic texts circulated under the name of Orpheus, such as the Derveni poem. Yet, in the absence of such an independent proof, we have to go back to the theogonies, and show the relevance of the One-Many problem in them. In what follows, I shall argue that the One-Many problem is implicit in the theogony of the Derveni poem in its cosmological and theological aspects. Its cosmological aspect is expressed in the attempt at giving a comprehensive and unified picture of the world, while the theological aspect is expressed in an essentially pantheistic view. Moreover, these two facets are not independent, but converge in the image of Zeus containing the whole world.

First, however, we should consider some mythological manifestations of the One-Many problem, which are *not* specific to the Orphic account. If by 'theogony' we mean a narrative 'of which the major part consists in an account of the gods from the beginning of the world to the present date,'[178] then we can say that such a narrative is intrinsically aimed at a systematic and comprehensive account of the history of the world. It lays stress on making its start from the very beginning, and then continues up till the birth of man.[179] A theogonic text thus tries to embrace, or rather to create, the widest possible temporal horizon. Furthermore, having identified the beginning, it systematically defines each generation, and preferably each member in each generation. This systematicity, which makes some parts of Hesiod hardly more than an inventory of divine names, is the sign of an endeavour to provide a comprehensive framework within which the place of all possible particular figures and events can be defined. It is the systematic elaboration of a comprehensive framework which distinguishes the temporality of the theogonies from the temporal totalisation expressed in the most general terms in the traditional phrase τά τ᾽ ἐόντα τά τ᾽ ἐccόμενα πρό τ᾽ ἐόντα.

[177] Finkelberg (1986) 321–35. [178] West (1983) 68; cf. 135 with n. 7 above.

[179] On the peculiar place of anthropogony in Hesiod, see Burkert (1999b) 101–2.

This feature of theogonic texts becomes even more evident when one compares Hesiod's *Theogony* with such collections of myths as Apollodorus' *Bibliotheca* or Ovid's *Metamorphoses*. In the latter, although the selection or arrangement may be governed by a thematic or other principle, the myths coexist without any attempt to establish a comprehensive system within which the individual stories would find their place.[180] Whereas the individual stories in these collections have their own time not in any way related to the time of other stories, the theogonies install a singular, unified, comprehensive temporality.[181]

Another aspect of the same attempt at totalisation is related to the political bearing of the theogonic myths. The backbone of the Babylonian and Greek theogonies is a myth which relates how, through clashes between subsequent generations, the ultimate king gains supremacy. This narrative pattern expresses an essentially monarchic political conception; the world, with all the beings in it, is under the rule of a single coercive power. In Mesopotamia, the New Year rituals apparently made an explicit link between the human ruler and the divine king Marduk, as the former enacted the latter's combat against the dragon (the Babylonian parallel to Typhon). The king thus mimed the fight which had led to the creation of the universe.[182] Although, due to the changes in the political structure, this layer of the succession myth had lost its immediate political relevance and its ritual representation, the cosmic aspect of it remained operant. It expresses the firm belief that it is one power which ultimately governs the world. In Hesiod, it is Zeus who, having once obtained royal power, sets the laws and allots the provinces or specific spheres of influence to the other gods.[183] Zeus' rules are the 'laws of everything',[184] coercive for gods and men alike. This single rule, however, is only the end-product of the succession story: it cannot effect the first developments.

As Burkert makes clear, the One-Many dialectic is implicit also in another general feature of theogonies. The logic of theogony implies a retrospective reduction of the number of entities. It requires that the author, by way of

[180] Lévi-Strauss introduced the distinction between 'mythologies' such as the Hesiodic systematisation, on the one hand, and the original plurality of 'myths' on the other. Note that the adjective 'original' does not necessarily involve chronological precedence.

[181] From a more philosophical, and less historical viewpoint, Richir identifies the symbolic emergence of the One, described as an 'implosion dans le système symbolique', or as a 'catastrophe symbolique' in the theogonies' attempt at totalisation. Cf. Richir (1994) 8–15 and (1995).

[182] The relationship between creation myth and rituals in Babylonia and elsewhere has been amply discussed, starting with the essays in Hooke (1933). Cornford (1950) tries to establish the connection with Hesiod. See also Vernant (1962) ch. 7; Burkert (1987) 23–4 etc.

[183] See Hes. *Thgn.* vv. 74–5: . . . εὖ δὲ ἕκαστα | ἀθανάτοις διέταξε νόμους καὶ ἐπέφραδε τιμάς.

[184] See Hes. *Thgn.* vv. 66–7: . . . πάντων τε νόμους καὶ ἤθεα κεδνὰ | ἀθανάτων. . . .

a systematic negation ('at the beginning there was no man, no mountains, no sea' etc.) removes the highest possible number of entities, to posit an economic starting point. The other side of this programme is the multiplication of entities by which, starting from the original economical state, we can explain the richness of the present world. The basic question is how can the one (or the few) become many.[185]

However, such an attempt at totalisation, intrinsic in all theogonic texts, is pushed considerably further with the episode of swallowing in the Orphic theogonies. It is stressed that by the act of swallowing everything, both the anthropomorphic gods and those divine creatures who personify different parts of the physical world become interiorised in Zeus. The result is that Zeus becomes the only one, or gets to be alone, μοῦνος ἔγεντο (col. 16.6). There is thus one viewpoint from which the totality and diversity of the world can be seen as a unity. It is not that the language of the Orphic poem is any more abstract at this point than Hesiod's. Both poets simply list the most important constituent entities of the world: the gods, and the major parts of nature.[186] The difference lies in the fact that in the Orphic poem there is one entity which can comprehend and represent the whole – and this one entity happens to be the supreme god.[187] It is difficult not to interpret this mythem as a mythical expression of the One-Many dialectic. Zeus is one but contains the many, which, in turn, is united in Zeus. Or, from a chronological point of view, the many is resolved into the one, just to become many again in the next moment.[188]

Furthermore, the same mythem has distinct theological overtones as well. On the one hand, the interiorisation of all the divine beings suggests that the different gods can get symbolically united or identified with Zeus. This conception presents the sphere of the divine as a unity. On the other hand, the insistence on the parts of the physical world being united in Zeus seems to propound a pantheistic view: the world is born from the one supreme god.

Clearly, both the One-Many problem and the correlated pantheistic view are amply testified in Presocratic philosophy. With all the differences

185 Burkert (1999b) 91–7. See also Komoróczy (1979), esp. 543–7. Komoróczy's formulations (published in Hungarian) often prefigure those of Burkert's. He argues for example that 'all cosmogonic conceptions are speculative retrospective deductions' (p. 546).

186 Cf. Hes. *Thgn.* vv. 108–10.

187 It has often been suggested that when Plato intimates that there were people even before Xenophanes who held the Eleatic thesis ἓν τὸ πᾶν (*Soph.* 242d4–7), he is referring to the Orphics. See Palmer (1998) 27 with bibliography in n. 53.

188 This is, in some sense, mirrored by Dionysos' story. Dionysos is torn apart, is eaten by the Titans – thus gets absorbed in the many – to be reborn united again.

acknowledged, both in the form of expression and in the religious character, the question of influence between the Orphic theogony and the Milesian type of monism arises. Not surprisingly, all possible scenarios have found propounders. According to some, Orphic lore prefigures Milesian philosophy;[189] while according to others the Orphic myth of the world unified in Zeus is a re-mythologisation of the philosophical monism and eventual pantheism of the Milesians. In my view, both its nature and the lack of datable evidence renders this question undecidable. The most likely is still Guthrie's view, according to which the pantheistic, monistic picture expressed in mythical terms by the Orphic theogonies and the cosmologies of the Milesians share a common world-view which was neither exclusively rational nor exclusively mystical.

The creationist model of cosmogony

Theogonies start with a restricted number of entities and then derive the multiplicity from the original unity, duality, or triad. There are several different models to account for the multiplication of entities. One way to classify these models is to distinguish between (i) the sexual type of generation, where the new entities are born from the intercourse of two previous entities of different sex; (ii) the developmentalist model where there is no sexual intercourse involved but from one entity a plurality of entities are born through internal differentiation; and (iii) the creationist model where a talented god creates the rest.[190]

In a way, (i) is the most obvious solution for a system in which we have a high number of anthropomorphic gods. In any case, it is by far the most common model for multiplication in Hesiod. We find, nevertheless, examples of the other models as well. For instance, it is explicitly stated, or even stressed, that Gaia gave birth to the sky, the mountains and the sea by parthenogenesis (vv. 132 and 213). This remark is customarily interpreted as indicating the view that the birth of these entities was conceived of as some sort of differentiation going on within Gaia, for which no husband was needed. This seems a plausible guess at least for the birth of mountains.

The relative importance of (i) and (ii) will be reversed in the Presocratic cosmogonies. In those accounts, the sexual model becomes far less important (although not completely forgotten), and the emergence of new entities

189 Most recently Palmer (1998) 27–30 argues that the Orphic Zeus encompassing the whole world is one of the most important sources of Xenophanes' conception of the cosmic god.

190 Other scholars use different terminology. Burkert (1999b) speaks about biomorphic and technomorphic models; see also Lloyd (1966) 203–5 and 207–8.

is most commonly accounted for by the differentiation of a previous, less articulated entity.

For (iii), the creationist model, pre-eminent in Genesis, we have very little evidence from Greek theogonies. There are some traces here and there in Hesiod, such as the fabrication of Pandora in the *Theogony* or the creation of the three races in the *Erga*.[191] Much more interesting is Pherecydes, where Zas makes Chthonie an embroidered gown for their wedding, the patterns of which will represent the topography of the earth as Chthonie becomes Ge.[192]

It is against this background that Guthrie declared: 'the conception which seems to me to have the best right to be called an Orphic idea is that of a creator.'[193] Guthrie was thinking here about the later versions, primarily the *Rhapsodies*, where there is the additional problem of two creators – first Phanes, who by an original demiurgic act created the whole physical world and everything in it, and then Zeus, who assimilated the original creator and created the world anew.[194] Although I do not think that we have two creators in the Derveni poem, Guthrie's point seems to apply to this version as well. Indeed, it is the act of creation which marks the most important difference from Hesiod, or for that matter, from all other known early Greek theogonies. The underlying intuition seems obvious, and very close to what has been said about the two senses of priority. The supreme ruler has to be source and cause of the world. His rule is more coercive and his power is more intense if he is not confined to the already given conditions, but can create his kingdom and can then steer a world fashioned according to his intents. This is the picture we get not only from Genesis, but also from some of the Babylonian theogonies.[195]

In this respect, there is a further notable feature of the story of second creation in the Derveni poem. For although Zeus assimilates the world by a simple swallowing, he does not create it anew by simply regurgitating what

[191] But Arist. *Metaph.* A 984b23–985a4 takes the Hesiodic Eros as the first approximation to the efficient cause.

[192] See Schibli (1990) 54–61.

[193] Guthrie (1952b) 106.

[194] The Neoplatonists thought that the first creation represented the creation of the intelligible world, while Zeus created the physical world.

[195] In the *Enuma Elis*, Marduk, the victorious god of the fourth generation, is the creator god. Having obtained the kingship, he cuts the primordial deity Tiamat into two. The upper part of Tiamat will be the sky, while the bottom part becomes the earth. Marduk then creates the stellar constellations and the other heavenly bodies, and defines the measures of time. The breast of Tiamat becomes the mountains, and the two eyes turn into the sources of the Tigris and the Euphrates respectively. When all this is ready, Marduk creates a shrine which will constitute the foundation of Babylon; then he creates man for the service of the gods. Incidentally, this account of creation, *mutatis mutandis*, is quite close to the creation effectuated by Phanes in the *Rhapsodies*.

has been swallowed. The extant text of the papyrus has preserved a verb to describe the creative act of Zeus. In col. 23, the Derveni author explains the way Okeanos and Achelous were born again. From this gloss, it seems certain that the verb used was μήcατο.[196] This verb, interestingly paralleled in Parmenides' *doxa*,[197] clearly indicates that Zeus did not simply spew up the previously swallowed entities. The verb implies instead that Zeus, as it were, processed the world inside himself, and reproduced it by a rational act of planning or contriving.[198] The characterisation of Zeus's demiurgic activity is a salient departure from the traditional Hesiodic account. This feature of the Derveni poem becomes crucial as we now proceed to the Derveni author's cosmogony and theology.

[196] On the reconstructed verse μήcατο δ' Ὠκεανοῖο μέγα cθένος εὐρὺ ῥέοντος see the discussion in the previous chapter on p. 129. For the use of μήcατο, see also ***L16*** = col. 25.14 as restored by Tsantsanoglou: [αὐτ]ὰ̣ρ̣ [ἐ]π̣εὶ δ[ὴ πάν]τα Διὸ[c φρὴμ μή]cατ̣[ο ἔ]ρ̣γα.

[197] Parm. B13 DK, where the *daimon* creates Eros: πρώτιcτον μὲν Ἔρωτα θεῶν μητίcατο πάντων. The *Rhapsodies* used the same verb for Phanes' demiurgic activity: μήcατο τ' ἄλλην γαῖαν ἀπείριτον, ἣν τε cελήνην κτλ. (*OF* 91 v.1 = Procl. *in Plat. Tim.* 32b). Cf. Tarán (1972) 407.

[198] The intellectual capacities of Zeus have already been stressed in the epithet μητίετα in ***L8*** = col. 15.6.

CHAPTER 5

The cosmic god

Having taken a stance on the structure, plot, and interpretation of the poem commented on in the Derveni papyrus, let us now turn to the views of the Derveni author. Certainly the reconstruction of the Derveni author's own theory is not any easier than the reconstruction of the poem, but it is perhaps not impossible, and undoubtedly worth trying.

Apart from the lacunose nature of the papyrus, what makes the reconstruction so difficult is that the Derveni author does not explain his theory in a linear way, but distributes the elements of it in his exegetical remarks. In other words, the exposition is not governed by the internal logic of the theory, but we learn it piecemeal, always as the Derveni author finds some part of his own doctrine explanatory of the verses under consideration. This means that we are obliged to hop forward and backward in the text in our efforts to reassemble the *disiecta membra* of the Derveni author's own views on the cosmos and on the works of the cosmic divinity.

Moreover, as long as the Derveni author remains unidentified – and I have to say in advance that I do not find any of the existing hypotheses concerning the identification of the author convincing, nor can I propose anything better – the text will continue to be in a frustrating interpretative vacuum, where no immediate external help can be expected. Even if we can often doubt the fidelity and exactness of the evidence provided by later doxographers, the set of doxographical reports – the 'A fragments' – always offers a convenient starting-point for the interpretation and discussion of the Presocratic philosophers. The doxographical tradition not only provides additional pieces of information to complete the evidence of the 'B fragments' (the verbatim quotations), but also presents an exegetical framework within which, or for that matter in contrast to which, one can explicate one's own reading of the text. But, alas, there is no 'A section' for the fragments of the Derveni text. There is no Diogenes Laertius or Clement of Alexandria to give us a hand in need.

In any case, the identification of the Derveni author is a slippery issue. On the one hand, one could assign to him some empty name, say Idaeus of Himera, about whom we know almost nothing except that together with Anaximenes, Diogenes of Apollonia and Archelaus he assigned the role of *arche* to the air.[1] Such a hypothesis would be just as difficult to verify as to falsify; and certainly, it would not lead us any further. On the other hand, the Derveni text could be assigned to an author we know more about from other sources, such as Diagoras in Janko's hypothesis, or Euthyphro known from the Platonic dialogues, as suggested by Kahn. Nevertheless, to test such a hypothesis we first need to reconstruct the views expounded in the papyrus itself so as to compare them in a second move with the information we possess about the suggested author from other sources. So we are back where we started: before speculating about the identity of the author, we should examine the text itself.

There seems to be a relative lack of interest in the views the Derveni author expresses in the course of his exegesis of the poem: scholars generally find it sufficient to assert that the Derveni author was an eclectic.[2] Once the author is characterised in this way, one can apparently dispense with a detailed reconstruction, and be content with pointing out, more arbitrarily than systematically, certain similarities with other early Greek philosophers. On these assumptions, and instead of proceeding more methodically, commentators tend to pick up some assertions or solitary terms in the papyrus and connect them with the language or specific doctrines of some Presocratic philosophers. There is no denying that significant points of contact have been discovered in this way, and I would certainly agree that such comparisons ultimately constitute an important means for the understanding of the Derveni author's views.

Yet again, eclecticism is another tricky issue. The Derveni author has been said to have been influenced by Anaxagoras, Diogenes of Apollonia, Democritus and Leucippus, Metrodorus of Lampsacus, Anaximenes and the Ionian philosophers in general, Parmenides and Empedocles and Pythagoreans such as Philolaus. The list is indeed long and heterogeneous. Nevertheless, I do not think that receiving influences – or eclecticism if one prefers this already pejorative term – necessarily precludes internal coherence and consistency, or that a new arrangement of elements from

[1] Sextus Empiricus *M* 9.360 = 63 DK.

[2] Burkert (1968) 99 and (1970); Reale (1974) 274; Boyancé (1974); Henry (1986); Edwards (1991).

different provenances cannot result in something original.[3] Also, Presocratic physics works within a set framework, with a limited number of key concepts. Hence authors coming later in time must necessarily reapply some of the concepts and patterns that have already come up in the doctrines of previous thinkers.[4] And even if someone comes with ground-breaking new ideas and the intention to shift the whole physical paradigm, as Plato does in the *Timaeus*, it will not be a *tabula rasa*; even the *Timaeus* shows the traces of doctrines of practically all the Presocratics mentioned in the above list.[5] All in all, the fact that the Derveni author is indebted to earlier thinkers gives us no dispensation from studying his own ideas as closely as possible.

Hence I shall start with the working hypothesis that the Derveni author did have a relatively coherent cosmological theory, no matter for the time being how original the particular elements of this doctrine were, and shall begin with an attempt at a primary, preliminary reconstruction of this doctrine based on a close reading of the remaining text of the papyrus (chapters 5 and 6). This will give a broad picture, with some points made relatively clear and a large number of questions left open. Then, in a second step, I shall undertake a series of comparisons with some prominent Greek philosophers in the subsequent chapters. In this manner, I shall shift the emphasis from the comparison of particular corresponding elements of different theories, viewed in isolation from the rest of the respective systems they originally belong to, towards an analysis which tries to examine the appropriation of certain doctrinal elements from one system to the other. Also, I shall try to see whether the idea borrowed has or has not changed its meaning in its new context. Just to give an example, it is not sufficient to show that both Diogenes of Apollonia and the Derveni author had an intelligent air, as this point does not have the same significance in the framework of a monistic doctrine, where air is the archetypal form of matter and the other elements are ultimately made of air, and in another, pluralistic system.

I shall start the exposition of the Derveni author's own theory by delineating his views on the identity of the cosmic god, to turn in the next chapter to his cosmology which relates the cosmic works of this god.

[3] Cf. Betegh (1993) 27–8. This point has also been emphasised by Laks (1997) – notably the author of the standard monograph on Diogenes of Apollonia, another so-called 'eclectic'.

[4] The same point is made by Laks (1983) xxxiii with regard to Diogenes of Apollonia. In fact, almost all the problems raised by Laks on the question of the eclecticism of Diogenes of Apollonia can apply *mutatis mutandis* to the Derveni author.

[5] The J. Burnet–A. E. Taylor hypothesis on Plato trying to reproduce a fifth-century physical system is improbable.

WHO'S WHO IN THE WORLD OF THE DERVENI AUTHOR

The Derveni author's understanding of the Orphic poem is marked by the fact that he identifies several different divinities – for the most part presumably mentioned in the poem itself – with each other, on the one hand, and with some more abstract concepts, both physical and psychological, on the other. Let us now try to explore this complex web of identifications.

Kronos and Mind

The first clear case of such an identification can be found in column 14, where the author identifies Kronos, son of Ouranos and father of Zeus, with Mind:

> He says that this Kronos was born from the sun to the earth because he became the cause through the sun that they were struck against each other. For this reason he says: '(He) who did a great deed.' And after this:
>
> *Ouranos son of Night, who first of all ruled*
>
> Naming Mind that strikes (? the beings) against each other Kronos, he says that it did a great deed to Ouranos; for (he says that) he (sc. Ouranos) was deprived of his kingdom (col. 14.2–9).

Thus, with the help of the Derveni author, and on the basis of the etymology κρούων νοῦς = Kronos, we discover that it was Mind that received the name 'Kronos' in so far as the things were struck or thrust (κρούεσθαι) against each other by it. As regards the Derveni author's conception of Mind, we can learn from this passage that Mind is something active. It can exercise its power over other beings, and can effect physical change by triggering off some kind of motion which leads to the collision of chunks of matter. Mind is the cause of or responsible for (αἰτίαν ἔσχε) this collision. Furthermore, the sun is somehow implicated in the action of Mind, playing the role of an intermediary in the causation started by the Mind (διὰ τὸν ἥλιον).

Zeus and Mind

As we follow the thread marked by the cosmic Mind, we find a little later that not only Kronos, but also Zeus is identified with Mind. In col. 16 we read the following:

> It has been made clear above [that] he called the sun a phallus. Since the beings that are now come to be from the already subsistent he says:

[with?] the phallus of the first-born king, onto which all
The immortals grew (or: clung fast to) blessed gods and goddesses
And rivers and lovely springs and everything else
That had been born then; and he himself became solitary.

In these (verses) he indicates that the beings always subsisted, and the beings that are now come to be from (*or*: out of) subsisting things. And as to (the phrase): 'and he himself became solitary', by saying this, he makes clear that the Mind itself, being alone, is worth everything, as if the others were nothing. For it would not be possible for the subsisting things to be such without the Mind. And in the following verse after this he said that the Mind is worth everything:

Now he is king of all and will always be.

. . .: Mind and . . . (col. 16.1–15)

As we have already seen in the preceding chapters, the lemmata quoted in this passage refer to the moment when – as a result of swallowing Ouranos' phallus (or, alternatively, Phanes) and with that the entire world – Zeus became the only one and took the whole world inside him. Consequently, the gloss 'And as to (the phrase): "and he himself became solitary", by saying this, he makes clear that the Mind itself, being alone, is worth everything, as if the others were nothing' must explain the universal Zeus of the poem by equating him with Mind.[6]

It is nevertheless worth noting already at this point that there is an important shift of emphasis here. The poem implies, on the level of literal meaning at least, that by the act of swallowing, Zeus interiorised everything. Everything was inside Zeus, and thus there was nothing outside of him. He was alone. Surely, such a sentence could be a perfect prop for a strictly monistic allegorical reading of the poem. It could mean that there is something unified, which contains in itself the entire plurality of things.[7] The Derveni author, in contrast, does not say that there is nothing outside of, or apart from, the Mind. He only says that the Mind in itself 'is worth everything'. Furthermore, the counterfactual [ὥσπερ]εὶ μηδὲν τἆλλα εἴη implies that in fact 'the others' do exist. We have the Mind, on the one hand, and a plurality of things, on the other; and as long as we have the Mind, it does not really matter whether or not the others exist. It remains to see what this idea can mean.

At any rate, we have seen that the Derveni author maintains that both Kronos and Zeus are to be identified with the Mind. And it seems that

[6] The wording makes it highly probable that the Derveni author is dealing with the same Mind that he was speaking about in connection with Kronos.

[7] Cf. p. 178 above.

we have further evidence showing that, for the author, Kronos and Zeus are not two independent divinities but only two names for the same god. In the second half of the existing part of col. 15, the author is explaining ***L8***=col. 15.6:

And in the next verse,

> *From him in turn Kronos, and then wise Zeus*

he says that there has been this rule since he has been the king. But his rule gets the explanation that it is by striking the things that are against each other that he separated them into their current reconfiguration, not different (?ones) from different ones, but [. . .]. And the (expression) 'and then prudent Zeus' makes clear that not another one, but the same one. He indicates this:

> *Holding wisdom and royal honour over the blessed gods*
>
> (col. 15.5–13)

As I have argued above, ***L8***, in all probability, came in the poem immediately after ***L7***=col. 14.6: Οὐρανὸϲ Εὐφρονίδηϲ, ὃϲ πρώτιϲτοϲ βαϲίλευϲεν. Thus, the poet here summarises the succession of divine kings.[8] As I understand the text of col. 15,[9] the first sentence after ***L8*** explains the 'from him' (ἐκ τοῦ δή) of the lemma. The Derveni author echoes this expression (ἐκ τοῦδε), and claims that the present rule started with the reign of Ouranos. That the author had Ouranos in mind should be clear also from the reference to kingship (ἐξ ὅϲου βαϲιλεύει), which refers back to ***L7*** (βαϲίλευϲεν).[10]

On this reading, the whole passage is about the continuity of the rule from Ouranos through Kronos to Zeus. What might have been taken as three different rules and rulers is one and the same; and, as I take it, this is what the repeated denial of difference and corresponding affirmation of identity – 'not different (? ones) from different ones . . .'; 'not another one, but the same one' – tries to inculcate.[11]

[8] Cf. 124 above.

[9] The construction of the text in lines 7 and following is controversial. The meaning of ἀρχή is ambiguous, and it is not clear whether the sentence ends after βαϲιλεύει or ἥδε ἀρχή. L&M translates 'he says that the beginning dates from the time since this particular rule has been king.' Rusten (1985) 138 renders the text in the following way: 'This ("Kronos") is described as the beginning because it is by striking the elements against each other that he has separated them into their current configuration – not (forming) one set of things to replace another, but (forming) the same things in a different way.' I am closer to Janko's solution (2001) 24; he translates '(Orpheus) means that his rule has existed since he became king'.

[10] Note also the political undertones of the term μετάϲταϲιϲ. The author explains the reason why the divinity is called 'king' by Orpheus in col. 19.8–15.

[11] Cf. also the parallel construction in the passage establishing the identity of Zeus and Okeanos in col. 23.4: οὔκουν . . . ἕτεροϲ . . . ἀλλ' αὐτὸϲ. On this identification see 194 below.

Zeus's mother and Mind

Mind appears once more in the last column of the surviving text:

'[of?] mother' because the Mind is the mother of the others; and 'of his own (ἑᾶϲ)' because she is good. He makes clear in these verses too that it means good:

Hermes Diaktoros son of Maia, giver of goods (ἑάων).

He makes it clear in these as well:

For two urns are placed on Zeus's threshold,
Of gifts such as they give: of evils, and the other one of goods (ἑάων)

Those who do not understand the term think that it is '[of?] his own mother'. But if he had wished to show the god wishing to mingle in love of his own (ἑαυτοῦ) mother, it would have been possible for him by altering some letters to say of his own (ἑοῖο) mother. For thus it would have become his own (ἑαυτοῦ), and he would be her son. . . . her . . . clear that . . . both . . . good . . . (col. 26.1–15)

According to the reconstruction of the poem as presented in the preceding chapters, the Derveni author's explanation must refer to the phase of the theogonic plot when Zeus is about to rape his mother who is most probably called Rhea also in this mythological genealogy.[12] Not surprisingly, the Derveni author tries to argue away this horrid episode. Yet he does it in a rather extraordinary way. He does not start off by allegorising the act itself, but rather explicates the reference of the word 'mother' in such a way that it completely excludes the possibility of any incestuous act. Apart from the fact that what we would have taken as a reflexive personal pronoun (ἑᾶϲ) is simply an adjective expressing appreciation, the word 'mother' does not refer to the parent of Zeus. The word refers instead to Mind, in so far as everything else is born from Mind. Or, as the meticulously argued interpretation of ἑᾶϲ emphasises, everything else is born from the 'good' Mind. And if we compare the outcome of this reasoning with what we have learnt about Zeus in col. 16, we have to conclude that both the alleged rapist Zeus and the alleged victim Rhea are the same: Mind. In such an arrangement, the possibility of rape and incest is definitely ruled out. But we have learnt something more. Mind can be seen as the 'mother,' that is to say the origin or source, of the other beings. Again, it remains to discover in what way Mind can be said to be the origin of other beings. As we shall see, the answer is in the cosmogony.

[12] Cf. 130 and 164 above.

Ge, Meter, Rhea, Hera, Demeter, Hestia and Deio

The identification of Mind with Rhea *qua* mother leads us to col. 22:

So he (sc. Orpheus) named all things in the same way as finely as he could, knowing the nature of men, that not all of them have a similar nature nor do all want the same things. When they have the power, they say anything that occurs to each one's heart, whatever they happen to want, never the same things, through greed (*or:* arrogance), sometimes also through lack of understanding. Earth (Ge), Mother (Meter), Rhea and Hera is the same (*or:* are one and the same). She/it was called Earth (Ge) by convention; Mother, because all things are born from her (*or:* from this one). Ge and Gaia according to each one's dialect. And (she/it) was called Demeter as the Mother Earth (Ge Meter), one name from the two; for it was the same. (col. 22.1–11)

The avalanche of identifications has definitely gained impetus. We find Ge, Gaia, Demeter, Meter, Rhea and Hera equated or identified with each other. It is worth recalling at this point that according to *OF* 145 (=Procl. *in Plat. Cratyl.* 403e), Rhea received the name 'Demeter' in the role of Zeus' mother in the *Rhapsodies.*[13] On the surviving evidence, it is difficult to say what the exact relationship between the individual items of this list of divine names was in the Derveni poem. The extant text does not allow us to decide whether some or all of these divinities were already identified by the author of the Derveni poem, or whether, alternatively, the identification of Rhea and Demeter in the later versions of the Orphic theogony was just the re-appropriation of a former allegorical reading operating with identifications – such as the passage of the Derveni text we are just discussing – of an earlier theogonic narrative in which these goddesses were still separate figures. At all events, in the light of the reasoning of col. 26 (Mother (=Rhea)=Mind), the joint mention of Rhea and 'Mother,' together with the identical explication of the term 'mother' ('Mother, because all things are born from her'), makes the conclusion inevitable that in the Derveni author's interpretation the being described by Orpheus with the help of these names of goddesses is ultimately Mind.

The interpretation of the immediate sequel of the lines quoted from col. 22 is not completely certain, however:

And it is said in the *Hymns* too: 'Demeter Rhea Ge Meter Hestia Deio'. For (she/it) is also called Deio because she/it was torn (*or:* ravaged: ἐδηιώθη) in the mixing/sexual intercourse. He will make it clear when, according to the verses, she

[13] Cf. 142 above.

is born. . . . And (she/it) is called Rhea because many and . . . animals were born . . . from her. Rhea and . . . . (col. 22.11–15)

On the most natural reading of the text, we have to add also the names Hestia and Deio to the repertory of denominations for the same being, even though this latter list is apparently quoted from another poem. But given that this other poem is probably also attributed to Orpheus by the Derveni author,[14] he could easily claim that Orpheus elaborated on the same question, i.e., the names attributable to the same being, in some of his other poetical works. This interpretation is reinforced by the fact that the grammatical subject to be supplied for the passive ἐδηι[ώθ]η in the etymology of Deio must be the same as the being whose unity is argued for in the preceding sentences.

Incidentally, in the light of a parallel in Philodemus *De pietate* (pp. 63+23 Gomperz=PHerc. 1428 fr. 3.14–18) where it is said that 'in the *Hymns* Orpheus in Philochorus (says) that Ge and Demeter are the same as Hestia',[15] one has the growing feeling that the divinities were already identified in the *Hymns*, and not just by the Derveni author[16] – barring, of course, the non-negligible eventuality, argued for by Obbink,[17] that Philochorus, a fellow *mantis*, takes this view from the Derveni author himself.

Aphrodite Ourania, Zeus, Peitho and Harmonia

There are more names to be identified with each other. In col. 21, in the column immediately preceding the one we have just discussed, we read the following:

. . . nor the cold to the cold. By saying 'by mating', he (sc. Orpheus) makes clear that divided up into small (pieces) they were moving and mating in the air, and as they were mating, they got put together with each other. They were mating until each came to its like. Aphrodite Ourania, and Zeus, and to aphrodise, and to mate, and Peitho, and Harmonia are given as name to the same god. A man mingling with a woman is said by common usage to aphrodise. For (this reason) as the things that are now got mixed with one another, (this god) has got the name

[14] Apart from the wording of the passage, this hypothesis is borne out by the parallel in Philochorus: see next note.

[15] =Philochorus (*FGrHist* 328 F 185): κἀν] | τοῖς Ὕμνοις δ' Ὀρφ[εὺς | π]αρὰ Φιλοχόρωι Γῆν [κ]αὶ Δήμητρα τὴν || αὐτὴν Ἑςτίαι. Cf. Obbink (1994) with the new papyrus readings and his arguments to the effect that this sentence in Philodemus would prove that the late fourth-century Philochorus read and quoted the Derveni text.

[16] Calame (1997) 71 says that the names were probably already juxtaposed in the poem. It is hard to see on the other hand how far this paratactic enumeration of names meant identification originally.

[17] Obbink (1994) and (1997) 49 n. 16.

Aphrodite. (It has got the name) Peitho, because the beings yielded to one another; and to yield and to persuade is the same. (It has got the name) Harmonia, because many of the beings got fitted to one another. (col. 21.5–12)

Zeus provides the link with the passages discussed above. As Aphrodite Ourania, Peitho and Harmonia are identified with Zeus, we should conclude that, once again, all these names are denominations of the same polyonymous being with whom we have been dealing so far. It should be noted at the same time that the denominations listed are explicitly said to be names of a god. Thus the divine nature of the being named is not doubted, even though his/her identity is conceived of in a considerably different manner from what we are accustomed to within the framework of traditional Greek religion.

Scholars have suspected a textual problem in this passage. According to a conjecture by Merkelbach, as published in the *ZPE* edition of the papyrus, the two infinitives, to aphrodise and to jump/mate, were shifted from the next sentence to this place by a scribal mistake. Hence Merkelbach's correction for col. 21.5–7:

> Ἀφροδίτη οὐρανία
> καὶ Ζεὺϲ {καὶ ἀφροδιϲιάζειν ̣καὶ θόρνυϲθαι} καὶ Πειθὼ
> καὶ Ἁρμονία τῶι αὐτῶι θεῶ̣ι ὄ̣νομα κεῖται. ἀνὴρ
> γυναικὶ μιϲγό̣μενοϲ ἀφροδιϲιάζειν λέγετα̣ι κατὰ
> φάτιν <καὶ ἀφροδιϲιάζειν καὶ θόρνυϲθαι>.

The trouble is that this text is hardly, if at all, smoother than the original. What we might have gained in the first sentence, we have lost in the second: the redoubling of ἀφροδιϲιάζειν is hard to make sense of. If anything, καὶ θόρνυϲθαι should come after the first ἀφροδιϲιάζειν in line 7.[18] But apart from this point, the correction, as a whole, is gratuitous and unnecessary. It is apparently based on the assumption that although the Derveni author's god can be called by the names of various gods of the Greek pantheon – this one can accept from an allegorist – one cannot allow that the god receives denominations other than proper names.

There are, however, a few considerations which might mitigate the unusualness of the point made in the text as it stands in the papyrus. First of all, it should be clear that at the time of the composition of the Derveni text, there was no explicit, clear-cut distinction between common

[18] This way of emending the text has now been suggested by Janko (2002).

nouns and proper names,[19] or even verbs and nouns.[20] True, the lack of such an explicit distinction does not mean that they would have simply mixed the members of the two groups. On the other hand, the belief that proper names (or more exactly, what will later be called proper names) have descriptive content just as every other word has, is the very foundation of the etymological practices exercised on the names of gods. Furthermore, it is worth noting that all the etymologies of names in the Derveni papyrus are based on verbal, as opposed to nominal, forms. Thus, the proper names themselves are explained as captions indicating different actions or cosmic functions of the god. As regards the names of goddesses dealt with in col. 21, the god can be called by the name of Aphrodite precisely because he/she makes the chunks of matter to 'mate' or 'aphrodise' with one another. The god can be called by the traditional name of the goddess for the very reason that the verbal form ἀφροδιϲιάζειν applies to her/him. And similarly for Peitho and Harmonia. As a result, unusual though it may sound, it seems better to accept the text as it is, and allow that, for the Derveni author, verbs could figure among other names of the god.

Boyancé, on the other hand, has maintained that the identifications of Zeus with the three goddesses had already been present in the poem, and that the Derveni author only provided the physical allegory.[21] Boyancé's only argument for this hypothesis is that in the Hymn to Zeus at the end of the *De mundo* Zeus is presented as both masculine and feminine.[22] This parallel, however, is not conclusive for at least two reasons. First, the *De mundo* hymn refers to the Zeus who has just absorbed the rest of the world: the hymn praises the universal, all-embracing god. Accordingly, it is this universal, pantheistic character of Zeus, embodiment of a temporary total unification and suspension of all distinctions, that is made emphatic in the hymn. In such a context, the possession of both sexes is just one aspect of this divine *coincidentia oppositorum*, as Zeus is also first and last, earth and sky, sun and moon. The passage of the poem commented on in col. 21, in contrast, is in all probability already speaking about the next phase of the story, i.e. the 'second creation' effected by Zeus. As we have seen in the previous chapter, the Derveni author is here commenting on the passage describing the birth of Aphrodite, and possibly of the two other goddesses mentioned in the passage, Harmonia and Peitho. In this phase of the story,

[19] The distinction between ὄνομα κύριον and προϲηγορία is most probably a Stoic invention. Cf. Chrysippus in *SVF* 2.45.147.

[20] Although the *onoma–rhema* distinction is already present in Plato's *Cratylus* (425a1–2 and 431b5–6), it is still not used systematically, nor does *rhema* there appear to mean 'verb'.

[21] Boyancé (1974) 102.

[22] *OF* 21a v. 4: Ζεὺϲ ἄρϲην γένετο, Ζεὺϲ ἄμβροτοϲ ἔπλετο νύμφη.

the momentary unification is over and the plurality is being formed from the unity. It would be hard to imagine that the poet maintained the unity of Zeus, Aphrodite Ourania, etc. even at this stage.

Moreover, to say that Zeus is of both sexes is one thing, and to identify him with certain particular goddesses is another. It is not without reason that although the universal god depicted also in the *De mundo* hymn can receive all kinds of, and even opposing, characteristics, he/she is not equated specifically with any other god of the traditional pantheon. Even though Zeus contains in himself/herself all the other divine beings, it is Zeus who contains them and not, say, Peitho. Identification is a symmetrical relation, whereas the relationship between Zeus and the other gods inside him is not a symmetrical one. After all, the hymn is to Zeus, and not to any other arbitrarily chosen divine being absorbed by him/her. Thus despite the momentary suspension of all distinctions, the identity of Zeus is not denied, but rather strengthened by the absorption of the rest of the world. We should not forget that the main objective of the whole plan masterminded by Night was to reaffirm the authority of Zeus over the others. That Zeus's integrity and pre-eminence are reinforced and emphasised is also suggested by the almost fastidious repetition of Zeus' name throughout the hymn: 'Zeus is the head, Zeus the middle and all the beings were born from Zeus etc.' As opposed to this, the text in col. 21 of the papyrus suggests absolutely no privilege granted to Zeus, or to the name 'Zeus': 'Aphrodite Ourania and Zeus and to aphrodise and to mate and Peitho and Harmonia are given as name to the same god.' 'Zeus' is just one, and even not the first, in a series of names. Thus, the alleged parallel from the *De mundo* hymn referred to by Boyancé does not entitle us to think that the identifications of col. 21 of the papyrus had already been present in the Derveni poem.

The Derveni author justifies the identification made by him by referring to some kind of mixing and conjunction of chunks of matter conceived of on the analogy of sexual intercourse. Therefore, the identifications of col. 21 lead us again to physical processes.

Zeus, Okeanos and the air

The snowball grows further, as a highly significant point is made in col. 23. The now lost lemma from the poem could be something like μήϲατο δ' Ὠκεανοῖο μέγα ϲθένοϲ εὐρὺ ῥέοντοϲ.[23]

[23] Cf. 129 above.

This verse has been made misleading and it is unclear to the many, but to those who understand correctly it is clear that Okeanos is the air and air is Zeus. It is not the case that another Zeus contrived Zeus, but that the same one (contrived) for himself great strength. But those who do not understand think that Okeanos is a river because he (sc. Orpheus) added the epithet 'broadly flowing'. But he indicates his meaning in current and customary expressions. For they say that the very powerful among men 'flowed great'. (col. 23.1–10)

The expression 'has been made misleading' might appear problematic. παράγω generally means 'lead aside' and hence 'to mislead', but as a grammatical technical term it means 'to modify' or 'change slightly'. Neither of these meanings is entirely satisfying in this context. Choosing the second sense, it would be difficult to see what kind of modification the author would think of; and, indeed, who could have introduced that modification and as opposed to what? The first sense, on the other hand, may seem to imply that the Derveni author blames Orpheus for making the verse misleading. However, in view of the author's usual approach to the poet, this is unlikely: Orpheus is beyond any criticism. Therefore, the expression should be taken in the sense that, although this line has caused some misunderstanding, Orpheus expressed the truth in it, but in an enigmatic way: 'has been made misleading' (for the many, but clear to the initiated). πα̣[ρα]γωγὸμ πεπόηται then gets close to αἰνίζεται.[24]

We learn from this passage that both Zeus and Okeanos are the air. Furthermore, the second sentence implies that through the identification with air, Okeanos is considered to be identical with Zeus. These identifications are all the more noteworthy since they show that we are in the realm of physical allegory, where the divine beings of the epic poems are explained as physical elements of the cosmos.

As regards the first identification, the equation between Zeus and the air comes up in two other passages of the papyrus. First, we have the first half of col. 17:

. . . it existed before it was named. Then it was named. For air existed even before the things that are now were set together and always will exist. For it was not born, but existed. And the reason why air received its name has been made clear above.

[24] Similarly Burkert (1968) 95 n. 4 with reference to Plato *Phdr.* 262d and *Gorg.* 493a: 'offenbar ist der grammatische Terminus aus der allegorischen Interpretationsmethode hervorgegangen: man fragt nach der "tieferen Absicht" (ὑπόνοια) des Autors, der im Ausdruck παράγει.' See also Blank (1991) 25–6; and Kingsley (1995) 168–9, with nn. 75 and 76. Kingsley's contention that παράγω 'also has the standard, semi-technical sense of "deriving" a word from another through the use of etymology' is of no help; to support his point, he refers to Plato *Crat.* 398c–d; 400c and 416b, where the word simply means 'to change' or 'to modify' the phonetic form of the word; and to the passage we are just discussing (PDerv. col. 23.1) where there is no question of etymological derivation.

But it was thought that it was born, because it got the name Zeus, just as if it did not exist previously. And he said that this will be 'the last', because it was named Zeus, and it will continue to be his name until the things that are now have got set together into the same form in which they were floating as they were before. (col. 17.1–9)

As we have seen in chapter three,[25] the Derveni author is probably dealing here with the verse, also attested in the *De mundo*'s Hymn to Zeus, Ζεὺς πρῶτος γένετο, Ζεὺς ὕστατος ἀργικέραυνος, or something close to this. In the first half of the column, the Derveni author tries to explain – or explain away – the first half of the verse, according to which Zeus *was born* first. For in his view, things in general,[26] and the air in particular, exist eternally, and thus could not have been born at any moment. So Zeus is air, and the air received the name 'Zeus' at a given moment of the story. This shows that the existence of air is not coextensive with the span of time within which it is called Zeus: the latter has a beginning, whereas the former is eternal. The second half of the passage, explaining the adjective 'the last' applied to Zeus by the poet, shows that the name Zeus, given at a certain moment, is not applicable for the whole of the rest of time. Rather, the denomination Zeus applied to the air corresponds to a particular state of the cosmos, and the name is valid only during the period within which that specific feature of the cosmos designated by the name of Zeus remains undisturbed. When this state of affairs is over, or the specific characteristic has disappeared, the air will not be called Zeus any more.[27]

The air-Zeus identification appears again in col. 19:

. . . existing things have been called each single name by reason of what dominates (them); all things were called Zeus according to the same principle. For the air dominates all as far as it wishes. (col. 19.1–4)

The Derveni author is explaining here the reason why 'all things' can be called by the name 'Zeus'. The Derveni author sets out the principle that each thing is named according to what dominates it, and hence claims that since Zeus is the air and 'the air dominates all things', therefore 'all things' can be called Zeus.

The identification between Zeus and air in itself cannot cause surprise. Although, on the basis of the phonetic resemblance, it became more common to equate Hera with air, the Zeus = air equation is also well attested

[25] Cf. 126 above.

[26] Cf. col. 16.7–8: 'In these (verses) he indicates that the beings always subsisted, and the beings that are now come to be from (*or:* out of) subsisting things.'

[27] On the reading and interpretation of the phrase speaking about the end of the Zeus-age, see 257–9 below.

in all kinds of texts. To begin with, Zeus had always been connected with the sky, and in Homer he is explicitly allotted this 'airy' portion of the world when the cosmic regions were distributed among the sons of Kronos. Sharing the earth and Olympus in common, Poseidon received the sea, Hades the underworld, and Zeus 'the broad heavens in the aither and clouds'.[28]

On this basis, allegorists could effortlessly identify Zeus with the elemental air. There was, however, some confusion owing to the ambiguity and shifting semantic field of the word αἰθήρ, gradually becoming in later times a synonym for fire. Nevertheless, in tragic texts the meaning is quite clear. So, for example when Euripides says

ὁρᾷς τὸν ὑψοῦ τόνδ' ἄπειρον αἰθέρα
καὶ γῆν πέριξ ἔχονθ' ὑγραῖς ἐν ἀγκάλαις;
τοῦτον νόμιζε Ζῆνα, τόνδ' ἡγοῦ θεόν. (fr. 941 Nauck)

we can be fairly sure that the *aither* identified here with Zeus is not fire, and not even merely the upper region of air: it is just as much the atmospheric air around the earth.

The identification of Zeus and air appears in philosophical texts as well. Philodemus reports that Diogenes of Apollonia had praised Homer for speaking about the god not in a mythical way, but according to the truth, as, on Diogenes' reading, Homer had considered the air to be Zeus, for he, Homer, had said that Zeus knows everything.[29] Although the reasons given are different, Homer's alleged identification of Zeus as air is also attested in a scholium to *Od.* 1.63.[30] Furthermore, according to some commentators, Empedocles also used the name of Zeus for the element air – this point, however, despite the recent brisk argumentation by Kingsley, must remain

[28] *Il.* 15.189–93: τριχθὰ δὲ πάντα δέδασται, ἕκαστος δ' ἔμμορε τιμῆς· | ἤτοι ἐγὼν ἔλαχον πολιὴν ἅλα ναιέμεν αἰεὶ | παλλομένων, Ἀΐδης δ' ἔλαχε ζόφον ἠερόεντα, | Ζεὺς δ' ἔλαχ' οὐρανὸν εὐρὺν ἐν αἰθέρι καὶ νεφέλῃσι· | γαῖα δ' ἔτι ξυνὴ πάντων καὶ μακρὸς Ὄλυμπος.

[29] Philod. *De piet.* c. 6b p. 70 Gomperz (*Dox.* 536) = A8 DK = T6 Laks: Διο[γέ]νης ἐπαι[νεῖ] τὸν Ὅμηρον ὡς [οὐ] μυθικ[ῶς] ἀλλ' ἀληθῶς [ὑ]πὰρ τ[οῦ] θείου διειλε[γ]μένον. τὸν ἀέρα γὰρ αὐτὸν Δία νομίζειν φησίν, ἐπειδὴ πᾶν εἰδέναι τὸν Δία λέγει καὶ [. . . . On the import of this testimony on the Derveni text, cf. 309 below.

[30] *Schol. E ad Od.* 1.63: καλῶς τὸν Δία νεφεληγερέτην φησί. *Ζεὺς γὰρ λέγεται ὁ ἀήρ·* τῶν δὲ ἀναθυμιάσεων πυκνὸν μὲν ἐν τῷ ἀέρι ἐγείρονται αἱ νεφέλαι. καὶ διὰ τοῦτο νεφεληγερέτην τὸν Δία φησί. The same identification appears in the schol. *Venetus B ad Il.* 8.250, where the scholiast explains Zeus' epithet πανομφαῖος: πανομφαίῳ: τινὲς τῷ ὑπὸ πάντων φωνουμένῳ, παρόσον αὐτός ἐστιν ὁ ἀήρ. It is worth recalling at this point that πανομφεύουσα, a *hapax* cognate of this rare epithet, is used for Night in the Derveni poem and commented on by the Derveni author in col. 10 of the papyrus.

controversial.[31] It seems possible on the other hand that Cleanthes and Chrysippus equated Zeus – or Zeus among other divinities – with air.[32] At all events, the idea must have been widespread enough to be worth putting on stage in comedies from Aristophanes' *Clouds*[33] to Philemon's play produced some times during the first half of the third century.[34]

The other half of the identifications of col. 23, the equation of Okeanos and air, remains quite striking nonetheless. Okeanos, if not envisaged as an anthropomorphic god, was customarily conceived of in the mythological topography as the great river flowing around the earth.[35] Surely, Okeanos is meant to be a river-god also in the poem commented on by the Derveni author, as is clearly indicated by the epithet εὐρὺ ῥέοντος. This is the obvious, literal meaning of the poetic text. And this is the way in which, according to the Derveni author, the ignorant would read the verses of Orpheus: 'But those who do not understand think that Okeanos is a river because he (sc. Orpheus) added the epithet "broadly flowing"' (col. 23.5–7). Thus this is the interpretation the Derveni author wants to replace with a more ingenious physical allegory: Okeanos is the air.

With this allegorical interpretation, the Derveni author certainly swims against the stream. Because Okeanos was conceived as the great river flowing around the earth, or, according to others, as the world-sea, it was easy, or even natural, to interpret him as elemental water in a physical allegorical reading. Although the mythological equivalent of water is more often

31 Kingsley (1995) chapters 1–5, with the conclusion on 47: Zeus is air, Hera earth, Hades fire, Nestis water. For the bibliography on the nineteenth-century forerunners of this distribution (F. Knatz, Burnet, G. Thiele, Bodrero) see p. 14 n. 4. On Kingsley's treatment of the Theophrastean material, see the objections raised in Mansfeld (1995).

32 For Cleanthes, the evidence is admittedly meagre. It mainly consists of a dismissive gloss by Plutarch (*Quomodo adolesc.* 31d12–e6) with reference to Cleanthes' interpretation of the Homeric expression Ζεῦ ἄνα Δωδωναῖε: κελεύων ἀναγιγνώcκειν ὑφ' ἕν, ὡc τὸν ἐκ τῆc γῆc ἀναθυμιώμενον ἀέρα διὰ τὴν ἀνάδοcιν ἀναδωδωναῖον ὄντα. The case for Chrysippus is less ambiguous and at the same time more problematic, cf. *SVF* 2.1076 = Philodemus *De piet.* col. 5.25–6.6: Δία δὲ τὸν αἰθέρα· . . . κ]αὶ Δία μὲ[ν] ε̣ἶ̣ναι [τὸ]ν̣ π̣ε̣ρὶ τὴν̣ [γ]ῆ̣ν ἀέρα, [τ]ὸ<ν> δὲ cκο[τ]εινὸν Ἅιδ[ην] τὸν δὲ διὰ τῆc γῆ[c κ]αὶ θαλάττη[c] Ποc[ειδῶ] (Henrichs (1974) 16). This allegorical reading clearly reflects the 'distribution of the world' in *Il.* 15.185–99. On the *aither-aer* problem, see above. It is true, on the other hand, that the Stoics often equated *aither* with fire and not with air, cf. *SVF* 2.143.40 1; 180.10–11; 185.12–14.

33 Cf. e.g. Arist. *Clouds* 263–4. The influence of Diogenes of Apollonia on the presentation of Socrates in the *Clouds* has long been recognised. It has recently been claimed by Van der Waerdt (1994) that all the ideas put into the mouth of Socrates come from Diogenes of Apollonia. See also 307 below.

34 Philemon (the older) fr. 91 v. 4 (Kock): Ἀήρ, ὃν ἄν τις ὀνομάσειε καὶ Δία.

35 On the different topographical interpretations, see the article by Gisinger in *RE s.v.* 1; on the mythological interpretations, see Herter in *RE s.v.* 2. On the mythological, iconographical material on Okeanos, see Cahn (1994), Weiss (1984) with Isler (1990).

Poseidon, the Okeanos = water identification occurs in numerous allegorical commentaries.[36] It is attested already in Plato[37] and in Aristotle,[38] and, as Bruno Snell has demonstrated, the Platonic and Aristotelian treatment of the Homeric verses on Okeanos goes back to the sophist Hippias.[39] The claim that Thales' water is the same as Homer's Okeanos is repeated several times in later antiquity, and the 'first philosopher' has often been accused of plagiarism on this account.[40] Okeanos' connection with water was supposed to be so strong that when Damascius, for example, explains to us that the four *rivers* of the mythical topography in Plato's *Phaedo* represent the four elements, he chooses Okeanos among the four to identify with water.[41]

The Derveni author's interpretation of Okeanos is not completely isolated, however. The one other text, so far as I am aware, in which Okeanos and the air are equated is Hesychius' lexicon, where, somewhat surprisingly, Okeanos' first 'meaning' is air. Unfortunately, Hesychius does not give any further explanation for this claim. Yet the previous entry, 'Okeanos' way', suggests that the background of the identification was eschatological:

Ὠκεανοῖο πόρον· τὸν ἀέρα, εἰς ὃν αἱ ψυχαὶ τῶν τελευτώντων ἀποχωροῦσιν Ὠκεανός· ἀήρ. θάλασσα, καὶ ποταμὸς ὑπερμεγέθης. φασὶ δὲ καὶ ὁμώνυμον αὐτοῦ ἐν Κρήτῃ

The epic expression Ὠκεανοῖο πόρον occurs in Hesiod's *Theogony* v. 292, where the context does not seem eschatological. It is true on the other hand that in the *Odyssey* the realm of the dead is encircled or lies beyond Okeanos.[42] However, Circe's suggestion to Odysseus that one can reach Hades by Okeanos appears as a novelty, such that no one has yet taken this route.[43] My only guess then is that the idea preserved in Hesychius goes back to an allegorical attempt to harmonise the common epic notion

[36] For a typical example see a scholium to Eur. *Hipp*. 121: πᾶν ὕδωρ Ὠκεανὸς λέγεται.

[37] Plato *Crat*. 402a–c; *Tht*. 152d–e; cf. *Tht*. 160e, 179e, 180d. It is true that in the Platonic passages the emphasis is on the fluidness, and not on the wateriness of Okeanos; the two notions are nevertheless closely linked, just as in Heraclitus' river-fragment also adduced in this context.

[38] Arist. *Metaph*. 983b20–984a5.

[39] Snell (1944). Cf. Mansfeld (1985) and (1986); Classen (1965); Patzer (1986).

[40] Heraclitus, *Homeric Allegories*, ch. 22; Stobaius 1.10.26, Sextus Empiricus *Adv. math*. 10.313–18; Iraneus *Adver. haer*. 2.14; Theodoretus, *Graec. affect. cur*., 2.9–11. Cf. Buffière (1956) 87 with further references.

[41] Damascius *in Pl. Phaed*. 112e4–113c8:"Ὅτι οἱ τέτταρες ποταμοὶ τὰ τέτταρα στοιχεῖά ἐστι τὰ ἐν τῷ Ταρτάρῳ· ὁ μὲν Ὠκεανός, φησί, τὸ ὕδωρ, ὁ δὲ Κωκυτὸς ἤτοι Στύγιος ἡ γῆ, ὁ δὲ Πυριφλεγέθων τὸ πῦρ, ὁ δὲ Ἀχέρων ὁ ἀήρ.

[42] *Od*. 4.563–9; 10.508–12; and 11.13–22.

[43] Cf. Sourvinou-Inwood (1995) 60–1 discussing also the well-known problem that in epic cosmography Hades is both under the earth and at the western extremities of the horizon.

that the aery *psychai* fly to Hades with Hades' location at the other side of Okeanos.[44]

A further notable feature of the Derveni author's exegesis of Okeanos consists in the fact that – just as in the case of Zeus' adventure with Rhea, commented on in col. 26 – the Derveni author equates the subject and the object of an action, as both terms become equated with Mind in one case, and with the air in the other. It is not true that Zeus contrived Okeanos, for Okeanos is the air – yet the air *is* Zeus, and it is not true that one Zeus created another Zeus.

However, the physical allegory of the two terms of the action does not elucidate the meaning of the act of 'contriving' itself. And it will turn out that the understanding of the act must modify the primary sense of the identification of the terms. The phrase 'Okeanos is the air and air is Zeus' would seem to imply a complete equivalence between the two terms via the identification with the air. Henceforth the creation should mean the duplication of Zeus, an admittedly absurd idea. Or, with the application of the outcome of the allegorical equation, it should mean that the air 'contrived', or created, more air. But this is impossible also on the physical level, as we shall see further on that the two conceptually possible ways of increasing the amount of a given element are excluded. In conformity with the mainstream of Greek tradition, there is no creation *ex nihilo*. Moreover, it is probable that, in the physics of the Derveni author, another element cannot transform into air. Therefore, the amount of air cannot be increased, and this rules out the possibility that an original part of air, designated by the name of Zeus, created a surplus of air, designated by the name of Okeanos.

The transitivity of 'contriving' is not completely lost, however, for the Derveni author states in so many words that Zeus did create something: great strength for himself.[45] It has to be noted in turn that Okeanos is not the 'strength' itself, but strength can be predicated of Okeanos. As the epithet 'broadly flowing' means 'strong', at least according to the Derveni author's explanation, the 'broadly flowing Okeanos' should mean 'the strong air'. Ultimately, the logic of the gloss is thus the following: Zeus created Okeanos; both Zeus and Okeanos refer to the air, but the epithet 'broadly flowing' applied to Okeanos means 'strong', and in this way we end up with the affirmation that the air created a strong air. This explication can

[44] As far as I could see, the literature on Okeanos is extremely silent on Hesychius' unexpected entry. Gisinger mentions it right at the beginning of his *RE* article, but his remark is almost as enigmatic as Hesychius himself.

[45] Col. 23.4–5.

only mean that the air reinforced itself. What exactly this self-reinforcement consists in must remain an open question for the moment. However, at a later stage, the cosmology of the author will provide us with some clues for the understanding of the passage.

Moira, breath and wisdom

There is a further allegorical explanation which is clearly linked up with the assimilation into air of different divine beings mentioned in the poem. Moira is explained by the Derveni author as *pneuma* or breath in col. 18:

> . . . and those moving downwards. But speaking about [. . .] he means that the [?earth] and all the other things are in the air, it being breath. Now Orpheus named this breath Moira. But all other men according to the common usage say that Moira spun for them and that those things which the Moira has spun will be, on the one hand speaking correctly, but on the other hand not knowing either what Moira is or what spinning is. For Orpheus called wisdom Moira. This seemed to him to be the most suitable out of the names that all men have given. For before Zeus received his name, Moira was the wisdom of the god always and through everything. But since Zeus received his name, they think that he was born, even though he existed even before, but was not named. (col. 18.1–13)

To this, we can add the continuation of the argument of col. 19:

> . . . existing things have been called each single name by reason of what dominates (them); all things were called Zeus according to the same principle. For the air dominates all as far as it wishes. And when they say that the Moira spun they say that the wisdom of Zeus ordains how the things that are and the things that come to be and the things that are going to be must come to be and be and cease. (col. 19.1–7)

In this case we have thus a threefold identification: Moira = breath = wisdom. As I have indicated in chapter three, I cannot follow the hypothesis, introduced by Merkelbach (1967), developed by Boyancé (1974), and now taken over by Bernabé (2002), to the effect that the concept of breath, or πνεῦμα, discussed in the first passage comes from a lemma of which there is no trace in the papyrus, but which we should be able to retrieve from verse 5 of the *De mundo*'s Hymn to Zeus: Ζεὺς πνοιὴ πάντων, Ζεὺς ἀκαμάτου πυρὸς ὁρμή.[46] The wording of the sentence τοῦτ' οὖν τὸ πνεῦμα Ὀρφεὺς | ὠνόμασεμ Μοῖραν suggests rather that the concept of breath is part of the Derveni author's physical allegory, since he claims that for the entity πνεῦμα Orpheus used the denomination Moira. As the Derveni

[46] Cf. 126 above.

author explains in the next column, the reason behind Orpheus' apparently idiosyncratic usage lies in the consideration that the idea preserved in the everyday saying 'Moira has spun' expresses well the role Orpheus – at least on the author's reading – assigns to breath and wisdom.

These two passages were among the fragments already published in the first transcript by Kapsomenos in 1964. Accordingly, the first papers on the papyrus already discuss these identifications. Some of these studies seem to contain some misunderstanding about the nature of the relationship between Zeus and Moira which has perpetuated itself in the subsequent literature. For both Burkert and Boyancé take the passages in columns 18 and 19 as implying a complete identification between the being designated by the name Zeus on the one hand, and that which is designated by the name Moira, on the other. On this reading, the Derveni author would argue that the same divine being was first called Moira and then received the name Zeus at a later phase of the story.[47] Such a reading might be supported by a different construction of the sentences in col. 18.7–13:

Ὀρφεὺc γὰρ | τὴμ φρόνηc[ι]μ Μοῖραν ἐκάλεcεν· ἐφαίνετο γὰρ αὐτῶι | τοῦτο προcφερέcτατον ε[ἶ]ν̣α̣ι̣ ἐ̣ξ ὧν ἅπαντεc ἄνθρωποι | ὠνόμαcαμ. πρὶμ μὲγ γὰρ κληθῆ̣ναι Ζῆνα, ἦμ Μοῖρα | φρόνηcιc τοῦ θεοῦ ἀεί τε κα̣ὶ [δ]ιὰ παντόc· ἐπεὶ δ' ἐκλήθη | Ζεύc, γενέcθαι αὐτὸν ἐ̣[νομ]ί[cθ]η̣, ὄν̣τ̣α μὲγ καὶ πρόcθεν, | [ὀ]νομαζόμ[ε]νον δ' ο[ὔ.

As the double accusative with καλέω makes the sentence inherently ambiguous, the difficulty lies in deciding whether or not the subject of the subordinate clause πρὶμ μὲγ γὰρ κληθῆ̣ναι Ζῆνα is the same as Moira, that is the subject of the main clause ἦμ Μοῖρα φρόνηcιc τοῦ θεοῦ ἀεί τε κα̣ὶ [δ]ιὰ παν̣τόc. Depending on our answer to this question, the subordinate clause can be rendered either as 'For before (Moira) received the name "Zeus" etc.'[48] or 'For before Zeus received his name etc.'.

However, the relationship between the respective terms of the equations turns out to be more subtle on closer inspection; and this consideration disproves the total identification between the being designated by the name 'Moira' and that which receives the name 'Zeus'. First, it should be noted

[47] Cf. e.g. Burkert (1970) 445: 'Ce qui parut n'était qu'un nouveau nom: «Zeus fut dénommé». Auparavant il y avait Moira, nom plus ancien pour une réalité à laquelle ne pensent certainement pas les hommes, quand ils emploient ce nom. En réalité, Moira est le souffle, l'intelligence du dieu; et Zeus lui-même n'est que ce souffle intelligent.' Cf. Burkert (1997) 172 strangely claiming that '"the god" is evidently identical with . . . φρόνηcιc τοῦ θεοῦ.' Cf. also Boyancé (1974) 96; and Ricciardelli Apicella (1980) 121–2, who makes the identification complete with πάντα.

[48] This seems to be Boyancé's reading: 'Nous y voyons que Zeus est le nom donné par les hommes à Moira' (Boyancé (1974) 96).

that the Derveni author systematically uses the same term with the same genitive construction in defining Moira's identity in the two columns: 'the wisdom of the god' in col. 18,[49] and 'the wisdom of Zeus' in col. 19. What is more, neither is mind (*nous*)[50] the same as wisdom (*phronesis*) nor is breath (*pneuma*) the same as the air (*aer*) – even though the relationship is close in both pairs. From these pairs, psychological on the one hand, and physical on the other, we obtain the following well-balanced structure:

MYTHOLOGICAL NAME	Zeus, the god	Moira
INTELLECTUAL ASPECT	mind (*nous*)	wisdom (*phronesis*)
PHYSICAL ASPECT	air (*aer*)	breath (*pneuma*)

Wisdom (*phronesis*) is to mind (*nous*) as breath is to air. As *pneuma* represents the active, moving part of the air, *phronesis* represents the practical, executive part or function of the mind. Thus, yet again, the outcome of the Derveni author's allegorical explanation is not a total identification. Moira/*pneuma*/*phronesis* is not the same as Zeus/air/*nous*, but an aspect or function of it; Moira is the active, executive, practical aspect of the god, that which 'ordains how the things that are and the things that come to be and the things that are going to be must come to be and be and cease' (col. 19.5–7). Furthermore, Moira/*pneuma*/*phronesis* could operate even before the air/*nous* received the name 'Zeus'.

PROVISIONAL CONCLUSIONS

We have now gone through all those passages in which the surviving portion of the papyrus gives explicit allegorical identifications for the different gods and goddesses of the traditional Greek pantheon. Before turning to the less certain cases – i.e. to those divinities who, although mentioned in the lemmata, are not explicitly treated by the same identificatory method in the extant text of the papyrus – let us spell out the recurring, characteristic features of the Derveni author's method. The recognition of such features may help us in formulating conjectures for the less obvious cases to be treated in the next section. Conversely, some of the remaining passages may lead us to modify the general picture to be drawn now.

[49] On the sentence Ὀρφεὺϲ γὰρ | τὴμ φρόνηϲ[ι]μ Μοῖραν ἐκάλεϲεν· Boyancé (1974) 95–6 writes that 'Manifestement (. . .) l'idée est que la pensée s'identifie à un pneuma. Il est clair dès lors qu'à ce stade de l'exposé il ne peut être question que de la pensée dans l'homme . . .' But why should it be so obvious that φρόνηϲιϲ is 'la pensée dans l'homme', and even more so as in the next sentence the Moira is explicitly said to be the φρόνηϲιϲ τοῦ θεοῦ?

[50] On the relation between Zeus and Mind, see the interpretations of col. 16 above.

First it should be noted that the Derveni author's unifying procedure, by means of which a number of gods and goddesses, very different in character, merge into one another, is apparently quite unique in Greek allegorical exegetical practice. For it seems that in both physical and moral allegories, during the classical period as well as in Hellenistic times, the prevailing method consisted in establishing one-to-one correspondences between the gods and heroes of the epic poems on the one hand, and certain entities of the physical world and basic psychological concepts on the other.

In all probability the early grammarian Theagenes of Rhegium, allegedly the first to write on Homer, was already using the method of one-to-one correspondences.[51] Furthermore, provided that new readings of Philodemus' *De pietate* do not substantially modify the picture, Metrodorus of Lampsacus systematically identified the heroes of the *Iliad* with the major parts of the cosmos, equating for example Agamemnon with aither, Achilles with the sun, Helen with the earth, Paris with the air, and Hector with the moon. As to the gods, Metrodorus somewhat extravagantly assimilated them to the parts of the human body. Demeter was the liver for him, Dionysos the spleen, and Apollo the bile. So we have a set of individual entities in the physical world, and a set of divine characters in the epic poem, and the allegorist's task consists in mapping the two sets by establishing a connection between the respective individual items. The same strategy prevails through the centuries, down to Heraclitus the rhetor, who compiled his *Homeric Allegories* probably in the first century BC, and to the *Theology* of Cornutus. And if occasionally more than one traditional god is equated with the same element within a single allegorical interpretation, say Apollo, Helios and Hephaistos with the cosmic fire,[52] it is not because the author wanted to conflate these gods, but rather because there were more gods in Homer than physical elements and namable heavenly bodies in the cosmology of the commentator.

The Derveni author's main claim, in contrast, is that the plurality of divine names used by Orpheus does not refer to a plurality of beings. The plurality is reduced to unity, and the natural assumption that, broadly speaking, there is a one-to-one correspondence between names and beings is denied. Yet we do not end up with sheer synonyms. Although the different

[51] Schol. *Venetus* B to *Il.* 20.67, probably going back to Porphyry, after having presented an allegorical interpretation of the theomachy based on the method of identifications, claims that this was the way already Theagenes tried to save Homer (οὗτος μὲν οὖν <ὁ> τρόπος ἀπολογίας ἀρχαῖος ὢν πάνυ καὶ ἀπὸ Θεαγένους τοῦ Ῥηγίνου, ὃς πρῶτος ἔγραψε περὶ Ὁμήρου, τοιοῦτός ἐστιν ἀπὸ τῆς λέξεως). According to Tatianus 31 p. 31 Schw., Theagenes was a contemporary of Cambyses. Brisson (1996) 55 expresses some doubt about the value of the testimony of Porphyry.

[52] As in the schol. *Venetus B* on the theomachy in *Il.* 20 mentioned in the previous note.

names refer to the same being – the cosmic Mind/air – they reveal different aspects, functions or activities of the being named. It is not true that 'Aphrodite' simply means the same as 'Zeus', even though both are names of the same god. Aphrodite is the name of the god so far as the god causes the conjunction and mingling of the separate chunks of elements, and so forth for all the other names. In this way, the identifications are not total, and not totally interchangeable, in so far as different names have the same reference but different meaning. Saying that Zeus is air, and Okeanos is air, does not mean that the names Zeus and Okeanos are interchangeable without any further ado. Although I shall continue to use the term 'identification,' I would like to make it clear that, in the case of the Derveni author, allegorical identification means assigning a *qua*-type description for the cosmic god.

Furthermore, the technique of allegorical identifications does not have the same theological import in the Derveni papyrus as in other allegories. Allegorical interpretation was most often used as a means of saving Homer from charges of impiety. The allegations of the philosophers trying to put Homer in disrepute are mistaken, claims the allegorist, as Homer is not speaking about the gods, but about the physical elements or human passions.[53] Thus, the names we find in the epic poems do not refer to the gods we worship in cult. And, notably, the same interpretative method could be used by other allegorists with less pious aims. For it is quite likely that for example Metrodorus, friend of Anaxagoras – who according to our sources invited the philosopher to Lampsacus when Anaxagoras was put on trial and expelled from Athens on charges of impiety – was not in the business of giving an allegorical interpretation of Homer from the standpoint of traditional Greek religion. On the contrary, he tried to show that the Homeric texts were not dealing with gods at all not in order to save Homer, but to undercut an important possible source of information on the gods. Yet we end up with the same result in both cases: the allegorist denies that the poetic text has any religious significance and can be treated as a source of knowledge about the gods. As opposed to this, the Derveni author maintains that Orpheus, if rightly interpreted, is the ultimate source on the nature of the god. Orpheus is thus not de-theologised by means of physical allegory, but, on the contrary, it is through physical allegory that we can appropriate the true theological teaching of Orpheus.

[53] Cf. e.g. the scholiast's remarks on Theagenes in n. 50 above. On the apologetic objectives of allegorical interpretations in general, with an overview of the ancient material, see Buffière (1956), esp. ch. 1, 'La bataille autour d'Homère'.

There is, however, a notable parallel to the method applied by the Derveni author. Diogenes Laertius, reporting (probably early) Stoic views on the nature of the divinity, says:

They say that god is a living being which is immortal and rational or intelligent, perfect in happiness, not admitting of any evil, provident towards the world and its occupants, but not anthropormorphic. He is the creator of the whole and, as it were, the father of all, both generally and, in particular, that art of him which pervades all things, which is called by many appellations according to his powers. For they call him Zeus [Δία] as the cause [δι' ὄν] of all things; Zen in so far as he is responsible for, or pervades, life [ζῆν]; Athena because his commmanding-faculty stretches into the aether; Hera because it stretches into the air; Hephaistos since it stretches into the designing fire; Poseidon since it stretches to the sea; Demeter because it stretches to the earth. And similarly they render the other appellations too by clinging to a proper meaning. (D.L. 7.147 = *SVF* 2.1021; trans. based on Long and Sedley)

Here too there is a – preponderantly physical – allegory of divine names, where the different names are not explained as referring to different entities of the world, but as different powers (δυνάμεις) of the one supreme cosmic divinity. Here too physical allegory is not primarily about understanding the physical constitution of reality (as compared e.g. with the standard allegories of the love of Zeus and Hera on Mount Ida), but about the nature of god: how is god related to the physical universe? The one basic difference between this explanation and the one offered by the author of the papyrus is that in this passage we still connect the different gods – by way of etymology or by referring to the god's traditional domain – to a cosmic sphere or element, whereas the Derveni author prefers interpreting the divine names as referring to cosmic functions or activities of the god.

OTHER GODS AND GODDESSES: FURTHER POSSIBILITIES AND SOME DIFFICULTIES

All that has been said so far could almost inevitably lead us to the conclusion that the Derveni author systematically, or even mechanically, identified all the gods and goddesses he found in the poem with different aspects or functions of his single divine being. Let us see now whether this could be the fate also of the remaining divinities mentioned in the papyrus. There is a handful of them: Dike, the Erinyes and Eumenides in the first seven columns; and Ouranos, Night, Metis, Achelous, Hermes and Maia in the second part of the text, dealing with the poem itself.

Ouranos and Mind?

There is no explicit allegorical interpretation of Ouranos in the remaining text of the papyrus. However, as both his son Kronos, and his daughter Rhea, just as his grandson Zeus, get directly identified with Mind (not to mention all the other relatives identified with Mind via one of these three gods), one could expect the same identification for Ouranos as well.[54] Furthermore, as I have tried to argue above, the author's interpretation of ***L7*** and ***L8*** in coll. 14 and 15 suggests that he perceived the Ouranos-Kronos-Zeus succession as one continuous rule, with a single ruler.[55]

Tsantsanoglou's *exempli gratia* reconstruction of the badly fragmented bottom lines of col. 14 must also rely on a similar train of thought:

[τῶν ἐ]όντωγ γὰρ ἁπάντ[ω]ṿ [οὔπω κρουομέ]νων
[ὁ Νοῦϲ] ῳ̣ϲ ὁρ[ίζω]ν φύϲιν [τὴν ἐπωνυμίαν ἔϲχε]ν
[Οὐρανό]ϲ.

Yet as the key-words of the sentence are entirely or almost entirely supplemented, and as the supplements themselves are quite obviously based on considerations similar to those I have formulated above, it would certainly be circular to build any specific interpretation on this sentence. All the more so, as Tsantsanoglou himself proposes an alternative for [ὁ Νοῦϲ]: [οὗτοϲ]. Nevertheless, the etymology suspected by Tsantsanoglou in itself would strongly support the idea that Ouranos too was identified with Mind, since if in fact ὁρ[ίζω]ν was supposed to account for the first part of the name, it is quite likely that Νοῦϲ was the explication of the end of it.[56]

Metis?

As argued above, I do not think that the goddess Metis appeared in the (surviving part of the) poem. I consider nevertheless the alternative hypothesis maintained by other scholars that Metis comes up in ***L9*** = col. 15.13. The remaining part of the author's text does not give any explanation of the word μῆτιγ/Μῆτιγ. It is possible, however, that Metis – if she indeed appeared in ***L9*** – had already been something like a reverse allegorical or

[54] So already Rusten (1984) 135. The remark, without exact reference, in Graf (1993a) 184 'He [sc. the Derveni author] believed that Ouranos, for example, referred to *nous*' must be a slip.

[55] Cf. 187 above.

[56] We can add that, as I have mentioned in the discussion of ***l6*** and ***L7*** = col. 14.5 and 6, on p. 159 above, Rusten has tried to interpret the slightly problematic gloss on these lemmata, speaking about Kronos' castrating Ouranos, with the help of the assumption that son and father were identified here as well.

etymological explanation of Zeus' epithet μητίετα occurring in the immediately preceding lemma, and hence it would make no difficulty for the Derveni author to say something about the 'cunning' or 'wise' nature of the cosmic Mind in order to argue away the separate identity of Metis. At any rate, the identification of Metis and Zeus should not cause any problem even on the Hesiodic account, where the goddess is swallowed by the king of the gods. So if Metis appeared in the poem and received a specific allegorical interpretation by the commentator, it is not unlikely that she, too, was assimilated to the cosmic Mind. But this must remain guesswork.

Night

As we have seen in the reconstruction of the theogonic plot implicit in the Derveni poem, Night has an especially important role in the story. To begin with, her position in the first generation of gods already makes her a reverential figure. Furthermore, she is the one who hides and nurtures the child Zeus, while at a later stage of the story, she helps the future king with oracular advice on how to obtain power. Nurturing Zeus earns her the epithet 'the nurse' (τροφ[ὸν in col. 10.11), and the dispensation of oracles makes her 'the one who proclaims all' (πανομφεύουσαγ in col. 10.9). In this way, Night is not just a primitive goddess of high reverence, but something like an *éminence grise*, conducting the development of the story from the depth of her cave. Also, she is Zeus's major ally in the very endeavour which constitutes the subject, if not of the whole poem, at least of that section of it which receives interpretation in the surviving part of the papyrus.

This high importance allotted to the figure of Night in the poem makes it all the more interesting to see what the Derveni author has to say about her. The references to Night are clearly concentrated in one section ranging from col. 10 to col. 13. The line of thought of this portion of the text can be reconstructed with a fair amount of certainty. From the sequence of the lemmata found in these columns, Bernabé reconstructed the following continuous passage of the poem (fr. 6 B.):

> Ζεὺς μὲν
> [ἧστο] πανομφεύουσα [θεῶν] τροφὸς [ἀμβροσίη] Νύξ·
> . . . χρῆσαι . . . ἐξ ἀ̣[δύτοι]ο
> [ἣ δ'] ἔχρησεν ἅπαντα τά οἱ θέ[μις ἦν ἀνύσασ]θαι,
> ὡς ἂν ἔ̣[χοι κά]τα καλὸν ἕδος νιφόεντος Ὀλύμπου.

Even though the exact wording might have been slightly different (especially in the first words of the last verse), Bernabé's reconstruction provides a convenient starting point for the examination of the Derveni author's interpretation. The overall structure of the argument seems to be the following. The first half of col. 10, evidently following up the now lost bottom part of col. 9, discusses the epithet πανομφεύουσα. Then, the transition being marked by a very short line in col. 10.10, followed by a *paragraphos*, the Derveni author starts explaining the other epithet τροφός in line 11.

In col. 11, he turns to the interpretation of the expression ἐξ ἀδύτοιο, to continue by considering the verb χράω referring to the oracular activity of Night. A very short line again in col. 11.7, followed by a *paragraphos*, and the Derveni author continues by quoting in full the verse [ἣ δὲ] ἔχρησεν ἅπαντα τά οἱ θέ[μις ἦν ἀνύσας]θαι (***L3*** = col. 11.10) to elucidate it in some detail.

Column 12 in turn focuses on the content of the oracle, by explaining the real meaning of Olympus, the seat from where the new king would rule his kingdom – a self-contained argument to which we shall return later.

Finally, col. 13 returns to the theme of proclaiming and receiving oracles by pointing to a parallelism between the prophetic advice offered by Night and that given by Kronos. For the latter, the Derveni author adduces the full-verse lemma Ζεὺς μὲν ἐπεὶ δὴ πατρὸς ἑοῦ πάρα [θ]έσφατ' ἀκούσα[ς] (***L5*** = col. 13.1). Again, we have a *paragraphos* after col. 13.3, and the Derveni author quotes the verse that speaks about Zeus's swallowing of the αἰδοῖον, which shows that Zeus in fact followed the oracular advice. The Derveni author can now continue by telling us about the real meaning of the αἰδοῖον, and so forth. This overview of the general structure of the section treating Night shows that, despite the apparent heterogeneity of the themes invoked, the glosses in the section extending from col. 10 to col. 13 are to be treated as parts of a larger argument, the subject of which is the poet's claim that Night was instructing Zeus by her prophecies. From this perspective, we can see better the connection between the individual and prima facie contrasting elements of the Derveni author's argumentation.

In the first subsection, stretching from line 1 to line 10 of col. 10, the Derveni author explains, with the help of a chain-argument, that the epithet πανομφεύουσα 'proclaiming all' means the same, or is the same, as πάν[τα] διδά[σκουσαν 'teaching all'. So far, we have not departed from the literal meaning of the poem, for, also on our understanding of Orpheus' text, Night did indeed teach everything, or at least everything concerning the coup against Kronos.

On the other hand, the explication of τροφός in the remaining lines of col. 10 brings us back to the world of physics and meteorology. Accepting Tsantsanoglou's supplements for these badly damaged lines, we read the following (col. 10.11–13):

τροφ[ὸν δὲ λέγων αὐ]τ̣ὴ̣ν̣ α̣ἰ̣ν̣ί̣[ζε]ται ὅτι [ἅ]ccα | ὁ ἥλ̣ι̣[οc θερμαίνων δι]α̣λύει ταῦτα ἡ νὺξ ψύ[χουcα | cυ[νίcτηcιν ±18] ἅccα ὁ ἥλιοc ἐ̣θ̣έ̣ρ[μ . . .]

By saying that she is 'nurse', he (sc. Orpheus) expresses in riddling form that whatever the sun dissolves by heating, the night unites by cooling . . . those things which the sun heated . . .

Although other supplements are certainly possible, the preserved fragments show clearly that, in relation to the 'nurturing' capacity of Night, the Derveni author sets up an opposition between Night and the sun. Furthermore, the supplements ψύ[χουcα and ἐ̣θ̣έ̣ρ[μ seem to be safe enough for us to infer that what is emphasised in relation to the sun–night opposition is not the light–dark, but the hot–cold distinction. And if we accept the admittedly less secure supplements cυ[νίcτηcιν and δι]α̣λύει, unification versus dissolution, we can conclude that, once more, the Derveni author is speaking not simply about the characteristics of cosmic physical forces, but specifically about the influence they can exercise on the other beings. The terms in which the sun–night opposition is expressed make it likely that the Derveni author treated Night – in conformity with his other interpretations – not so much as a personal goddess who gives advice and hides the child-god in a cave, but rather as a physical, cosmological force.

This passage will become even more interesting when we discuss the cosmogonical works of Mind in the next chapter. As we shall see there, the fundamental role of Mind during the current phase of cosmic history consists in regulating the amount of heat which the fiery particles collected in the sun can communicate to the other beings. If there is too much heat, the other beings lose their individual form as by melting, but when the heat of the sun is under control, the other beings can coagulate. Notably, all the words in the papyrus beginning with the letters cυ- happen to refer to the unification and coagulation of chunks of matter during this cosmic phase, and under the influence of Mind.[57] This fact gives further weight to

[57] cυνίcταcθαι in col. 9.6 and cυμπαγῆναι in 9.8: the god withdraws the fiery particles in order to let the others coagulate; cυ̣cταθῆναι in col. 17.2 and cυνεcτάθη in 17.8: although the air existed even before, the era marked by the name of Zeus is characterised by the coagulation and unification of chunks of matter; c̣υ̣[cτάcεωc in col. 17.14 probably says that the beginning of the rule of Zeus means the beginning of the coagulation or constitution of entities; cυνεcτάθη in col. 21.3: particles

the hypothesis that cυ[νίcτηcιν, as contrasted with δι]ạλύει, also refers to a physical force put into play in the cosmogony.

These considerations make it probable that the cooling and perhaps coagulating force of the night, as described in the last lines of col. 10, should be seen in relation to the cosmogonic works of Mind. The cool night is definitely on the side of the sun-regulating Mind, restricting and counterbalancing the heating and melting force of the sun. Furthermore, it has to be said that even in the Mediterranean climate, there are after all not that many things which actually melt or dissolve during the day and then coagulate again during the night. Hence, it is perhaps not totally improbable that the night referred to in this passage is something more general than the dark hours of a normal day, and close to the long-duration cosmogonic function of air/Mind.

On the other hand, a more immediate connection, or even identification between night/Night and the cosmic air, also identified with Mind, depends on the physical characteristics the Derveni author assigns to elemental air. Keeping in mind that the characterisation of air is shifting during the Presocratic period, and is notoriously difficult to pin down in the individual texts, it is also not inconceivable that the Derveni author was using the more archaic concept of air implying coldness, darkness and wetness.[58] In this case, the physical characteristics attributed to Night/night, such as coldness and darkness, can be very close to the physical characteristics of elemental air.

The next bit of text related to prophetic utterances is far more problematic and difficult to understand. As every other word requires some comment, let us see the passage in Greek (col. 11.1–12):

> [τ]ῆc Νυκτόc. ἐξ ạ̓[δύτοι]ο δ' αὐτὴν [λέγει] χρῆcαι
> γνώμην ποιού[με]νοc ἄδυṭον ẹἶναι τὸ βάθοc
> τῆc νυκτόc· οὐ γ[ὰρ] δύνει ὥ[c]περ̣ τὸ φῶc, ἀλλά νιν
> ἐν τῶι αὐτῶι μέ[νο]ṿ αὐγὴ κạτạ[λ]αμμάνει.
> χρῆcαι δὲ καὶ ἀρ̣κέcαι ταὐτ̣ọ̀ [δύ]ναται.
> cκέψαcθαι δὲ χρὴ ἐφ' ὧι κεῖτạ[ι τὸ] ἀρκέcαι
> κ̣ạὶ τὸ χρῆcαι.
> ‾χρᾶν τὸνδε τὸν θεὸν νομίζọ[τεc ἔρ]χ̣ονται

of matter get together by collisions; cύνηθεc in col. 21.5: the 'like' particles of matter find each other by the end of the process originated by Mind and designated by the names of Zeus, Aphrodite, Peitho and Harmonia; cυνίηι in col. 25.7, cυνέλθοι in 25.8 and cυνεcτάθη in 25.9 refer, on the one hand, to Mind's gathering the fiery particles in order to form the sun and, on the other hand, to the natural but controlled tendency of the residual fire, i.e. the stars scattered through the sky, to come together in one mass.

[58] Cf. Kahn (1994) 143–9.

[π]ευϲόμενοι ἄϲϲα ποῶϲι· τὰ δ' [ἐπὶ τούτ]ῳι λέγει·
[ἥ δὲ] ἔχρηϲεν ἅπαντα τά οἱ θέ[μιϲ ἦν ἀνύϲαϲ]θαι
[ἐν τούτ]οιϲ ἐδήλωϲεν ὅ[τ]ι ο[
[±12 τὰ] ἐόντα ι[

The first words guarantee that the Derveni author is still concentrating on Night. In the next two sentences the author explains by an etymology what he takes the *adyton*, the oracular shrine of Night, to mean. This time he agrees with modern linguists in analysing the word into the privative *a-* and the verb δύνω or δύω. For the rest, he offers another proof of his ingenuity. For the Derveni author, the word δύω does not refer to the innermost shrine in the sense that 'there is no entering' into it, but signifies that the night does not set as the sun does, but remains stationary while the daylight lays hold of it.[59]

Again, this image becomes considerably more powerful if we substitute night/Night with the air. For the corresponding physical reality is not the night, as such, but the air which remains at the same place in the morning. The air remains where it was during the night but the daylight, communicated by the sun, gradually overtakes it. The author is still concentrating on the night–sun antithesis, but has moved from the question of thermic difference to the discussion of the light–dark opposition.

The next sentence, connected with the following one by a δέ, passes on to the explanation of the word χρῆϲαι, 'proclaiming oracles', somewhat abruptly. As the word ἔχρηϲεν occurs in the next lemma [ἥ δὲ] ἔχρηϲεν ἅπαντα τά οἱ θέ[μιϲ ἦν ἀνύϲαϲ]θαι (***L3*** = col. 11.10) quoted a few lines later, this explanation seems to pave the way for the exegesis of the forthcoming lemma. Thus, we are back to the theme of oracles. The Derveni author contends first that χρῆϲαι means the same, or has the same power (ταὐτὸ [δύ]ναται), as ἀρκέϲαι. Having established the equivalence of the two terms, he announces the need for further investigation:

'furnish/proclaim oracles' (χρῆσαι) and 'assist' (ἀρκέϲαι) mean the same. But one has to examine to what 'assist' and 'furnish/proclaim oracles' apply (as a name).[60] (col. 11.5–7)

[59] As Tsantsanoglou remarks (*apud* Laks and Most (1997) 14 n. 20), the expression ἐν τῶι αὐτῶι μέ[νο]ν can also mean 'remaining in the same condition'. Nevertheless this would be less natural in opposition to a verb signifying locomotion. At another level, the Derveni author's contention about Night/night remaining in the same place might be taken as an answer to Hesiod. In *Thgn.* 748–57 Hesiod relates that Night and Day leave and go back to the House of Night in turns; the Derveni author would say that only the Day moves, the Night is stationary.

[60] This is how I understand the phrase ἐφ' ὧι κεῖτα[ι. ὄνομα would certainly be welcome however, as we have it e.g. in col. 21.7. But κεῖμαι with the dative and without ὄνομα is attested in Plat. *Crat.* 392b8–9: Τὸν "Ἀϲτυάνακτα" ἄρα ὀρθότερον ᾤετο κεῖϲθαι τῷ παιδὶ ἢ τὸν "Ϲκαμάνδριον".

It is somewhat odd that the Derveni author first establishes that the two words have the same meaning, and then tries to find out what the common reference is. We would expect that one has to know the common reference prior to establishing the connection between the two terms, as it is only the common reference that can provide the link. As the statement about the equality of the two terms is not just arbitrary but also stands without any kind of explanation, the likeliest conjecture is that not only χρῆcαι, but also ἀρκέcαι, probably in the form of ἤρκεcε, comes from the poem.[61]

The author's procedure here is comparable to the one we have encountered, for instance, in col. 21, where he first established that the different names (Aphrodite Ourania, Zeus, ἀφροδιcιάζειν, θόρνυcθαι, Peitho and Harmonia) refer to the same god, to turn to substantiating this claim by describing the physical processes expressed by these names. The difference is, however, that the arguments in the former case consisted in showing why and how the different denomination can refer to the same being, and not in examining what they refer to.

Apart from this problem, the exact meaning of ἀρκέcαι is far from evident. The semantic field of the word is relatively wide, and the context in the papyrus does not offer much help in deciding which meaning we should choose. The different relevant senses fall into two larger groups: 'ward off', 'defend' and 'assist' on the one hand, and 'be strong enough', 'suffice' and 'endure', on the other. L&M translates 'preventing harm', whereas Janko suggests 'suffice'. My guess is that the author is trying to play on the first meaning of χράω 'to furnish (what is needed)' and take it to be equivalent with 'to help' or 'to assist'. The original idea expressed in the poem, i.e. Night giving oracular advice to Zeus, definitely supports this reading. Yet, in view of the Derveni author's basic aim, i.e. to collapse the different divine characters of the poem into one god, it cannot be excluded that, by exploiting the polysemy of the word, he was about to shift the literal meaning of χρῆcαι, 'proclaiming oracles', by making the connection with ἀρκέcαι. The second group of meanings of ἀρκέcαι would certainly be welcome for this purpose, because they could lead the way towards an interpretation where the dispenser and the recipient of the oracle are not separate entities. Zeus is not in need of external help, but is self-sufficient, 'strong enough'. In this case, we would have a rather close parallel with the interpretation of Okeanos in col. 23, where the separate identity of the contriver and the

[61] Compare the reasoning of col. 10, where the equality of 'uttering', 'saying' and 'teaching' is painstakingly argued for.

contrived is denied, and we end up with a self-strengthening air. In any event, the passage is quite enigmatic, even by the standards of the papyrus.

This line of interpretation receives further support from the last part of the section under investigation. At this stage, the Derveni author connects the oracular advice given by Night with the prophecy Zeus receives from his father. Having quoted the lemma Ζεὺϲ μὲν ἐπεὶ δὴ̣ π̣ατρ̣ὸϲ ἑοῦ πάρ̣α̣ [θ]ἔ̣ϲφατ᾽ ἀκούϲα[ϲ] (***L5***) in the first line of col. 13, the Derveni author apparently denies that Zeus was following any advice, oracular or otherwise:

For neither did he hear this time[62] – but it has been made clear in what sense he heard – nor does Night command (this time).[63] (col. 13.2–3)

It is explicitly claimed here that Zeus is not receiving oracular advice from his father. As regards the parenthetical reference to an explanation of the interpretation or modality of 'having heard', the best bet is that it points back to column 8:[64]

And how he (sc. Zeus) begins, he makes clear in this:

> *And when Zeus took from his father the prophesied rule*
> *And the strength in his hands and the glorious daimon*

They ignore the fact that these words are transposed. They are to be taken as follows: 'Zeus when he took the strength from his father and the glorious daimon.' Since this is so he (sc. Orpheus) does not maintain that Zeus hears [from his father] but that he takes the strength [from him]. If it is taken the other way, [it would seem] that he [took it] contrary to the divine decrees . . . (col. 8.3–11)

Admittedly, the translation in the second part of the passage is quite conjectural. But let's try to reconstruct the author's reasoning from the beginning. Clearly, as usual, he contrasts two possible constructions of the text; he argues for one and rejects the other. He also explains that the reason why people tend to misunderstand the text is that they do not notice a *hyperbaton*. But where is the *hyperbaton*? Rusten suggests that the Derveni

[62] τότ̣ε is preferable to the anaphoric τόδ̣ε as suggested by Tsantsanoglou and translated by L&M. There is nothing in the immediate context to which the τόδ̣ε could refer to. On the other hand, the temporal reference is in place, since the author refers back to another alleged case of prophecy received by Zeus.

[63] Although the Derveni author continues by saying that 'But he makes clear saying as follows' (ἀλλὰ δηλοῖ ὧδε λέγων·), to quote ***L6***, I cannot see in what way it is supposed to clarify the sentence we are dealing with.

[64] Although the Derveni author customarily reserves the different forms of δηλόω for referring to Orpheus' own way of expression, the expresssion δεδήλωται ἐν τοῖϲ προτέροιϲ in col. 17.4 probably refers to a previous passage of the Derveni author's own text. West (1983) 85–6; Rusten (1985) 125–6 already connects col. 13 with col. 8 on account of the similarity of the two hexameters, the one in col. 13.1 and the other in col. 8.4. On their treatment of the question see 116 above.

author wilfully separated an enjambement which had originally connected the lemmata in col. 8.5 and col. 13.4: . . . καὶ δαìμονα κυδρὸν | αἰδοῖογ κατέπινεν. . . .[65] The *hyperbaton* would then refer to the separation of the two lines.[66] The question of αἰδοῖον set aside, there is a much more plausible understanding of the author's reference to a *hyperbaton*. The technical sense of the term is transposition of words. Now if we compare the two contrasting interpretations as presented by the Derveni author, there is in fact one word which changes its place in such a way that it ḍoes make a difference for the construction of the line:

[ἔc]τιν δὲ ὧδ' ἔχοντα· Ζεὺc μὲν ἐπεị τ[ὴν ἀλ]κὴν
[πα]ρὰ πατρὸc ἑοῦ ἔλαβεγ καὶ δαίμονα̣ [κυδρ]ọ́ν

ἄλλωc δ' ἔ]χοντα π̣αρὰ̣ θέcφατα δ[όξειεν ἂν λαβεῖ]ν

The author's problem, then, is the position of παρά in the expression πατρὸc ἑοῦ παρα θέcφατα (I am not marking the accent of παρά to retain the ambiguity): does it go with θέcφατα, or is it in postposition, to be construed with πατρὸc ἑοῦ? Grammatically both are possible.[67] And this time the Derveni author opts for the construction we would choose. It is another question who could ignore that παρά is in postposition.

But why is the other possibility to be discarded so vehemently? The author stresses that with the interpretation he proposes, the verse does not imply that 'Zeus hears' something from his father.[68] If παρά is not in postposition and goes with θέcφατα, there is necessarily some kind of prophecy or divine decree implied. Moreover, πατρὸc ἑοῦ could also be understood as a possessive genitive referring to θέcφατα. The meaning of the words would then be that Zeus does something 'on account of' or 'contrary to the prophecy of his father'. If, on the other hand, παρά goes with πατρὸc ἑοῦ, then θέcφατα can be taken as an adjective with the more general sense of 'divine', so common in epic poetry. Hence there is no prophecy issued by the father implied, and Zeus does not need to hear

[65] On my view on this point see 116–17 above. [66] Rusten (1985) 125 with n. 10.

[67] In the former case, πατρὸc ἑοῦ could probably be taken as a possessive.

[68] I take οὐκ ἀκούειν τὸν Ζᾶ̣[να an *acc. cum inf.* with Tsantsanoglou (L&M 12 n. 11), but rather with a verb of saying such as ἀμφιcβη]τεῖ which seems to fit the lacuna exactly. It is indeed tempting to take ἀκούειν in the sense of 'understand' with L&M and Janko. But in this way we would lose the denial of Zeus hearing an oracle which, I believe, is strongly suggested both by the author's argument about the alleged *hyperbaton* and by the denial of hearing in col. 13 with the backward reference. Moreover, if we take τὸν Ζᾶ̣[να to be the object and not the subject of ἀκούειν then we are in a difficulty with]τει at the end of the line. Apparently this is why Janko tried to suggest δυναc]τεί|αν which is unlikely in view of the fact that the scribe avoids separating words in such a position and 40 characters would not be too long for a line. For the two separated words in the whole papyrus see 61 above.

anything, but can simply take over the signs of power from his father. And the Derveni author can continue by explaining how one should understand the phrase 'from his father' in the next column.

At any rate, the act of hearing seems to be denied here, and the same point is repeated in col. 13: 'for neither did he hear this time'. I have to admit, however, that this reconstruction is still unable to explain what the expression 'but it has been made clear in what sense he heard' refers to. My only guess is that the last two legible words of col. 8, καὶ μαθών were part of the explanation.

But why is it so important for the Derveni author to show that Zeus did not hear any prophecy from his father, or at least not in the literal sense? The answer must be that if Kronos and Zeus are two names, or two aspects, of the same divinity, one cannot hear an oracle delivered by the other. It would be just as ridiculous as to say, by misunderstanding of Orpheus' point about Okeanos, that 'another Zeus contrived Zeus'.[69]

In any case, the advisory function of Night is also explicitly denied in col. 13, and in parallel to the denial of Zeus's receiving advice from Kronos. Provided that the above explanation of the Derveni author's gloss on Kronos' prophecies has some validity, the parallel treatment of Night's oracle suggests that Night cannot 'command' Zeus for the same reason that Zeus cannot listen to Kronos in the ordinary way: the source of the advice is the same as the recipient of it.

At the end of this lengthy treatment of the Derveni author's interpretation of Night, I would conclude that there are some strong indications that Night too was identified with the supreme air/Mind. She is no more the source of external help, as her advisory, commanding function, so important in the plot of the poem, is explicitly denied. On the physical level, Night/night is allotted the cosmogonical function of the air, i.e. counterbalancing the heat of the sun and hence allowing the other bits of matter to coagulate.

Achelous and water?

Some further improvements on the reading of the last lines of col. 23, apparently developing an interpretation of Achelous, might turn out to be of great importance. Let us see the Greek first, because we have to discuss some possible alternative supplements. At the current state of affairs the

[69] Cf. col. 23.4.

text is as follows, leaving out for the time being a crucial supplement in line 12:

> τὸ δ' ἐχόμενον·
> ¯ἶναϲ δ' ἐγκα̣[τέλε]ξ̣' Ἀχελώιου ἀργυ̣[ρ]οδίνε̣[ω]
> ¯τῶ[ι] ὕδα[τι . . τίθη]ϲ̣ι Ἀχελῶιον ὄν̣ομ[α . .] δ̣ . [ ]
> τὰ[ϲ] δ' ἶνα[ϲ ἐγκαταλ]έξαι ἐϲτ[ὶν τ]ὸ̣ ἐγγε̣[νέϲθ]αι
> τη .[ ]. .[±11]. ων αυ[±13].
> ἑκαϲ̣[τ ±12]δεβουλ[
> . .ν[±13]ο̣ντο̣[(col. 23.10–16)

The crucial part is τῶ[ι] ὕδα[τι . . τίθη]ϲ̣ι Ἀχελῶιον ὄν̣ομ[α. If the very attractive supplement τίθη]ϲ̣ι (or alternatively δίδω]ϲ̣ι as suggested by Tsantsanoglou) is correct, then we have to envisage the possibility that the Derveni author's method is not as systematic as it has appeared thus far. For it seems to imply that the author thinks that this divine name does not stand for Mind or air, but for water. Achelous, son of Okeanos and Tethys, is the god of fertile and life-giving waters, and outside the sphere of mythological imagination Achelous is the geographical name of more than one actual Greek river.[70] On all accounts, Achelous' identification with water would be utterly natural in the case of another allegorist,[71] but it is unexpected for two intimately connected reasons in the Derveni author's text. First, it would mean that the Derveni author does after all resort to the more usual procedure of physical allegory, distinguishing different parts of the cosmos – elements, celestial bodies etc. – and identifying them with different gods of the mythological narrative. Hence not only the air but also the other elements could receive the name of traditional gods. The identification of other elements with gods would furthermore imply that some kind of higher, divine status is granted also to the elements other then air, and this would certainly complicate the theology of the author.

This identification, if in fact present in the text, becomes even more startling when we consider that, as we shall see in the next chapter, fire has an incomparably more important role in the cosmology of the Derveni author than water. The theme of fire comes up often enough in the remaining part of the text for us to expect a god to be identified with it – if the divine identifications apply to the elements other than air. Water, on the other hand, has no recognisable function in the author's cosmology. Notably, the

[70] On the mythological, iconographical material see the comprehensive study by Isler (1970), cf. also Isler (1981) in *LIMC* I.1 and Weiss (1984). For Achelous as the geographical name of different rivers, cf. Herod. 2.10.3; Thuc. 2.102.3; Strabo 8.342; 9.434; Paus. 8.38.9–10.

[71] Achelous is treated as a synonym for water e.g. in Eur. *Andr.* 167; *Bacchae* 625; Ar. *Lys.* 381. For further references, see Isler (1970).

Derveni author does not seem to be interested in the watery theme even when he quotes the poet singing about 'rivers and lovely springs' in col. 16. And if he identifies Achelous with water, why does he not identify Okeanos with it as well?

Moreover, as water is unlikely to be further identified with air, the poor Achelous – and only he, as far as the surviving text allows us to see – gets excluded from the company of his fellow-divinities unified in the cosmogonic works of the divine Mind/air. Surely, this would be unfair. Thus, on all accounts, we would be much less surprised if the water-god Achelous were identified with air, and much better off if we could find in the last lines of col. 23 that it is only the 'ignorant' who take Achelous to refer to water. Indeed, I think the author starts his exegesis by discarding the customary but erroneous identification of Achelous with water. This is why I suggest a negation (οὐ as opposed to e.g. ὡς suggested by Burkert) before the verb.

Hermes and Maia

Hermes and Maia appear only in the penultimate full-verse lemma of the remaining text in ***L17*** = col. 26.4 ≈ *Od.* 8.335: 'Hermes Diaktoros son of Maia, giver of goods'. The context of the quotation makes it almost certain that these two gods are not glossed by the Derveni author. But the fact that Hermes and Maia do not receive any allegorical treatment should not cause surprise. As we have seen in the examination of the lemmata, ***L17*** is adduced only as an illustration in connection with the interpretation of the word ἑᾶς, an apparent reflexive possessive pronoun which could suggest not only Zeus's incestuous desire but also the separate identity of Zeus and Rhea. Naturally, the Derveni author does his best to eliminate this option, and he does so by claiming that ἑᾶς is an adjective meaning 'good', assimilating it to the epic word ἐύς. Thus, irrespective of the question whether the Derveni author considered ***L17***, the verse mentioning Hermes and Maia, to be by Homer or thought he was quoting Orpheus, the only important word in the lemma is ἑάων, and the rest does not require interpretation at this point.[72]

[72] One could object that I treated the Derveni author's quotation from the *Hymns* differently, inserting Hestia and Deio into the system of identifications. Yet the parallel does not hold, as in that case the point of importance was precisely the identity of the goddesses mentioned together with Hestia and Deio.

Divine names in the first six columns: Dike, Erinyes and Eumenides

These names certainly cannot be treated on the same footing as the previous ones, as they do not come up in relation to the Derveni poem itself. Dike is mentioned only in the Heraclitus quotation in col. 4, and hence, similarly to what we have seen in the case of Hermes and Maia, does not necessarily require specific interpretation from the Derveni author. She might have nevertheless been the object of exegesis in the second part of the column, as the single remaining word δίκης, or for that matter Δίκης, in line 12 suggests. Yet the physical condition of the papyrus excludes any conjectures on the content of this gloss.[73]

On the other hand, the Erinyes, who figure in Heraclitus' text as the helpers of Dike, are referred to repeatedly and with emphasis in the Derveni author's own text in the first six columns. Thus the Erinyes evidently constitute part of the author's own eschatological views as developed in this section of the papyrus. If so, we have to see if there is any clue in the text about the relationship of the cosmic divinity and the Erinyes, as, despite the strong thematic divide between the first six columns and the rest of the text, it is natural to suppose that the eschatological views of the author are not unrelated to the cosmic theology he develops through the exegesis of the poem in the second part of the text.

To begin with, one would certainly expect the Erinyes to be identified with the Eumenides, also mentioned in col. 6. On the other hand, we cannot take this point for granted, for we have seen numerous instances where it turned out to be misleading to approach the Derveni author with presuppositions based on the traditional Greek understanding of the identity of such divine beings. He could just as well say that only 'the ignorant' think that the Eumenides and the Erinyes are the same. The Eumenides, for their part, are explicitly identified with souls in col. 6.[74] Yet the rather desperate scraps of lines 4 and 5 of col. 2 make it possible that the Erinyes too were identified with souls.[75] To complicate things further, we have the daimones mentioned repeatedly in columns 3 and 6; and it is not altogether clear how the daimones mentioned in this context relate to the Eumenides/souls and Erinyes.

There is, however, one conspicuous feature of the treatment of these divine or semi-divine beings which clearly sets them apart from the traditional gods dealt with in the second part of the papyrus. As we have seen

[73] Cf. Tsantsanoglou (1997) 97. [74] Col. 6.9–10: Εὐμενίδες γὰρ | ψυχαί εἰσιν.

[75] Col. 2.4–5. See Tsantsanoglou's comments ad loc., (1997) 96.

above, the most striking characteristic of the Derveni author's procedure in the exegesis of Orpheus' poem is that all his techniques of allegoresis are geared to the objective of reducing the plurality of traditional gods to a unity. In contrast to this, he is not at all bothered by the plurality of Erinyes, Eumenides, souls and daimones. It follows that these beings are not on the same level as the cosmic god whose features are explained through the identifications in the second part of the papyrus.[76] The elucidation of the identity of Erinyes and Eumenides is governed by other principles and has other aims.[77]

CONCLUSION

The most important result of this survey is our understanding that one of the Derveni author's primary concerns in his commentary is to fuse the gods and goddesses of the poem so as to come up with a single cosmic god. Prima facie, this is an outright violation of the meaning of the poem. The poem speaks about the birth of gods, the way they interact with one another and how the social organisation of their world evolves – how they marry each other, give birth to their offspring, gain and lose power, castrate, help, rape, and occasionally swallow each other. All this must disappear as the personal identity of the individual gods is absorbed into the monumental figure of the one cosmic Mind. Furthermore, since the divinity is explained as a physical element, air in this case, and the absolute generation of physical elements is explicitly denied, the cosmic god can have no birth. Theogony is ruled out. Yet, paradoxically, on a deeper level all this is quite in line with the theological import of the Orphic poem. It is time to cash in our own interpretation of the Derveni poem.

As I have tried to show in the previous chapter, one major theoretical objection that could be, and was in fact, raised against the Hesiodic narrative was the unresolved tension between the two types of priority: the supreme ruler Zeus was temporally posterior to other gods of lesser importance.

The ultimate solution to this problem is to say that the supreme divinity is not born at all.[78] This step cannot be taken within the framework of the traditional mythologic–theogonic discourse whose organising principle and main explanatory tool is to relate the generations of gods, and

[76] So Tsantsanoglou (1997) 99.

[77] On further conjectures on this point, see 85–9 above and 346 below.

[78] Indeed, already Xenophanes had declared that those who say that the gods were born are just as impious as those who say they die (see 21 B11 DK = Sextus Emp. *adv. math.* 9.193 and Arist. *Rhet.* 1399b5–9).

where the 'backbone' of the story is the succession myth. The solution was probably first proposed by Pherecydes of Syros who posited that his three ultimate divinities were not born at all: 'Zas and Chronos always were and Chthonie.'[79] But this is exactly the point that entitles Aristotle to say that Pherecydes (together with the magi) is already mid-way between mythological and philosophical discourse.[80] The Derveni author clearly sides with the philosophers in this respect, for he says explicitly that his cosmic god, air/Mind, was not born.

Yet I have also tried to show that the Orphic theogony has made a serious effort to offer a solution to the problem of priority within the limits of the mythological discourse. One of the characteristic novelties of the Orphic theogony, as compared to the Hesiodic one, is the story of Zeus swallowing the world. As a result of this episode, Zeus can, in a certain sense, be temporally prior to the other beings: 'Ζεὺϲ πρῶτοϲ γένετο . . .'. The cosmic ruler is nevertheless said to have been born, as the genetic aspect of theogony still prevails, but he can be considered older than all the other characters of the story. Zeus was there before all the others. The Derveni author concurs with this, since he says that the divine air, also identified with Zeus, was there even before the things that are now were formed – but he brings the issue one step further by asserting that it was not ever born.[81]

A further notable theological implication is that, as a result of the episode of swallowing, there can be a considerably clearer distinction between Zeus and the other gods when we compare the Orphic narrative to the Hesiodic representation. For a mythical moment everything is in Zeus, and then Zeus brings to light, creates anew the whole world. To express it in Platonic terms, Zeus rises to a higher ontological level, for he is not just the supreme king, but becomes the ultimate source of all other beings, including all the other gods: '. . . Διὸϲ δ' ἐκ πάντα τέτυκται'. Although the content of the affirmation will become clear only after the examination of the Derveni author's cosmogony, we have already encountered several times the claim that the Mind is the 'mother' and source of all other beings. Thus, unlike Hesiod's Zeus, but very much like the Zeus of the Orphic poem, the Derveni author's cosmic god is responsible for the birth of the other beings.

Perhaps even more important is the issue of plurality versus unity. As we have seen at the end of the last chapter, the story of swallowing is probably

[79] D.L. 1.119 = 7 A1, B1 DK: Ζὰϲ μὲν καὶ Χρόνον ἦϲαν ἀεὶ καὶ Χθονίη. Cf. Damasc. *De princ.* 1.321.3–4 Ruelle = 7 A8 DK.

[80] Arist. *Metaph.* 1091a30–b7. The passage is discussed in Schibli (1990) 18 n. 11.

[81] Col. 17.2–3: 'For air existed even before the things that are now were set together and always will exist. For it was not born, but existed.'

the most important contribution from the side of the mythological discourse to this question – which is, at least according to some, the major theme during the sixth and fifth centuries. Under this heading, the swallowing episode offers the striking image of the one Zeus containing the entire plurality of all the other beings. In the sphere of theology it shows that the universal Zeus can 'physically' contain all the divine beings of the world, and thus the other gods can symbolically be identified with Zeus. The individual identities of the other gods are temporarily suspended, since they have been absorbed in the unique cosmic god: '. . . αὐτὸς δ' ἄρα μοῦνος ἔγεντο'.

This pantheistic and monistic moment of the Orphic narrative (everything is in Zeus and Zeus is alone) is, once again, an approximation of the philosopher's understanding of the nature of the divine,[82] but still from within the framework of the mythological mode of expression. Seen from this perspective, the Derveni author's system of identification is nothing more than an elaboration on the same theme. He picks out and makes absolute the most important and novel element of Orpheus' narrative and explains in detail in what sense all the gods and goddesses mentioned by Orpheus can be identified with each other and ultimately with the cosmic god.

In view of the above considerations, I would maintain that the Derveni author's interpretation abuses the poem (in the sense of intentionally misrepresenting its meaning) far less than is usually held to be the case. There is no denying that the interpretation of the individual lemmata is far from obvious. Nevertheless, the outcome of his exegesis does bring out some of the most conspicuous elements of the Orphic poem's possible theological implications, by making explicit and absolute what is already there in embryo and implicit in the crucial episode of the swallowing. The Orphic poet, with his conception of the temporally prior, generative and unificatory Zeus, had made an important move towards a more abstract conception of divinity, in some respects similar to the one which becomes *à la mode* in Presocratic circles. As he is explaining the mythological narrative, the Derveni author, already acquainted with the later philosophical developments and armed with the toolbox of exegetical techniques, makes a serious attempt to bridge the remaining distance between the two sides.

Besides, there is, I think, a viewpoint from which the author's attempt to interpret the different divine figures of the poem as different names of the one cosmic god becomes, if not less far-fetched for us, at least defensible

[82] The most ready parallel is Xenophanes. See Palmer (1998).

in a particular religious and ritual context. We have seen that the author probably treats the Orphic poem as a hymn;[83] I have argued, however, that the poem does not readily fit into this category. One aspect of this problem is the fact that, even though the poem focuses on the figure of Zeus more directly than Hesiod's *Theogony*, it still speaks about more gods than a hymn *qua* prayer, i.e. a song directed to one god (or a closed group of gods).[84] On the other hand, it is well known that a prayer *casu quo* a hymn invokes the god it is directed to by calling him or her by his or her different names and titles. Moreover, together with the cult-names, patronyms or metronyms, and names referring to the habitual residence of the god, the different names indicating different *functions* of the god could have an important role in the invocations.[85]

It is just as important in this respect that the Orphic hymns brought the multiplication of the god's denominations to the extreme and were often reduced to a mere succession of *epicleseis*. A further specific feature of these hymns is that in the lists of denominations the authors do not stick to those names which traditionally refer to the same god (e.g. Pallas and Tritogeneia as names of Athena) but, as Jean Rudhardt has put it, 'Ils interpellent un dieu sous le nom d'un autre, traitant comme des synonymes les noms de plusieurs divinités que les traditions mythiques et cultuelles distinguent entre elles.'[86] Although the majority of the Orphic hymns are certainly of a later date, the quotation from the *Hymns* ('Demeter Rhea Ge Meter Hestia Deio' in col. 22.12) shows that this practice was present, at least in embryo, already when the Derveni text was composed.[87]

Therefore, in merging the different characters of the poem into the figure of the one cosmic god, the author, as I tried to show above, radicalises theological tendencies already present in the Orphic narrative. Furthermore, in the course of this he relies on a characteristic feature of the Orphic hymns, i.e. that though the hymn is to one god, this god can be called and characterised by names which were traditionally treated as belonging to independent entities. Thus, the Derveni author could treat the poem as a hymnic prayer in which the god, the one cosmic god, is called by different names, and these names indicate the god's different functions in organising the cosmos.

[83] Cf. col. 7.2, with 137 above. [84] See 137–8 above.
[85] Cf. Bremer (1981) 194–5. [86] Rudhardt (1991) 272.
[87] As noted above, it is possible or even probable that some gods were identified with each other already in the text commented on by the author (e.g. Rhea and Demeter). Nonetheless, I find it completely implausible that all the identifications argued for by the author were already present in the Orphic poem.

We have seen, however, that the image of the all-containing Zeus of the Orphic narrative could have yielded to an interpretation which takes the ultimate step on the road of unifications. As the supreme god contained within his belly not only all 'the blessed gods and goddesses', but also the 'rivers and lovely springs and everything else . . .',[88] this could have been readily interpreted as an expression of a pantheistic view. Yet we have seen that the Derveni author resisted this option by keeping apart the Mind/air and the 'other beings'.[89] These two sides of the author's world must remain separate, since the other things will display contrasting forces and, eventually, the order imposed by the cosmic god. This is what we will learn through the Derveni author's cosmogony.

[88] Cf. ***L10*** = col. 16.3–6. [89] Cf. 186 above.

CHAPTER 6

Cosmology

In the interpretation of the Derveni author, the divine characters, actors in the plot of the Derveni poem, have been absorbed into the monumental figure of the cosmic god. The narrative – that is, the temporal sequence of events – is not, however, completely lost. What was related as a series of interactions between the different gods in the poem has become a series of interactions between the cosmic god and the 'things that are' in the story of the Derveni author. The theogony has been transformed into cosmogony.

In what follows I shall attempt a reconstruction of the Derveni author's cosmology and physics. First, I shall discuss the author's doctrine on the formation of the cosmos, and then in a second move I shall take a closer look at the ontology and physics implied in the cosmogony.[1]

COSMOGONY

The outlines of the Derveni author's cosmogonical theory are relatively clear. He distinguishes the present state of the cosmos from a previous, pre-cosmic state. Before the current cosmic configuration emerged, there had been no independent entities with definite contours, but a cosmic mixture containing all the matter out of which the presently existing entities would later form. Then the cosmic Mind separated the fiery particles, for he knew that it was the excessive heat of fire which did not allow independent entities to form. He fashioned the sun out of these fiery particles, and placed it at a convenient location. Because the sun was still perilously large he dispersed the surplus fiery particles and fixed them in the sky, thereby creating the stars. After the excess of heat had been removed the sun could

[1] Thus I reverse the usual order. The reason for this is that the physical elements and the forces operating on them are more readily understandable once we have seen the underlying logic of the story.

still communicate sufficient motive force in the form of heat to the rest to set different blocks of matter in motion: entities could now separate, collide and coagulate. This is the process through which the world as we know it received its shape.[2] Let us now consider the successive phases of this story more in detail.

BEING AND GENERATION

As we examine the Derveni author's ideas on the generation of the cosmos, the first thing to note is that he explicitly excludes the possibility of generation in an absolute sense. In col. 16 we read the following:

It has been made clear above [that] he called the sun a phallus. Since the beings that are now come to be from the already subsistent he says:

[with?] the phallus of the first-born king, onto which all
The immortals grew (or: clung fast), blessed gods and goddesses
And rivers and lovely springs and everything else
That had been born then; and he himself became solitary.

In these (verses) he indicates that the beings always subsisted, and the beings that are now come to be from (*or*: out of) subsisting things. (col. 16.1–8)

At first sight it is not easy to see in what way the lemma and the Derveni author's interpretation are related, even though the passage is apparently meant to constitute a self-contained argument.[3] At any rate, the Derveni author is making here an explicit ontological claim. His point is that the things which exist now were not created out of nothing, but were formed out of already existing things.

Even though a cosmogony involving a genuine *creatio ex nihilo* had probably never been a real option in Greek philosophy, the urge to express the underlying ontological principle in an explicit way was, in all likelihood, an effect of Parmenides' *aletheia*. After the Parmenidean ontological challenge, philosophers found it necessary to make clear that what appears as birth is not generation in an absolute sense, but that the same constitutive elements can be reorganised, appear in new structures and thus form new entities.

[2] Cf. Betegh (1993) 29–40; Laks (1997) 129–34; Most (1997) 118. For a short summary with different conclusions, see Rusten (1985) 127–8.

[3] The preceding sentence, of which only the last part is conserved in the first line of the column, speaks about the identification of the sun and the phallus. The ὅτι δέ, with which the passage I have quoted is introduced, marks the beginning of a new argument. The sentence after the passage, beginning with a τὸ δέ, speaks about the Mind which 'became solitary'.

The ontological population[4] has to be constant and the birth of a new entity is explained as the rearrangement of already existing ingredients.[5]

The warning that the word 'birth' can be misleading, so one has to make sure that it is rightly understood as merely the emergence of a new configuration out of the previously existing entities, also appears in the Derveni author's text:

> For they existed even before, but were spoken of as 'being born' since they were separated out.[6] (col. 21.13–14)

Surely, by the time of writing, the author's insistence on the conservation of matter is hardly more than a truism within the confines of the philosophical paradigm. On the other hand, such an emphatic denial of generation in an absolute sense is quite significant when one is discussing a mythological-theogonical text. For the biological, sexual model for the generation of new entities, necessarily prevailing in the mythological mode of expression, might actually give the impression that there is an actual, genuine multiplication of beings. The birth of new gods is certainly not balanced by the decomposition of some older divinities, and thus the society of the blessed, deathless gods shows a constantly and exponentially rising demographic curve.[7]

It is worth noting then that the Derveni author does not need to quote philosophical arguments, or even to put another cunning allegorical explanation into play to support the ontological principle he will use in his physics. Instead, he feels entitled to quote these four verses from the poem as sufficient to demonstrate Orpheus' commitment to the principle of the constant preservation of matter through cosmic changes. His move is warranted in that the verses he cites from the poem say that Zeus did not create the world out of nothing, that the entities of the new world order had already existed before they were engulfed, and that they were then

[4] I take this expression from André Laks.

[5] The *locus classicus* is Anaxagoras B17 DK; cf. also Empedocles B17.1–13 DK. For a clear exposition of the view that these were responses to Parmenides, see Furley (1989) ch. 6. On the difference between Anaxagoras' and Empedocles' response, see *ibid.* 48–51.

[6] With L&M and Janko, I take the subject to be 'the things that are'; but 'the air' is not impossible either.

[7] By pressing the point, one could say that in Hesiod, when Gaia gives birth to the mountains by herself (vv. 126–8), or when the different river-gods are born from Okeanos and Tethys (vv. 337–70), it is the differentiational, and not the biological, sexual model, which is at work. As a result, the sum total of earth and water could, at least theoretically, remain constant. But even this reasoning, not quite relevant to Hesiod's discourse, would only be applicable to a very limited number of cases in the *Theogony*.

created anew by the king of gods. There is no birth in the absolute sense, only a new order of things.

Although the point made is clear, the author's terminology is somewhat loose. He repeatedly uses three terms, τὰ ὑπάρχοντα, τὰ ὄντα and τὰ νῦν ὄντα.[8] The assertion that τὰ νῦν ὄντα are born from τῶν ὑπαρχόντων seems to imply that τὰ νῦν ὄντα refers to the currently existing composite entities, whereas τὰ ὑπάρχοντα designates the ingredients which are stable throughout the changes.[9] However, τὰ ὄντα are said to have always subsisted (ὑπῆ[ρ]χεν ἀεί), which must mean that τὰ ὄντα are not the composite entities with a limited lifetime, but the imperishable components themselves. And this is what we shall find in other parts of the text as well. The expression τὰ νῦν ὄντα (or τὰ νῦν ἐόντα) – implicitly or explicitly contrasted with those which existed previously[10] – refers to the presently existing entities that have come into being at a certain moment in time and which, presumably, will cease to exist in this particular form sometime in the future. τὰ ὄντα, on the other hand, without the temporal designator, can refer either to the everlasting components as such, or to the composite entities which exist during a limited span of time.

THE PREVIOUS STATE

In line with a well-established tradition of Greek cosmogonical speculation, ranging from Anaximander through Anaxagoras to Plato's *Timaeus*,[11] the initial cosmic situation is expressed not so much in positive terms, but as a negation of the presently existing entities and the overall cosmic arrangement. As matter was already there in some form or other, what was lacking was the shape and structure of the individual beings existing now on the one hand, and the general spatial constitution of the cosmos on the other.

8 The author freely alternates between ὄντα and ἐόντα, so τὰ νῦν ἐόντα in line 8 is unproblematic.

9 It has to be noted that ὑπάρχειν in the sense of 'originally or already existing' is not limited to the technical vocabulary of philosophy. LSJ *s.v.* B2, for example, refers to Herodotus 7.144 already with this meaning: αὗται αἱ νῆες τοῖςι Ἀθηναίοιςι ὑπῆρχον where these ships are opposed to those which the Athenians were about to build. On the other hand, τὰ ὑπάρχοντα cannot mean 'things which existed before but not any more'. See, however, Sider (1997) 138: 'the Commentator's use of ὑπάρχοντα in the sense "enduring things" in a philosophical context is unexampled in the fifth century, and in fact does not show up until Aristotle. I wonder whether this indicates that the year 400, which is usually given as a terminus ante quem, may not be too early.'

10 Cf. e.g. col. 17.2 and 8–9.

11 Ignoring now the question whether or not Timaeus' cosmogonical narrative represents a temporal sequence. For the point I am about to make, 'initial' can be taken in a wider sense, encompassing also structural or explanatory priority, from which the present situation can be derived through a series of steps.

A reference to air's continuous existence through cosmic changes makes this point explicit:

For air existed even before the things that are now were set together and always will exist. (col. 17.2–3)

As we shall see in a moment, 'setting together' (cυcταθῆναι) is the author's key concept in describing the birth of the presently existing entities, and, correlatively, this is what is denied for matter in the previous state.

Col. 9 provides us with a more precise idea about the state of affairs during the initial, pre-cosmic situation:

He made the strength to belong to the strongest, just as a son belongs to the father. Those who do not understand the meaning of the things said think that Zeus takes the strength and the daimon from his father. Now, knowing that fire, [in as much as] it is mixed with the others, agitates the things that are and hinders them from getting set together because of fomenting, he removed it to an adequate distance, so that once it is removed, it does not hinder the things that are from coagulating. For whatever is kindled is dominated (sc. by fire), and when dominated, it mixes with the other things. But (as to the expression) that 'he took in his hand', he gave a riddling meaning just like the other [. . .] knows firmest [. . .] strong, he said that Zeus [. . .] the daimon . . . just as. . . . strong. (col. 9.1–13)

We learn from this passage that it was not just sheer accident that things did not form during the previous state, but that they were actually not able to do so. The reason for this is also given: fire was mixed with them, and did not let them compound or coagulate. Thus the initial situtation can be described as a mixture without entities with definite shapes, and this mixture includes fire. Furthermore, the verb ταράccω, with the sense 'agitate', 'stir up', and 'cause confusion', entails not only that this situation was disorderly, but also that fire communicated some kind of motive force to the others, and that the excess of this force was to be blamed for the lack of more or less stable, independent entities. Therefore, we have to assume that the mixture was in a state of constant chaotic motion. And, as we shall see, the passages describing the cosmic function of the sun, made out of fire, also lay stress on the motive function of fire.[12] Moreover, the use of cυμπαγῆνα[ι], 'coagulate' – which will be echoed by πήγνυμι in col. 15.4 – might suggest that this universal blend is to be imagined as fluid like some kind of hot magma.

Incidentally, it has to be noted that the substantive form θάλψιc, which describes the immediate effect of fire, is a technical term of the doctors.

[12] This idea is also present in Hippocratic texts. See 273 n. 136 below.

According to the *TLG*, the word does not occur outside the Hippocratic corpus in pre-Hellenistic times.[13]

The last two relatively well-preserved sentences of the passage might provide further information on the initial state of affairs. As we encounter here some difficulties caused by verbal ambiguity, we have to discuss the Greek first:

ὅcα δ' ἂ[ν] ἁφθῆι ἐπικρα[τεῖται, ἐπικ]ρατηθὲν δὲ μίcγεται | τοῖc ἄλ[λο]ιc. (col. 9.9–10)

The first problem is the meaning of ἁφθῆι in line 9. Does it mean 'to touch' or 'to kindle, to set on fire', as both meanings seem possible in the context? L&M and Janko in their respective translations adopt the latter (indicating the alternative in a footnote), whereas Rusten chooses a version of the former.[14] At any rate, the ambiguity of the sentence is aggravated by the fact that – partly due to the Derveni author's predilection for passive constructions – neither the subject nor the object is clearly stated. And, conversely, both the subject and the object will be different depending on our interpretation of the verb.

The main advantage of Rusten's interpretation is that it establishes a better connection with the lemma, as ἅπτω in the sense of 'to grasp',[15] could in this way be taken as the Derveni author's variant for the ἐγ χείρεccι ἔ[λ]αβ[εγ of the lemma.[16] On this reading, then, the agent of the passive verb is Mind, who grasps fire, takes it under his control, and thus commences a new reign; just as when the Zeus of the poem takes from his father the power and the signs of divine rule in his hands. On the physical level it would express the thesis that a new cosmic order begins as the air comes to dominate (ἐπικρατεῖ) fire. This picture would be paralleled with the air dominating all other things in general (col. 19.3–4) and the fiery corpuscules in particular (col. 25) during the present cosmic order.

However, the last clause (ἐπικ]ρατηθὲν δὲ μίcγεται τοῖc ἄλ[λο]ιc) must imply on this reading that, as an effect of being ruled, fire gets mixed with the others. Therefore, this construal of the sentence is only possible if one accepts the other part of Rusten's interpretation too, namely that fire has to be mixed with the other things to enable the other things to coagulate. But, as we shall see, this is difficult to reconcile with other parts of the

[13] *De dieta in morb. ac.* 7.8; *Aphorisms* 3.1.3; *De humor.* 3.7 and 5.14; *De humid. usu* 1.20. In later times too, θάλψιc remains a primarily medical term.

[14] Rusten (1985) 129 translates 'Whatever is "grasped" is "mastered" and, in its subordinate status, it is mixed with the other things.'

[15] As a matter of fact, I cannot find an exact parallel for this meaning in LSJ *s.v.* ἅπτω.

[16] Rusten (1985) 129–30.

text. Mind's aim is not to mix fire with the others but, on the contrary, to separate it from the rest.

Tempting as it may have appeared at first, the ἅπτω = ἐγ χείρεςςι ἔ[λ]αβ[εγ version must therefore fall. On the other hand, given that the Derveni author is considering fire and heat in this passage, the 'kindle' option can be just as attractive. As we can rule out the possibility that air/Mind ignites things under his control, fire must be the agent of both of the passive verbs in the first sentence: whatever is kindled is dominated by fire. And whatever is dominated by fire mixes with the others. In this case, μίςγεται,[17] caused by fire, must be the opposite of cυνίςταςθαι and cυμπαγῆναι; and indeed, μίςγεται is an appropriate word to refer to the original jumbled situation in which separate entites could not form. Hence the function of this sentence is to explain further why it is necessary to remove fire if one wants the things to coagulate.[18]

Thus, in the initial situation fire 'dominates' the others, and as a consequence of this, the world is an ignited, agitated mixture of all matter. But this passage of col. 9 already brings us to the cosmogonical activity of the divinity.

THE CREATION OF THE SUN

The creation of the sun is undoubtedly the crucial moment of the story, triggering off the whole cosmogonical process. The author states this in the plainest possible terms:

> If the god had not wished that the things which are now should exist, he would not have made the sun. But he made it of such a sort and of such a size as is explained in the beginning of the account. (col. 25.9–12)

Although it is not immediately clear which previous passage the author had in mind, I shall argue later on that he refers to ***L2*** of the poem.[19] Anyhow, we can certainly learn some further details about this important cosmogonical act from the passage explaining ***L2***. Indeed, the lines I quoted in the previous section from col. 9 constitute the most explicit statement

[17] As LSJ indicates, μίςγω is the dominant variant for μίγνυμι in Homer and Herodotus, but also attested in Attic prose.

[18] One could object that in col. 21, when the author explains the name of Aphrodite as referring to the *mixis* of elements, conceived on the analogy of sexual intercourse, it is much more probable that the text speaks about the unification and composition of separate entities and not about the pre-cosmic formless mixture. But even if μίςγεται is appropriate for the pre-cosmic state, one should not infer that there is no *mixis* in the cosmos as well; cf. the parallel case in Anaxagoras' cosmogony.

[19] For further arguments, see 327–8 below.

about the beginnings of the cosmic rearrangement effected by the divine Mind, while the repetition of words and the interpretation attributed to the ignorant ensures that the Derveni author is still dealing with ***L2***, as quoted in col. 8.4–5:

> Ζεὺϲ μὲν ἐπεὶ δὴ π̣α̣[τρὸϲ ἑο]ῦ πάρα θέ[ϲ]φατον ἀρχὴν
> [ἀ]λκήν τ' ἐγ χείρεϲϲι ἔ[λ]αβ[εγ κ]α̣ὶ̣ δαίμον̣[α] κυδρόν

This lemma, as I have argued above, was, in all likelihood, at the very beginning of the narrative part of the poem, launching the story *in medias res*, describing the way Zeus obtained royal power.[20] This is reassuring in so far as the Derveni author, in his turn, also commences his story not with an 'In the beginning . . .' but immediately at the crucial moment of the change of cosmic regime. So far so good.

Yet the interpretation of ***L2*** in col. 9 is nonetheless full of ambiguities, partly resulting from the physical condition of the papyrus, partly stemming from the equivocal nature of some of the terms used by the author. Thus, before drawing any conclusions regarding the cosmogonical activity described in these lines, let us first take a stance on these textual and interpretative problems.

ε̣ἶ̣ν̣α̣ι̣. τ̣ὴ̣[ν ἀλ]κ̣ὴν οὖν τοῦ ἰ̣ϲχυρ̣[ο]τάτου ἐπόη̣[ϲεν] | εἶναι ὡϲ̣[περ]ε̣ὶ παῖδα πατρ̣ό̣ϲ. οἱ δὲ οὐ γινώϲκον[τεϲ] | τὰ λεγό[μεν]α δοκοῦϲι̣ τ̣ὸ̣ν̣ Ζᾶνα παρὰ τοῦ αὑτο[ῦ] | πατρὸϲ [τὴν] ἀλκήν [τε καὶ] τὸν δαίμονα λαμμά[νειν]. | γιγνώϲκ[ων] οὖ̣ν τὸ πῦρ [ἅτε] μ̣εμειγμένον τοῖϲ | ἄλλοιϲ ὅτι ταρ̣άϲϲοι καὶ κ[ωλ]ύ̣οι τὰ ὄντα ϲυνίϲταϲθαι | διὰ τὴν θάλψιν, ἐξή̣λλαξ[ε ὅϲ]ο̣ν̣ γ̣ε ἱκανόν ἐϲτιν | ἐξαλλαχθὲμ μὴ κωλύ[ειν τὰ] ὄντα ϲυμπαγῆναι. | ὅϲα δ' ἅ[ν] ἁ̣φθῆι ἐπικρα[τεῖται, ἐπικ]ρ̣ατηθὲν δὲ μίϲγεται | τοῖϲ ἄλ̣[λο]ιϲ. ὅτι δ' ἐγ χείρ[εϲϲιν ἔλαβ]ε̣ν ἠινίζετο | ὥϲπε[ρ τ]ἄλλα τὰ π̣[±14]. ομεν [. . . .]α | [βε]βαιότατα νοε̣ι̣[±18]ν ἰϲχῦρὸϲ | ἔφη τὸν Ζᾶνα τ̣.[±16]ν δαίμονα | [ὥ]ϲ̣περεὶ ε[ἰϲχ]υ̣ρ̣ο̣ῦ̣ (col. 9.1–13)

The first problem concerning the passage is that the subjects of the first and the third sentences are missing. Furthermore, as the second sentence is a dismissive aside about the misguided interpretation of the ignorant – i.e. the Derveni author gives here the literal, surface meaning of the lemma – we would expect that the subject to be supplied in the third sentence is identical with the subject of the fragmented first sentence. Prima facie, it would seem likely that the subject of ἐπόη̣[ϲεν in the first sentence is the poet, assuming that the Derveni author is trying to explain here that Orpheus was speaking about the possession of strength and its relation to the Kronos–Zeus relationship. The author's explanation seems to

[20] See 110 above.

imply that we should take the father–son relationship only as an analogy (ὡς̣[περ]ε̣ί).[21] This reading would be in concord with our expectations, namely that the Derveni author wants to argue away the genealogical relations, which presuppose the independent identity of the two gods. In this way the argument would be analogous to the one explaining the relationship, or more precisely the lack of it, between Zeus and Rhea in col. 26.[22]

On the other hand, the third sentence clearly describes a cosmological activity, involving fire, the setting together of entities and so forth. Therefore, it is reasonable to suppose that the subject of the main verb, which must be ἐξή̣λλαξ[ε,[23] is the supreme cosmic agent. Hence, this sentence probably describes what Mind/Zeus does, the reasons for his action, and that – *nomen est omen* – he introduces these cosmic changes on rational grounds (γιγνώσκ[ων] οὖν).[24] We have to suppose then that there is a change of subject between the first and the third sentence: in the first the poet is the subject, whereas in the third, the cosmic god.

Furthermore, we learn from this passage about fire's fundamental role in the cosmogonical process.[25] As we have seen above, fire is the major hindrance to the combination and coagulation of entities.[26] From this it follows that the supreme agent has to do something with fire if it wants

[21] This seems to be the interpretation governing the translation of L&M: 'So he made him come from the strongest thing, just like a son from his father' and also the one suggested by Tsantsanoglou ad loc.: 'So he made the power be (in relation) to the strongest just like a child to a father.' But I think Janko gives the best translation ('So (Orpheus) made the rule belong to the strongest, as a son (belongs) to the father'), and my translation follows his quite closely.

[22] So already Rusten (1985) 128 n. 13.

[23] The *ZPE* text gives ἐξα̣λλας̣[. As the aorist is clearly preferable here, I tentatively change the two dotted letters and supplement ἐξή̣λλαξ[ε. Rusten (1985) 129 with n. 15, adopts ἐξα̣λλας̣[cει], but assumes that it is a historical present. This would, however, be the only historical present in the papyrus. On ὅ]ς̣ο̣ν̣ γε, see the apparatus of my text.

[24] Rusten (1985) 129 translates 'Since he (Orpheus – or perhaps the creator himself) knew that etc.' Calame (1997) 77, on the other hand, apparently construes the sentence on the basis of the οἱ δὲ οὐ γινώσκον[τες]–γιγνώσκ[ων] οὖν antithesis and gives the following paraphrase: 'So too, when Zeus gathers in his hands his own father's power, the ignorant give a literal interpretation of the text, thinking (. . .) that Zeus inherits his father's divine power, whereas he who knows (col. 9.5) knows that this power corresponds to the capacity of mixing the elements.' Although the parallel is indeed tempting, 'not knowing' has its object in τὰ λεγό[μεν]α, whereas the content of 'knowing' is directly cosmological, prompting a cosmological action. Janko also takes Zeus as subject.

[25] Fortunately, the important addition by Tsantsanoglou to the L&M translation of the column to the effect that in line 5 'πῦρ is read in the papyrus', obviates a lengthy discussion on the question what exactly makes this trouble for the other beings by agitating and heating them.

[26] Rusten (1985) gives the opposite interpretation. On his understanding of the text, the passage is not about the negative, but about the positive cosmogonic role of heat. Thus, having written before Tsantsanoglou made it public that in line 5 'πῦρ is read in the papyrus', he supplemented 'heat' in the lacuna indicated in *ZPE*, but with a negation [θερμὸν μὴ]. Thus the general sense of the passage would be that fire has to be mixed with the other things, for otherwise the things cannot combine. Although as far as the number of characters go, the negation is not excluded with πῦρ either, this would be difficult to square with what follows, becauses it would entail that fire distributes *less* heat

things to take shape. But what exactly does the supreme agent do with fire? The required action is described by the verb ἐξαλλάττω. The trouble is that this verb has two connected, but distinctly different meanings, both of which seem possible in the present context. ἐξαλλάττω can either have the more general meaning 'to change, to change utterly', albeit implying mainly qualitative change; or alternatively, it can signify change regarding the location of something, hence 'to remove' or 'to withdraw'. L&M and Janko prefer the first meaning, although L&M indicate the alternative 'changes its place' in a footnote ad loc. These translations convey, I take it, that the subject of the sentence, be it the Mind or Orpheus, initiates some kind of qualitative change on fire.[27]

I would maintain nonetheless that the spatial 'remove' gives better sense, since other passages indicate that the principal cosmogonical activity of the cosmic divinity consists in the creation of the sun and the stars. And this could be done by relocating the fiery particles. This should become evident when one considers the description given in col. 9 in the light of a passage in col. 15 explicitly speaking about the creation of the sun:

> . . . striking[28] them against each other and, if he made the [su]n separate, (the result is that) the things which are stood apart from one another.[29] For as the sun got separated and encircled,[30] he coagulated and held fast both the things that are above and those which are below the sun. (col. 15.1–5)

With most commentators, I assume that this paragraph looks back at the same cosmogonical action which has been adumbrated in col. 9. Moreover, just as in cols. 13. and 25, it is stated here that the creation of the sun is prerequisite for the coagulation of entities; or in other words that Mind created the sun in order to obtain the generation by coagulation of the

to the others when it is mixed with them than when it is taken apart. All the more so, as the crucial demiurgic activity of the god is the creation of the sun (cf. col. 25.10–11), and this act does not consist in mixing the fire with the others but, as col. 15.1–5 makes explicit, in separating it from the rest.

27 Rusten (1985) 129–30 is ambiguous on the point. He translates 'he altered it', which implies qualitative change to me, whereas he renders it as 'distributing it among the other elements', which seems to imply change of location, in his interpretative paraphrase.

28 As Rusten (1985) 135 n. 34 suggests, κρ[ο]ύεν in the papyrus must be an infinitive; Tsantsanoglou in L&M 16 n. 35 has the same suggestion. L&M, however, translates 'they were striking against one another'. Apart from the question of the tense of the verb, this is unlikely, since the subject with the active verb must be the cause of the collision; when he speaks about the entities being thrust against each other, the Derveni author ordinarily uses the middle. I think with Janko that one has to supplement an iota after the epsilon.

29 This is how I supplement and understand the sentence; for the alternative suggestions, see the apparatus ad loc. in the text. The supplement τὸ[ν ἥλι]ον not only gives a satisfactory meaning, but is borne out by the repetition, highly characteristic of the author's style: τὸ[ν ἥλι]ον | χωριcθέντα . . . χωρ[ι]ζομένου γὰρ τοῦ ἡλίου.

30 On this rendering, see my arguments at the end of the next section, 242–3 below.

currently existing entities. Yet the verbs used in this passage state unequivocally that the creation of the sun involves a change of location. And indeed, with this point in mind, col. 9 and this passage constitute a nice parallel: cυνίcταcθαι and cυμπαγῆνα[ι] of col. 9 are echoed here by πήξαc, whereas χωρίζω and ἀπολαμβάνω in this passage answer the repeated ἐξαλλάττω of col. 9. Hence this parallel also strongly suggests that what the Derveni author states in col. 9 is not that Mind effected some kind of qualitative change on fire, but that he changed the location of it, so as to remove the fiery particles from their previous position in which they were mixed with the remainder of matter. This removal of fiery particles is actually the creation of the sun (and the stars) and the beginning of the cosmogonical process.

Although the sun is not explicitly mentioned in the surviving text of this passage, it is reasonably certain that the creation of the sun is explained in the first fragmentary sentence of col. 14.

> [?he made] to leap the brightest and hottest having separated it from himself. (col. 14.1–2)

The phrase ἐ]{κ}χθόρηι τὸ{ν} λαμπροτάτον τε [καὶ θε]ρμό[τ]ατον | χωριcθὲν ἀφ' ἑωυτοῦ shows obvious parallels with the end of the previous lemma ὃc αἰθέρα ἔχθορε πρῶτοc. It seems to me, however, that it is easier to come to grips with the sentence starting from χωριcθὲν ἀφ' ἑωυτοῦ. For this is manifestly a reference to the act of castration, 'the great deed' mentioned a few lines below. So the myth says that Kronos castrated Ouranos. But we have just seen that the castrated phallus is the sun; besides, Kronos and Ouranos are avatars of the same god; therefore the castration must mean that the god separated (from himself) the sun. Note also that the author will use the same verb in the next column, which I have just discussed, in connection with the sun: χωρ[ι]ζομένου γὰρ τοῦ ἡλίου (col. 15.3). The interpretation proposed is reinforced by the fact that τὸ{ν} λαμπροτάτον τε [καὶ θε]ρμό[τ]ατον must refer either directly to the sun or at least to fire, and, in view of the parallel, it must be the author's interpretative paraphrase of αἰθέρα. But what can he make of the verb ἔχθορε? My guess, reflected in the translation, is that he took it as a transitive verb with αἰθέρα as direct object. It is, by the way, quite understandable in view of the fact that the ἔχθορε + acc. construction has caused some trouble also for modern commentators. If so, the author takes the phrase ὃc αἰθέρα ἔχθορε πρῶτοc to mean that Mind made the matter of the sun (αἰθέρα = τὸ{ν} λαμπροτάτον τε [καὶ θε]ρμό[τ]ατον) separate from the rest by making it 'to leap out'. Moreover, the first part of the same lemma, αἰδοῖογ κατ̣έπινεν, could well refer to the act by which the Mind/air encircles or envelops the newly created sun. In this way, the whole lemma αἰδοῖογ κατ̣έπινεν, ὃϲ αἰθέρα ἔχθορε πρῶτοϲ can encapsulate the two moments of the creation of the sun – separation from the rest, and enveloping it in air – which is taken up by the phrase χωρ̣[ι]ζομένου γ̣ὰρ τοῦ ἡλίου καὶ ἀπολαμβανομένου | ἐμ μ̣έϲωι, 'as the sun got separated and encircled' in the next column.

THE POSITION OF THE SUN?

A standard rendering of the phrase χωρ[ι]ζομένου γ̣ὰρ τοῦ ἡλίου καὶ ἀπολαμβανομένου | ἐμ μ̣έϲωι in the passage just quoted from col. 15 (col. 15.3–4) offers us information about the place to which Mind removes the fiery particles, and, moreover, on the geometrical structure of the author's cosmos. Laks and Most, for example, translate the phrase as follows: 'For when the sun is separated and confined in the middle', while Janko translates it as 'For as the sun was being separated and isolated in the centre'.[31] The expression ἀπολαμβανομένου | ἐμ μ̣έϲωι requires particular consideration as it could be our most important clue to the reconstruction of the structure of the author's cosmos. Eventually, I shall argue that the phrase does not mean that the sun was placed in the middle, but that it was *encircled*. In other words, I shall maintain that the sentence does not speak about the cosmographical location of the sun, but, rather, stresses the fact that once the fiery particles are removed, the newly created aggregate of them (i.e. the sun) is encircled by the air, and in this way delimited, confined and bridled. But since this is a dissenting interpretation of the phrase, let us first see what the standard rendering would imply. We shall need to consider this point in the context of the history of planetary theories.

First and foremost, we have to note that if the sun is in the centre (leaving open for the time being the question of the centre of what, and in what sense it is a centre), this is clearly an arrangement based on functional considerations and has apparently nothing to do with astronomical observations and calculations. The sun has to be in the middle because, for teleological considerations, this is the most appropriate location for it. At this place

[31] Janko (2002). Cf. also 'For while the sun was removed and separated in the center': Rusten (1985) 137; 'En effet, lorsque le Soleil fut séparé et retenu au centre': Brisson (1996) 53; 'For when the Sun is separated and cut off in the middle': Janko (1997) 64; 'For as the sun was being separated and cut off in the middle': Janko (2001).

it is far enough, but not too far from the earth: it does not communicate too much heat to the earth and thus separate entities can form, but it still emits the heat needed for terrestrial life. Thus, the position ascribed to the sun fulfils both functional and also aesthetic requirements. By granting this central place to the sun, cosmic balance and symmetry are guaranteed.

However, the locution 'in the middle' is vague and can be understood in a number of different ways. In fact, the sun is allotted a central position, presumably on similar grounds, in quite different astronomical schemes. The sun acquires a central position, for example, in the well-known planetary system which became dominant in later Hellenistic times, and in astrology. In this arrangement the sun is the *middle* of the seven planets, flanked by three of them on each side (Venus, Mercury and the moon in the direction of the central earth, and Mars, Jupiter and Saturn towards the periphery). By this ordering, the sun could be said to be μέϲη with regard both to its position and to its function.[32] On this account, however, the sun's position relative to the planets is more important than its distance from the centre and from the periphery.[33]

Moreover, this scheme could easily integrate elements from the theory of proportions in such a way that the respective distances of the celestial bodies from the centre constitute a geometrical or harmonic series. One possible result is the mixed scheme reported by Plutarch (where the above mentioned distribution is contaminated with the Philolaic system of a central fire and counter-earth), in which the sun's orbit is 729 units from the centre, 720 units from the earth and 58,320 units from the sphere of the fixed stars, yet it could still be said to be 'in the middle' between the earth and the periphery.[34] This is surely not what the Derveni author had in mind.

Another, less well-known, manifestation of this pattern can be found in the Hippocratic *De victu*. In this text too the sun's orbit is in the middle, but the two terms in this case are the sphere of the fixed stars on the one hand, and the orbit of the moon on the other.[35] This conception could be all the more relevant to the papyrus in that the fire concentrated in the

[32] For the evidence for and interpretation of this arrangement, see e.g. Burkert (1961) 28–43.

[33] I do not want to enter here into the thorny issue of dating. Some ancient and modern authors attribute this system to Pythagoreans before Plato, and notably to Philolaus. Burkert (1972) 318–20, on the other hand, argued that it is a late development of which there is no trace before Archimedes.

[34] Plut. *De animae procr.* 1028b.

[35] *De victu* 1.10: ἐν δὲ τούτῳ ἐποιήϲατο <τὸ> πῦρ περιόδουϲ τριϲϲάϲ, περαινούϲαϲ πρὸϲ ἀλλήλαϲ καὶ ἔϲω καὶ ἔξω· αἱ μὲν πρὸϲ τὰ κοῖλα τῶν ὑγρῶν, ϲελήνηϲ δύναμιν, αἱ δὲ [πρὸϲ τὴν ἔξω περιφορήν,] πρὸϲ τὸν περιέχοντα πάγον, ἄϲτρων δύναμιν, αἱ δὲ μέϲαι καὶ ἔϲω καὶ ἔξω περαίνουϲαι <ἡλίου δύναμιν ἔχουϲιν>. Cf. 4.89: ἄϲτρων μὲν οὖν ἡ ἔξω περίοδοϲ, ἡλίου δὲ ἡ μέϲη, ϲελήνηϲ δὲ ἡ πρὸϲ τὰ κοῖλα.

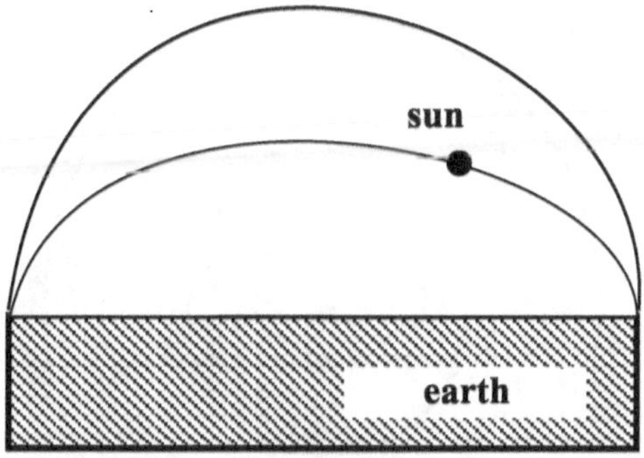

Figure 6.1.

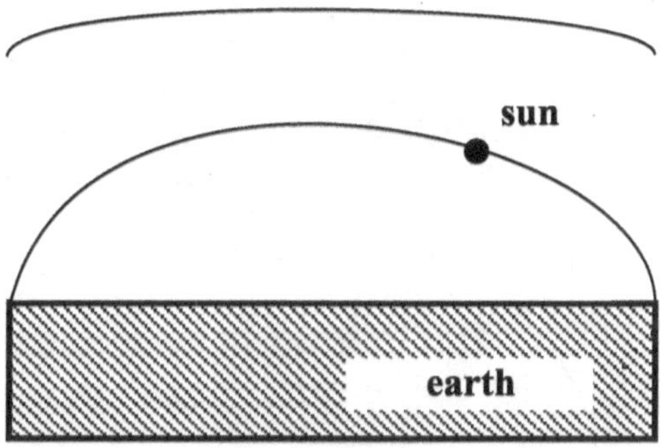

Figure 6.2.

sun is declared by the Hippocratic author to govern or dominate the other things, and, more important, to have a pre-eminent role in structuring the cosmos and, on the analogy of the cosmos, the human body.[36]

In the formulation of col. 15, as it is usually translated, the terms in respect of which the sun has been located are not stated. Moreover, the author will speak about the creation of the moon and the stars only at a later stage (in col. 25). Accordingly it seems probable that the meaning of 'middle' should be more self-evident than in the planetary systems mentioned above.

In a more primary sense, and ignoring for the time being external, historical considerations, the sun can be said to be 'in the middle' in the arrangements shown in figures 6.1–6.4.

[36] 1.10: ἐνὶ δὲ λόγῳ πάντα διεκοσμήσατο κατὰ τρόπον αὐτὸ ἑωυτῷ τὰ ἐν τῷ σώματι τὸ πῦρ, ἀπομίμησιν τοῦ ὅλου, μικρὰ πρὸς μεγάλα καὶ μεγάλα πρὸς μικρά and τὸ θερμότατον καὶ ἰσχυρότατον πῦρ, ὅπερ πάντων κρατεῖ (ἐπικρατεέται M) κτλ.

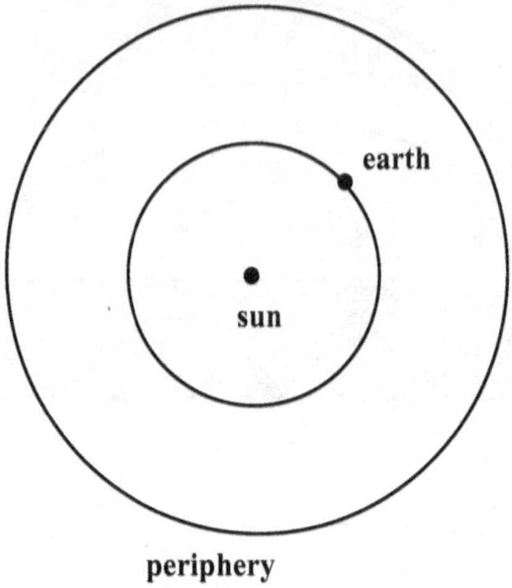

Figure 6.3.

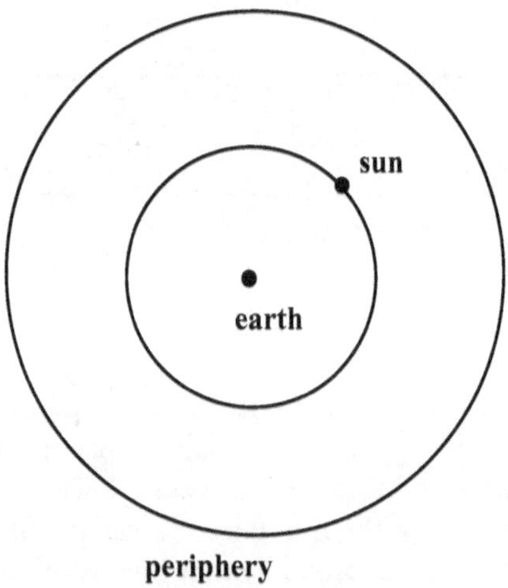

Figure 6.4.

Figures 6.1 and 6.2 are *exempli gratia* representations of the least accurate and cosmologically most primitive interpretations of the sun being 'in the middle'. The difference between the two is simply that the model in fig. 6.1 entails a spherical or hemispherical cosmos, whereas the one in fig. 6.2 represents a world built up by stratified more or less horizontal layers. On this reading the author would only mean that, approximately and on average, the sun is somewhere halfway between the 'top' of the firmament and the horizon, i.e., that we see the sun at noon roughly at 45°.[37] Obviously, it does not say anything about either the orbit of the sun between sunset and sunrise, about the difference of its position in winter and summer, or the shape and dimensions of the flat earth, or whether or not the earth is encircled by water. Although these views cannot be completely excluded, some considerations make them unlikely. Apart from the questionable observational support for such views, the above models with their low level of abstraction, naively earthbound perspective, and rudimentary geometrical imagery would be more likely in a sixth-century or even earlier context.[38] It seems quite improbable that the Derveni author, being well-versed in the cosmological theories of the last generations of the Presocratics and trying to be both up to date and original,[39] would espouse such a crude and antiquated world view. Furthermore, the author, as we shall shortly see, employs geometrical notions and vocabulary with a high level of consciousness,[40] which is completly lacking in these models.

On internal grounds, the model depicted in fig. 6.3 is much more attractive. To begin with, once we have granted the author a spherical cosmos – which is a far better bet than the hemispherical or the flat and layered ones – this appears the most natural reading of the text. All the more so, as in col. 24.1 μέϲον means the geometrical centre of a circle. On this reading, we would have a spherical universe, the sun being in the middle of the sphere, and the earth revolving around it at some unspecified distance. The great advantage of this representation would be that it can best account for the teleological aspect of the sun's position: it is from this central position that the sun can help the Mind most effectively to coagulate and hold fast everything in the cosmos, and conversely, the air completely encircling the sun could make sure that it remains at the place determined for it. Also, as the stability of the whole cosmic order depends on the stability of the sun, it is perhaps more assuring to have a fixed sun than one which is constantly on the move.[41] Furthermore, it has to be noted that most of the other early cosmogonies commence with some kind of change in the centre, and then the dynamics of this change prompt the whole process. True, this could hold true for the Derveni author's cosmogony, although with less force, also if the fire was *taken away* from the centre.

[37] This model was suggested to me by Reviel Netz.

[38] Perhaps Hesiod and Anaximenes would be the closest parallels.

[39] The next chapter should substantiate this claim.

[40] Cf. col. 24.1–2 with comments on 248 below.

[41] Note that the verb ἀπολαμβάνω also has the strong meaning of 'to arrest' and 'to stop'. Cf. LSJ *s.v.* IV.

It also has to be noted that the crucial cosmogonical and cosmological position ascribed by the author to fire and, more specifically, to the sun could well account for a central position also in the geometrical structuring of the cosmos. As a matter of fact, an often quoted passage of Aristotle testifies that there were people, Pythagoreans or others, who had ventured to place the fire in the middle of the cosmos, precisely for these a priori reasons:

There are many others who might agree that it is not right to allocate the central position to the earth – people who look for proof to theoretical argument rather than to phenomena. For they believe that what is most worthy of honour deserves the most honourable place; and as fire is more honourable than earth, a bounding position more honourable than an intermediary one, and the outermost limit and the centre are both boundaries, they arrive by way of conclusion at the opinion that fire, and not the earth, lies at the centre of the spherical universe.[42]

To be sure, Aristotle is apparently speaking about a central fire and not specifically about a central sun. But it shows that there were attempts to build spherical and 'pyrocentric' models.

It is noteworthy that this is the model adopted for the Derveni author's cosmology by Rusten who in his 'Interim Notes' is giving a close reading of certain passages of the text, without considering the historical context.[43] Yet the historical context constitutes a major problem for this interpretation. It would be more than surprising to find our author coming up with a heliocentric cosmology about a century before Aristarchus of Samos.[44]

Therefore, even though I have tried to make the best case for the cosmological model in fig. 6.3, the geocentric model in fig. 6.4 remains more likely for historical reasons. We have again a spherical cosmos, and the earth, in accordance with the mainstream cosmologies, is in the centre, while the sun revolves halfway between the centre and the periphery.[45] The symmetry of this picture is provided by the fact that, all through its path, the sun keeps an equal distance from the central earth and the periphery.

A further consequence of this reading would be that when the author says 'above and below the sun', he would mean between the sun and the

[42] Arist. *De caelo* 293a27–b1.

[43] Cf. Rusten (1985) 130 n. 18: 'This is why the sun had to be kept separate in the center of the universe when the creator combined the rest of the elements above and below it . . .'

[44] On the other hand, even if one were able to prove that the Derveni author did place the sun in the centre, this would still not make him a more ancient Copernicus than Aristarchus, precisely for the reason given by Aristotle: the considerations behind this bold move are purely a priori and not in any way based on mathematical, astronomical arguments.

[45] The Hippocratic *De hebdomadibus* explicitly attributes this position to the moon.

firmament and between the sun and the earth respectively. Hence the language used by the author would suggest that 'up' and 'down' stand here in the more sophisticated geometrical meaning, i.e., 'towards the periphery' versus 'towards the centre'; a usage which becomes explicit in the *Timaeus*, but is perhaps already there in Parmenides in an implicit form.[46]

Obviously, even if the author had had this concept of up and down, it would still not follow that he had a genuine centrifocal dynamics. Indeed, he would certainly resist the idea that a spherical earth in the centre of the finite cosmos will not fall simply on account of 'the likeness of the heaven itself to itself everywhere',[47] as this consideration would decrease the mighty tasks of the divine air. It is more fitting to have the air encircle the earth and maintain it in its place. But this is all quite speculative, as, even on this reading, the author might well have used the phrase 'above and below the sun' more carelessly, with an absolute notion of up and down in mind, ignoring the fact that, strictly speaking, this language conveys the original idea only at noontime, when an absolute 'above' and an absolute 'below' happen to correspond with 'between the sun and the periphery' and 'between the sun and the earth' respectively.

Nonetheless, it is worth while recalling that no matter which of the four models we eventually opt for, any of these constructions would vindicate a relatively high degree of originality for the author. For, as far as I can see, none of the other known early cosmologists places the sun, or the sun's orbit, explicitly 'in the middle', be this 'middle' between earth and sky, between the centre and the periphery, or in the geometrical centre of a spherical cosmos.

Furthermore, regardless of our eventual choice for any of the above models, the expression 'in the middle' only makes sense if we think in terms of a finite cosmos. If we take the author to say that the god placed the sun in the middle in the pre-cosmic situation, then, barring that 'middle' is used here in the sense of 'what was to become the middle',[48] we have to suppose that the matter out of which the cosmos was to be formed was

[46] *Tim.* 62c3–63e8. For a modified version, see Arist. *De caelo* 268b20–2. On the centrifocal world picture implicitly present in Parmenides, see Furley (1989) 24–6. The Hippocratic *De hebdomadibus* also has this concept of up and down (cf. ch. 2), but, as Mansfeld (1971) argues, the treatise is of later date. Incidentally, if we accept West's supplement on the basis of ps.-Galen, the Hippocratic author uses the same construction as the Derveni author but with reference to the moon: ἡ δὲ cελήνη μέcη οὖcα <πάντων τῶν ἄλλων μερέων, τῶν μὲν ἀνωτέρων, τῶν δὲ κατωτέρων,> cυναρμόζει αὐτὴ τἄλλα πάντα κτλ.

[47] Plato *Phd.* 109a2–4.

[48] Anaxagoras' Mind started the rotation at an unspecified point of the infinite pre-cosmic mass, but this point was to become the middle of the spherical cosmos.

itself limited. True, it would still be possible for this finite matter to be encircled by unlimited void, as we have it in Stoic cosmology.

Yet all the above speculations on the cosmological position of the sun and the geometrical structure of the author's cosmos can become futile if we opt for a different interpretation of the phrase ἀπολαμβανομένου | ἐμ μέϲωι, and I shall argue now that this is the right solution. For, even though this meaning is not listed in LSJ or the other standard dictionaries, the results of a *TLG* search show that ἀπολαμβάνω ἐν μέϲῳ was a set phrase with the fixed meaning 'to encircle' or 'to enclose', especially in the military sense.[49] This latter is clearly the meaning of the phrase e.g. in Polybius[50] and in Flavius Arrianus.[51] In a more general sense, the phrase can express that something is or gets surrounded by something else.[52] The only passage where the expression could mean that something is placed in the (geometrical) centre is in Diodorus Siculus, where a baggage train is said to be arranged in the centre of the troops. But here too, the accent is on the fact that the baggage train thus gets surrounded and in this way protected by the troops.[53] Another possible but closely connected meaning in the passive is 'to be caught between two opposing forces', as e.g. in Flavius

[49] A full-corpus search showed the following passages: Flav. Arr. *Alex. anab.* 4.27.4; Flav. Joseph. *Hist. antiq. Jud.* 15.212.1; Polyb. *Hist.* 1.76. 2; Diod. Sic. *Hist.* 2.13.3 and 19.39.4; Ioan. Chrys. *De virg.* 34, *In Genes.* vol. 53. p. 196, and *Ad pop. Antioch.* vol. 49. p. 65; Basil. *Epist.* 242.1; Paus. 3.5.5.8; ; Plotinus 3.6.8.16; Polyaenus *Stratagemata* 8.23.7.13; Theodoret. *Int. in Psalmos* 80.1085.1

[50] Polyb. *Hist.* 1.76.2: οἱ δὲ περὶ τὸν Ϲπένδιον ϲυνέντεϲ τὸ γεγονὸϲ ἀπήντων εἰϲ τὸ πεδίον καὶ παρεβοήθουν ἀλλήλοιϲ, οἱ μὲν ἐκ τῆϲ περὶ τὴν γέφυραν πόλεωϲ ὄντεϲ οὐκ ἐλάττουϲ μυρίων, οἱ δ᾽ἀπὸ τῆϲ Ἰτύκηϲ ὑπὲρ τοὺϲ μυρίουϲ καὶ πεντακιϲχιλίουϲ. ἐπεὶ δ᾽εἰϲ ϲύνοπτον ἧκον ἀλλήλοιϲ, *νομίϲαντεϲ ἐν μέϲῳ τοὺϲ Καρχηδονίουϲ ἀπειληφέναι*, ϲπουδῇ παρηγγύων ἅμα παρακαλοῦντεϲ ϲφᾶϲ αὐτοὺϲ καὶ ϲυνῆπτον τοῖϲ πολεμίοιϲ.

[51] Flav. Ar. *Alex. anab.* 4.27. 4: καὶ ταῦτα ὡϲ ἐξηγγέλθη Ἀλεξάνδρῳ, περιϲτήϲαϲ τῆϲ νυκτὸϲ τῷ γηλόφῳ τὴν ϲτρατιὰν πᾶϲαν κατακόπτει *τοὺϲ Ἰνδοὺϲ ἐν μέϲῳ ἀπολαβών*, τήν τε πόλιν αἱρεῖ κατὰ κράτοϲ ἐρημωθεῖϲαν τῶν προμαχομένων, καὶ τὴν μητέρα τὴν Ἀϲϲακάνου καὶ τὴν παῖδα ἔλαβεν.

[52] E.g. Basil. *Epist.* 242.1. where we are said to be surrounded by a sea of toils (εἰ καὶ ἐν μέϲῳ πελάγει κακῶν ἀπελήφθημεν).

[53] Diod. Sic. *Hist.* 19.39.4: ἐπιφανέντων δὲ πρότερον τῶν Ἀντιγόνου ϲτρατιωτῶν οἱ τῶν ἐλεφάντων ἡγεμόνεϲ τάξαντεϲ εἰϲ πλινθίον τὰ θηρία προῆγον, *ἐν μέϲῳ μὲν ἀπειληφότεϲ* τὰ ϲκευοφόρα, κατὰ δὲ τὴν οὐραγίαν ἔχοντεϲ ϲυναγωνιζομένουϲ ἱππεῖϲ οὐ πλείουϲ τετρακοϲίων. In Diod. Sic. *Hist.* 2.13.3 the phrase appears in the following context: ἐντεῦθεν δ᾽ ἀναζεύξαϲα καὶ παραγενομένη πρὸϲ Χαύονα πόλιν τῆϲ Μηδίαϲ κατενόηϲεν ἔν τινι μετεώρῳ πεδίῳ πέτραν τῷ τε ὕψει καὶ τῷ μεγέθει καταπληκτικήν. ἐνταῦθ᾽ οὖν ἕτερον παράδειϲον ὑπερμεγέθη κατεϲκεύαϲεν, *ἐν μέϲῳ τὴν πέτραν ἀπολαβοῦϲα*, καθ᾽ ἣν οἰκοδομήματα πολυτελῆ πρὸϲ τρυφὴν ἐποίηϲεν κτλ. Here the park is built around the rock, and thus the rock got encircled by it. I do not think that the Loeb translator's 'putting the rock in the middle of it (sc. the park)' is correct, for I do not think that first they created the park and then placed that immense rock in the centre of it, but, the other way around, they built the park around the rock.

Josephus and Pausanias; but this usage has nothing to do either with the geometrical sense of the centre.[54]

On the strength of these passages, I infer that by the phrase χωρ[ι]ζομένου γὰρ τοῦ ἡλίου καὶ ἀπολαμβανομένου | ἐμ μέσωι the author simply means that the Mind/air first separated the fiery particles and then encircled or enveloped them, so that in this way, enveloped in the air, this cluster of fire, i.e. the sun, gets bridled and confined to a certain position. This reading is not incompatible with the geometrical interpretation developed above, but certainly does not compel it. That is, ἐμ μέσωι could be intended as mathematically precise, but need not be.[55] Moreover, we should not ignore that all the other examples are relatively late, so it is possible that the phrase had a more precise meaning at the time when the Derveni author was composing his text.

THE STARS

Even though the creation of the sun is certainly the crucial element in Mind's manoeuvres to dispose of fire, and consequently in setting the general *diakosmesis* in action, we learn somewhat later, almost at the end of the remaining part of the papyrus, that Mind's primary cosmogonical tasks were not completed with the creation of the sun:

> And there are other things now in the air, floating far away from each other. But during the day they are invisible being dominated by the sun, whereas during the night it is visible that they exist. They are dominated because of their smallness. Each of them floats in necessity, so that they do not come together with one another; for otherwise all those which have the same characteristics[56] as those out of which the sun was set together would come together in one mass. If the god had not wished that the things which are now should exist, he would not have made the sun. But he made it of such a sort and of such a size as is explained in the beginning of the account. (col. 25.3–11)

So the stars are of the same kind of matter as that out of which the sun is composed. Furthermore, as the last sentence makes clear, the sun's size is of primary importance for the maintenance of the present world order designed by the god. Yet there is more of the fiery matter than is needed for

[54] Joseph. *Hist. antiq. Jud.*; 15.212.1 οὕτως οὖν *ἐν μέσῳ τοῦ στυγεῖν καὶ στέργειν ἀποληφθείς*; Paus. 3.5.5.8: ἅτε γὰρ εὖ εἰδὼς ὁ Παυσανίας ὡς τὰ σφάλματα ἀεὶ Λακεδαιμονίοις γίνονται *ἐν μέσῳ πολεμίων ἀποληφθεῖσι* κτλ.

[55] As if one were to say 'When visiting Bosnia, he was caught in the middle of a war', where normally one would not mean 'exactly at the half-way point of the war', but in the right context one might.

[56] Probably, but not necessarily, the word δύναμις is to be taken in the more terminological sense, referring to the basic causally relevant physical characteristics, such as heat, light etc.

the composition of the sun; the surplus has to be kept under control. This is why the superfluous fiery matter is dispersed in the air in small pieces, and kept there by force.[57]

Incidentally, this passage about the stars offers further support for the interpretation I have suggested for the meaning of ἐξαλλάττω in col. 9. For had Mind been able to change the stellar particles qualitatively, converting them into something which does not jeopardise the cosmic order, this procedure would have been a much more convenient means of dealing with this unwanted surplus of hot matter. Had this option been open to Mind, he surely would not have employed the much less definitive solution of keeping the stars in confinement, apart from each other, by a continuous effort. But the solution he did opt for is basically the same as in the case of the sun on my interpretation of col. 9: he regulated the fiery particles by means of a deliberate spatial distribution of them.

The author does not tell us anything about the exact location of the stars. The only clues we have are the claim that the stars are 'floating in the air', and that in view of the force of 'like to like', it is a matter of importance that the stars are 'floating *far away* from each other'. From this it follows that – if the author's cosmos was finite – it is better if they are closer to the periphery, because in that way they can be further both from each other and from the sun.[58]

THE MOON

Prima facie, it seems reasonable to assume that, as in most other cosmologies, the moon was created in the same phase of cosmogony together with the sun and the stars. Yet the creation of the moon, in the case of the Derveni author's cosmogony, cannot have been prompted by the same reasons as motivated the creation of the other celestial bodies. In col. 25, immediately before the passage just quoted about the stars and their position, the Derveni author says

> . . . and brightness. Those things out of which the moon is (composed) are the whitest of all, divided according to the same principle (*or*: measure), but they are not hot. (col. 25.1–3)

As the next sentence will speak about the stars, and the particles of the moon are opposed to something which is hot, it is reasonable to suppose with

[57] On the nature of this force see 271–2 below.

[58] Burkert (1997) advanced the hypothesis that the author is speaking specifically about the Milky Way. As this suggestion is based on an alleged parallel with Anaxagoras, I shall consider it later in 301–2 below.

Burkert that the previous sentence, of which only the last two words are preserved here, was speaking about the solar matter, describing the physical characteristics of it.[59] At any rate, we learn now that the particles out of which the moon was composed are not hot. This entails that it is not the removal of the hot fiery particles which necessitated the formation of the moon.

The phrase 'divided according to the same principle' makes things even more complicated, as it apparently shows some kind of correspondence between the sun and the moon. μεμεριcμένα can be read in two different ways here. One possibility is that μεμεριcμένα refers to the lunar particles individually. This usage is paralleled in col. 21 where we hear about particles moving in the air κατὰ μικρὰ μεμεριcμένα. Thus, κατὰ τὸν αὐτὸν λόγον μεμεριcμένα in col. 25 would probably mean that although the lunar particles have different physical characteristics than the solar particles, their sizes are equal.

Another possibility is that the expression means that the lunar particles taken collectively are divided from the rest of matter 'according to the same principle'.[60] As the other term of the 'same' most probably refers to the sun, the phrase would mean that there is an underlying common principle governing the creation of both sun and moon. But what would this principle be? As the primary cosmogonical rationale effective in the creation of the sun is excluded, I would conjecture that the author refers to the fact, well known in antiquity, that the apparent sizes of the sun and the moon are equal.[61]

On either reading, it is still not clear whether or not there was a specific cosmogonical reason for the creation of the moon. It may be that the moon is not treated on the same footing with the sun and the stars because the physical characteristics of lunar matter are cosmologically less important. It is possible that the formation of the moon actually belongs to the next cosmogonical phase, during which other varieties of matter form into independent entities as a result of the disposal of fire.[62] It would mean that,

[59] Burkert (1997) 168 suggests τὰ μὲν οὖν ἐξ ὧν ὁ ἥλιοc cυνεcτάθη, ὑπερβάλλοντά ἐcτιν θερμότητα] for the missing first part of the sentence. So also Janko, with a different wording for the end of the supplement.

[60] The translation in Burkert (1997) 168 'separated according to the same principle' suggests this option. However, on the next page he speaks about the corpuscularism of the author in connection with this verb.

[61] On this reading λόγοc can just as well mean 'measure' (cf. Heraclitus B31 DK . . . εἰc τὸν αὐτὸν λόγον . . .).

[62] The fact that the poem speaks about the creation of the moon at the end of the 'second creation' by Zeus (cf. ***L15*** = col. 24.3) can give further support to this interpretation. It occurs thus considerably later than the episode of swallowing, which the Derveni author takes to speak about the creation of the sun.

in contradistinction to the other celestial bodies, but similarly to the other beings, the moon came into being not by a direct intervention of the cosmic god, but as an end-product of the mechanical chain-reaction triggered off by the creation of the sun and the stars.

There is a further cosmologically relevant question to be discussed with regard to the moon. Where does her light come from?[63] Does she have her own light, or does she borrow it from the sun? The answer must reside in the physical characteristics of the constitutive matter of the moon. Now these are described as [λ]ε̣υκότατα (...) τῶν ἄλλωγ, but not hot.[64] As the Derveni author uses λαμπρός and its cognates in describing the brightness of the sun,[65] we can restate the question in terms of the relationship between λευκός and λαμπρός.

λευκός, together with the majority of Greek terms for colours, is ambiguous. Broadly speaking, it can mean (i) 'light', 'bright' and 'brilliant', (ii) 'white' but again ranging from snow-white to grey, and (iii) 'pale' or 'wan'.[66] (i) is unlikely as it would imply that the moon is the brightest of all, that is, brighter even than the sun.[67] On the face of it, 'pale' or 'wan' would be possible with regard to the light of the moon, especially in a comparison with the brightness of the sun. It would simply mean that although the moon also emits light, this light is less brilliant than that of the sun.[68] However, the superlative closes this option: [λ]ε̣υκότατα μ̣ὲν τῶν ἄλλωγ, preceded by κ̣αὶ λαμπρό[τ]η̣τα, would be quite an unnatural way to express the claim that the moon is less brilliant than the sun.

Hence (ii) seems the best bet. But, as opposed to (i) and (iii), it does not say anything specific about the luminosity of the moon.[69] It only states, more or less in agreement with our visual experience, that the moon appears the whitest among the heavenly bodies, without pointing out whether it is reflected light, or emitted by the moon herself.[70]

[63] On ancient theories of lunar light, see the recent studies by Panchenko (2002) and Graham (2002).

[64] Col. 25.1–2. [65] Col. 14.1; 25.1. [66] Cf. LSJ *s.v.*

[67] Empedocles in B 21.3 DK uses λευκός in this sense, but applies the adjective to the sun: ἠέλιον μὲν λευκὸν ὁρᾶν καὶ θερμὸν ἁπάντηι.

[68] Cf. e.g. Heracl. A12 DK: λαμπρότερον μὲν τὸν ἥλιον· . . . τὴν δὲ σελήνην . . . ἀμαυροτέραν φαίνεσθαι.

[69] This is also the way Janko translates the word, but he emends the superlative into a comparative.

[70] Burkert (1997) 169, on different grounds, arrives at a similarly unresolved conclusion: 'It seems strange that the author appears to have forgotten Anaxagoras' brilliant insight that the Moon gets her light from the Sun: ἥλιος σελήνῃ ἐντίθησι τὸ λαμπρόν (B18 DK); but he might have thought or said that even in order to function as a mirror the Moon has to be 'white' and not dark.' To be sure, the whiteness of the moon was not a commonplace in antiquity.

We can nonetheless expect some help from the preceding column, which focuses on the moon, and more specifically on moonlight. The text, however, is dauntingly obscure:

. . . are equal measured from the centre, but those which are not round-shaped cannot be of equal limbs. This (verse) makes it clear:

which shines for many articulate-speaking humans on the boundless earth

Someone might think that this verse is said wrongly, namely that when she is at her utmost, the things that are show up more than before she is at her utmost. But he does not mean this (by saying that) she shows, for if he had meant this, he would not have said that she shows for many, but that for all at the same time, both for those who work the land, and for those who sail when they have to sail, and for the former the seasons. For if there were no moon, people could not have discovered the counting either of the seasons, or of the winds . . . and all the others . . . (col. 24.1–12)

Presumably, the first sentence concerned with geometry is connected to the rest through the hapax ἰcομελῆ, which was, in all probability, an adjective applied to the moon in the poem. It is also clear that the author tries to infer from ἰcομελῆ that the moon is circular. Barring the possibility that the expression 'This (verse) makes it clear' is no more than a formulaic phrase to lead up to the next, independent bit of the argument, implying no logical connection whatsoever,[71] the subsequent lemma, together with the author's gloss on it, should underpin the point on circularity. It is, however, far from obvious in what way the next passage could give any support for the argument about circularity.

Commentators take it that the author's claim about the circularity of the moon is trivial, assuming that ἰcομελήc, 'equal-limbed' must have meant 'circular' already in the poem.[72] It should be noted, however, that this hapax adjective is a curious coinage to express circularity. As a matter of fact, the compounds with -μελήc, such as ἀρτιμελήc, αἰγομελήc, λυcιμελήc, μικρομελήc, πηρομελήc, ὑγρομελήc etc., in accordance with the standard meaning of μέλοc, tell something about the limbs, that is the arms, legs and other projecting members of a being.[73] In this context, ἰcομελήc is all the more surprising as a circular or spherical being would be more aptly described by saying that it has no limbs at all.[74] As far as I can see, the only

[71] δηλοῖ in coll. 8.3; 16.12; 21.1; 26.2; 26.5 does indicate a logical connection; in col. 13.3, however, the logical connection is not entirely clear.

[72] Cf. e.g., Bernabé (forthcoming) ad fr. 15 B. *Crit.*: 'μεccόθεν ἰcομελὴc πάντηι (i.e. sphaerica)', clearly construed on the basis of Parm. B8.44 DK μεccόθεν ἰcοπαλὲc πάντηι.

[73] Cf. LSJ. *s.vv.*

[74] Cf. Plato *Tim.* 33d2–34a7.

other instance of a -μελήc word used with reference to a rounded entity is the οὐλομελήc in Parmenides B8.4, which, one should remember, is a highly vexed locus.[75]

Besides, had the obvious literal meaning of ἰcομελήc been 'circular', the author would not have employed a geometrical argument in order to explain it. For, indeed, we are not accustomed to seeing him make such trivial, and – without playing on words – circular arguments.[76] As far as the adjective goes, I cannot see why it could not just as well refer to the equal horns of the new moon.

On the other hand, the Derveni author is apparently grafting in technical geometrical vocabulary. According to the *TLG*, κυκλοειδήc is a late word, the first attested occurrence of which is in Euclid's *Optics*.[77]

The gloss after the lemma is also problematic, as the syntax is even more garbled than usual. Furthermore, contrary to the Derveni author's general practice, the imaginary opponent's interpretation does not seem to correspond to the literal meaning of the lemma in this case.

ὑπερβάληι causes some extra difficulty. The verb usually suggests that something exceeds or transgresses its measures.[78] It is used in this sense in the Heraclitus quotation in col. 4, with reference to the sun's natural size, not outgrowing its measure. For such a transgression would lead to a thermic cataclysm. In our passage ὑπερβάλλω is not in the negative, yet there is no suggestion of a cosmic catastrophe. As far as I can see, the closest, somewhat isolated parallel is in Euripides, *Hel.* 1367, ὑπέρβαλε cελάνα, which however is part of a *locus desperatus*, daggered by most editors. The reference might be to the full moon.[79] I tentatively suggest then that the line of reasoning behind the interpretation discarded by the author is the following. The author first establishes that Orpheus speaks about the full moon by the argument about ἰcομελήc, either in agreement or contrary to the literal meaning of the poem. Then he shifts the attention to the word πολλοῖc taken also from the lemma, taking it in a comparative sense, to

[75] It is true, on the other hand, that *sphairos* has 'limbs' in Emp. B30 DK: αὐτὰρ ἐπεὶ μέγα νεῖκοc ἐνὶ μελέεccιν ἐθρέφθη κτλ.

[76] Cf. the *exempli gratia* supplement (without commitment regarding the meaning of ἰcομελήc) by Burkert (1968) 95: '["gleichgliedrig" heißt das runde, weil es] von der Mitte aus gemessen ist; was nicht rund ist, kann nicht "gleichgliedrig" sein.' This is followed now also by Janko.

[77] Eucl. *Opt.* 36.

[78] The word was previously taken in this sense by Janko. His new translation has 'waxes full'.

[79] Cf. also Herod. 4.184.2: οὗτοι τῷ ἡλίῳ ὑπερβάλλοντι καταρῶνται καὶ πρὸc τούτοιcι πάντα τὰ αἰcχρὰ λοιδορέονται, ὅτι cφέαc καίων ἐπιτρίβει αὐτούc τε τοὺc ἀνθρώπουc καὶ τὴν χώρην αὐτῶν. This passage is of little help because it does not specify the meteorological or astronomical condition inciting the Atarantes to verbally abuse the sun. The cursing could happen every noontime, or just on some exceedingly hot summer days.

conclude that the full moon emits more light, and thus makes things more visible to people than other phases of the moon.

But, anyway, this is just a hypothetical and incorrect interpretation for the Derveni author. For his part, he urges that φαίνειν is used here in the transitive sense 'to show', and not in the intransitive 'to shine'. For if Orpheus wanted to say that the moon shines, he would have said that it shines for all the people.[80] But Orpheus said that it shows or gives indication for many, that is, not for everyone but for those for whom it is of special importance. The moon indicates, for instance, to agricultural workers and sailors the appropriate time for their activities.[81]

The author's insistence on the point that the moon does not shine, but 'shows', strengthens the hypothesis – without of course proving it conclusively – that the moon does not have light of her own in this cosmological theory.

TIME AND OLYMPUS

Before we turn to the advent of independent entities, I have to discuss what I consider, in a certain sense, the most perplexing argument in the whole papyrus. I treat it at this point because if my admittedly highly conjectural interpretation is valid, the passage logically belongs in this part of the story. In the remaining part of col. 12 we read the following:

> And the next line goes like this:
>
> *so that he may rule on the lovely abode of snowcapped Olympus*
>
> Olympus and time are the same. Those who think that Olympus and the heaven are the same are entirely mistaken, for they do not know that the heaven cannot be longer rather than wider; but if someone were to call time long, he would not be wrong at all. And whenever he (sc. Orpheus) wanted to speak about heaven, he added the epithet 'wide', whereas whenever (he wanted to talk) about Olympus, on the contrary, he never (added the epithet) 'wide', but 'long'. By saying that it is 'snow-capped', the power . . . snowy . . . snowy . . . white . . . bright . . . grey . . . and . . . (col. 12.1–15)

[80] The author might have in mind passages like the one in Hesiod's *Theogony* 371–3: Θεία δ' Ἠέλιόν τε μέγαν λαμπράντε Cελήνην | Ἠῶ θ' ἣ πάντεccιν ἐπιχθονίοιcιν φαείνει | ἀθανάτοιc τε θεοῖcι κτλ.

[81] The construction of the text is difficult. In line 8 the papyrus has ἅμα. It was corrected by the *ZPE* editor to ἀλλά, probably on the assumption that this must be the counterpart of the οὐ τοῦτο in line 6 (with another οὐκ–ἀλλά pair in the parenthetical phrase in between). If we retain ἅμα, it can be construed either with πᾶcιν or with the subsequent ṭοῖc τε (. . .) καὶ ṭοῖc̣. Moreover, without ἀλλά, the reference to the peasants and sailors might seem to be not an antithesis but an explanation of πᾶcιν. In this case the argument is even less intelligible.

Luc Brisson in his contribution to the L&M volume focuses exclusively on this column, and gives a highly instructive close reading of the text. He discusses the philological problems of the passage, and also explains the logic underlying the argument.[82] So let me just briefly indicate the major points here.

The author assumes that if φ can be predicated of *a* but not of *b*, then *a* and *b* cannot be identical.[83] On the other hand, if φ can be predicated of both *a* and *c*, *a* and *c* can be identical.[84] On the basis of this reasoning, and contrary to the contention of some unnamed rivals, Olympus and heaven (or Ouranos) cannot be identical.[85] But, the author urges, Olympus is identical with time, and indeed both can be qualified μακρόc, 'long' or 'high'. Again the author comes up with an eccentric option. Olympus and heaven are sometimes conflated already in the epic;[86] but for that matter we can also recall the curious statement in Parmenides B11 about the celestial layer of the 'outermost Olympus'.[87] As to the relationship between the sky and time, Aristotle informs us that there were people who identified time with the celestial sphere.[88] But who else would identify Olympus with time?

The major problem consists in determining what the author means by 'time' (χρόνοc) here. Brisson argues convincingly that the χρόνοc spoken of here is not the primordial god Chronos who some centuries later will become a primordial divinity in the *Rhapsodies*. The question Brisson does not ask is why the author wants to convince us that Olympus and time are the same. What is the role of this identification in the system of the Derveni author?

[82] Brisson (1997), especially 150–9. [83] Cf. Arist. *Top.* 7.1.

[84] This is a point where I do not follow Brisson. On p. 155 he says that 'The reason the commentator gives here to prove that Olympus is identical to heaven [this must be a slip for 'time'] (both are qualified by the same epithet) is the same given by the allegorist Heraclitus a few centuries later to prove that, even in Homer, Apollo and the sun are identical'. I do not think the language of the column entitles us to say that this part of the argument is actually to *prove* the identity of the two entities. It only says that the objection just raised against the identification of Olympus and heaven is not effective here. Janko, on the other hand, supplements 'time' in line 9 instead of Olympus. But, as far as I can see, the argument from the same predicate cannot work in this way: we need Orpheus to qualify Olympus 'long' if we want to establish that he meant 'time' by saying Olympus.

[85] Funghi (1983) notes a striking parallel from *Schol. Il.* 1.402.: ὅτι μακρὸν τὸν Ὄλυμπον ὡc ὄροc· τὸν δὲ οὐρανὸν εὐρύν.

[86] *Od.* 6.41–7. For further occurrences see Funghi (1983) 11 n. 1 and Brisson (1997) 154.

[87] Parm. B11.2–3 DK. Cf. also Empedocles B44 DK; and Philolaus A16 DK (judged spurious by Burkert (1972) 243–6 followed by Huffman (1993) 396–400 with further bibliography); Hipp. *De hebdom.* 2.10. See also [Plato] *Epin.* 977b1–3: ἐὰν γὰρ ἴῃ τιc ἐπὶ θεωρίαν ὀρθὴν τὴν τοῦδε, εἴτε κόcμον εἴτε ὅλυμπον εἴτε οὐρανὸν ἐν ἡδονῇ τῳ λέγειν κτλ.

[88] Arist. *Phys.* 218a32–b1.

Let us first see its context in the poem. In all probability, the lemma discussed here refers to the oracular advice Night gave to Zeus: what Zeus should do in order to establish his supremacy over the other gods. This is what the poet expresses by the phrase 'that he may rule on the lovely abode of snowcapped Olympus'. We have also seen that the content of the oracle was that Zeus has to swallow the phallus of Ouranos (or Phanes), and this was explained by the commentator as the creation of the sun.

I would tentatively suggest that the Derveni author connects the creation of the sun – closely connected also to the creation of the other celestial bodies – to the question of time. I have the following pieces of evidence for this claim (none of them, admittedly, conclusive). The most important clue is that however desperately damaged the last lines of the column are, we can still see that, after having established the identity of Olympus and time, the author immediately starts speaking about Olympus' epithet 'snowcapped'. It is also clear that he conducts the discussion towards the physical characteristics implied in the adjective. In a first move, he shifts from νιφόεις to νιφετώδης. The former is epic and often used with respect to mountains, especially Olympus, whereas the other is rare, technical, and apparently late. A *TLG* full corpus search shows only 13 occurrences. The first chronologically is in Aristotle's *Meteor.* 364b21; and before the beginning of the Christian era it is used only by Theophrastus, Posidonius, Theopompus, Strabo and Polybius, once by each of these authors.[89] One really wonders where the Derveni author could have taken this word from.

At any rate, νιφετώδης also expresses that the entity in question is 'like snow'. This can help the author to show that Orpheus is just using a simile.[90] Also, it paves an easy way towards focusing on the visual, and, conspicuously, not the thermal characteristics of snow. The remaining words of the column are all in this register: λ̣αμπ̣[ρός, λευκός and πολιός (which means grey or grizzly, mostly used for the colour of hair, but it can simply be a synonym of λευκός). It is important to note at this point that every other time the author uses any of these adjectives, he is describing the heavenly bodies and their constitutive matter.[91] It seems at least conceivable then that here too he is leading the discussion in this direction: he wants to connect time with the sun and the other heavenly bodies.

The other piece of indirect evidence is from col. 24, where the author, as we have just seen, underlines the calendrical importance of the moon:

[89] Polybius, *Hist.* 3.72.3; Strabo, *Geogr.* 4.5.2; Posid. fr. 338b; Theophr. fr. 6; Theopomp. fr. 130.
[90] Tsantsanoglou *apud* Brisson (1997) suggests the supplement ε[ἰκάζει in line 11.
[91] Cf. coll. 14.1; 25.1

For if there were no moon, people could not have discovered the counting either of the seasons, or of the winds . . . (col. 24.10–12)

Yet even if we conclude that in col. 12 the author speaks about time in connection with the cosmic god's creating the heavenly bodies, any further interpretative steps would certainly be far too speculative. The most obvious parallel would certainly be the *Timaeus*: the heavenly bodies are the 'instruments of time'.

THE FORMATION OF INDEPENDENT ENTITIES

In view of the description of the initial state, one would expect the emergence of independent entities out of the original mixture to be described in terms of cooling and cessation of movement. This is not entirely so. We find the only passage which apparently speaks about this aspect of the process in the last, badly damaged lines of col. 10 – thus closely following on col. 9 which, as we have seen, gives the outlines of the cosmogony. In the bottom part of col. 10 the author starts glossing Night's epithet 'the nurse':

By saying that she is 'nurse', he (sc. Orpheus) expresses in riddling form that whatever the sun dissolves by heating, the night unites by cooling . . . those things which the sun heated . . . (col. 10.9–12)

I have already discussed this part of the text in another context.[92] I shall now complete the account given there with a few additional points. First it should be noted that we already have the sun here. This means that the author must have made it clear in the now lost part of col. 9 that the removal of the surplus of fire resulted in the creation of the sun.

Further, the antithetical structure of the sentence might suggest at first sight that there is a war going on here. The sun endeavours to destroy the newly emerged separate entities by dissolving them through its heat, pushing them back to the formless mixture they have just emerged from, whereas the night (which, as I have argued, might be just another name for air) tries her best to cool them down and thus keep them in shape.[93] On this reading δι]ạλύει would mean 'dissolves' as by melting. Yet in view of the author's other statements about the cosmogonical role of the sun, this picture cannot hold true. For in these other passages – we shall see in a

[92] Cf. 209 above on the identification of Night.

[93] διαλύω and συνίστημι are contrasted in this sense in Aristotle's testimony on Leucippus: συνιστάμενα μὲν γένεσιν ποιεῖν, διαλυόμενα δὲ φθοράν (Arist. *GC* 325a32 = 67 A7 DK).

moment – as soon as the fire is separated and encircled, in a word, bridled by the air, the sun plays an active and positive part in the emergence of independent entities.

In view of this, the sentence under consideration must mean that the sun and the night are co-operating. The sun breaks up the original shapeless mass into smaller chunks of matter, while the night cools down and moulds these already manageable portions of stuff. So the relevant sense of διαλύω is not 'to dissolve' or to thaw something into a shapeless mass, but to break down something into parts.[94] It will turn out that this is exactly the force of the verb κρούω, explicitly described as the effect of the sun.

Let us follow the thread of the cosmogonical role of the sun. In col. 13, in explaining the lemma about the swallowing of the phallus, the author stresses yet again the crucial role of the sun in the formation of the entities:

But he makes this clear by saying as follows:

He swallowed the phallus of [. . .], who sprang from the aither first.

Since in his whole poetry he speaks about facts enigmatically, one has to speak about each word in turn. Seeing that people consider that generation is dependent upon the genitalia, and that without the genitals there is no becoming, he used this (word), likening the sun to a phallus. For without the sun the things that are could not have become such . . . things that are . . . the sun everything . . . (col. 13.3–12)

Although the word order in these sentences is contorted and the formulation clumsy, the meaning is reasonably clear. The aorist of γε̣ν̣[έcθαι in line 11 suggests that the generation claimed to be dependent on the existence of the sun is not the generation we can observe day by day, but rather the original process of cosmogony. Therefore, it is expressed here that the sun is prerequisite for the generation of the particular entities we have in this world order.

The next column is apparently still discussing the same lemma, and further elaborating on the cosmogonical role of the sun:

[?he made] to leap the brightest and hottest having separated from himself. He says that this Kronos was born from the sun to the earth because he became the cause through the sun that they were struck against each other. For this reason he says: '(He) who did a great deed.' And after this:

Ouranos son of Night, who first of all ruled

[94] A nice parallel is Plato *Tim.* 68d ὅτι θεὸc μὲν τὰ πολλὰ εἰc ἓν cυγκεραννύναι καὶ πάλιν ἐξ ἑνὸc εἰc πολλὰ διαλύειν κτλ.

Naming Mind that strikes (? the beings) against each other Kronos, he says that it did a great deed to Ouranos; for (he says that) he (sc. Ouranos) was deprived of his kingdom. He named Kronos himself from his action and the others too according to the same principle. For of all the things that are . . . nature . . . that he got deprived of his kingdom . . . the things that are . . . (col. 14.1–14)

As I argued above, the first incomplete sentence must speak about the creation of the sun.[95] The next sentence indicates the result of the formation of the sun: the things started to be 'struck' against each other.[96] Although the immediate cause of this state of collision is the sun, ultimately the god is responsible for it, as he himself has created the sun. This is the 'great deed' marking the dawn of a new cosmological era. And on account of this deed, the god can now be called Kronos, the Mind who makes things collide (κρούων νοῦϲ).

The cosmogonical significance of collisions becomes more lucid in the next column, the first part of which we have already considered above:

. . . striking them against each other and, if he made the [su]n separate, (the result is that) the things which are stood apart from one another. For as the sun got separated and encircled, he coagulated and held fast both the things that are above and those which are below the sun. And in the next verse,

From him in turn Kronos, and then wise Zeus

he says that there has been this rule since he has been the king. But his rule gets the explanation that it is by striking the things that are against each other that he separated them into their current reconfiguration, not different (?ones) from different ones, but [. . .]. And the (expression) 'and then prudent Zeus' makes clear that not another one, but the same one. (col. 15.1–13)

Although colliding is beyond any doubt a key concept in the author's technical vocabulary, it is not altogether easy to retrieve the precise value of this term. It has to be noted that the collisions, κρούεϲθαι, are related not so much to the setting together of entities, which will be the next phase of cosmogony, but rather to the preliminary division of the original mass into smaller parts. One would expect that the separation is prerequisite for collisions and not the other way around. The only way I can picture the process is that due to the sun's heating, smaller parts of the mixture are now gaining motive energy on their own, independently of the whole. Then, the whole is split or torn apart as a result of this growing internal tension. One might wonder, however, whether κρούεϲθαι is the most appropriate

[95] See 234 above.

[96] Rusten (1985) 127 suggests that this refers to an intermediate state where the sun is already in existence but is still in the proximity of the other entities.

term to describe this action. Or was this term only dictated by the fact that it yields a nice etymology for the name of Kronos?

To make things more complicated, one step of the story is apparently missing. We have seen that the original mixture was not just red-hot, but probably also stirring. But what happened to it after the fire got extracted and before the newly created sun could start making things collide? In other words, is the mixture still in the same agitated state, or has it already more or less, or even totally, calmed and cooled down by the time the sun commences to exercise its influence on it? So does κρούεσθαι mean the addition of a new dimension to the motion of the whole, or does it mean that the sun sets the by now immobile mixture into motion again? And yet again, we can envisage a more complex dynamics, to the effect that the original movement of the mixture was gradually transformed into the seemingly chaotic, albeit eventually purposeful, motion of κρούεσθαι. It is all the more regrettable that we do not have an answer to this point as it would make some difference also in the comparisons with other cosmogonies.

In the next column (col. 16) we have the statement of the conservation of matter which we have already discussed. Then come three columns elaborating on the different aspects of the priority, domination, and power of the air, which is also said to determine the generation and destruction of other beings (coll. 17–19), while col. 20 is an excursus on initiatory rites, apparently little to do with cosmogony. It is only after all this that the author gets back to the formation of distinct entities in col. 21:

> . . . nor the cold to the cold. By saying 'by mating', he (sc. Orpheus) makes clear that divided up into small (pieces) they were moving and mating in the air, and as they were mating, they all got put together with each other. They were mating until each came to its like. Aphrodite Ourania, and Zeus, and to aphrodise, and to mate, and Peitho, and Harmonia are given as name to the same god. A man mingling with a woman is said by common usage to aphrodise. For (this reason) as the things that are now got mixed with one another, (this god) has got the name Aphrodite. (It has got the name) Peitho, because the beings yielded to one another; and to yield and to persuade is the same. (It has got the name) Harmonia, because many of the beings got fitted to one another. For they existed even before, but were spoken of as 'being born' since they were separated out. . . . to be separated out makes clear that . . . now . . . (col. 21.1–16)

The negation in the first fragmented sentence possibly refers to the initial state, where the force of 'like to like' could not yet be operative. In the next sentence, providing the explanation of θόρνηι, the process gets started. I assume that the phrase 'divided up into small (pieces) they were moving and mating in the air' picks up the story roughly at the stage of collisions. We can see that, presumably as a result of the collisions, matter is already portioned into small pieces, and these pieces are now milling about in the air. While κρούεσθαι designated the movement necessary for the division of matter, θόρνυσθαι concerns the movement paving the way for the conjunction of smaller pieces into larger entities.[97] We have every reason to assume that θόρνυσθαι, just as well as κρούεσθαι, is effected by the sun. Besides, both suggest disorderly motion, yet both are parts of the great cosmogonic project of Mind.

The motion described by θόρνυσθαι makes it possible for the lumps of matter to meet each other, and thus render the force of 'like to like' effective. These encounters are haphazard but when the lumps of matter fit one another (σύνηθες), they stay together.[98] The point is notable since it shows that the lumps of matter, split from the original mass by the force of κρούεσθαι, have their own, different characteristics.

The author also exploits the sexual associations of θόρνυμαι, as the word can mean 'to mount' or 'to impregnate'. He puts it into parallel with ἀφροδισιάζειν, which in turn receives its physical value via μίγνυμι. The latter verb, as also in col. 9.5 and 9, signifies the 'togetherness' or conjunction of gobs of matter.[99]

Finally, the explanation of the remaining two divine names, Peitho and Harmonia, offers additional colouring to the description of the process. The lumps of matter 'yield' to each other, and so finally they can get 'fitted' to each other.

Let us now make an inventory of the terms describing the different phases of the emergence of independent entities. (The asterisks indicate those verbs which are apparently more immediately dictated by the lemmata.)

(i) separation from the rest:
  διαλύω (col. 10.12)
  διίστημι (col. 15.2 and 9)
  διακρίνω (col. 21.14 twice)
  χωρίζομαι (coll. 14.2; 15.2 and 3 – always with regard to the sun)
  κρούω* (coll. 14.4 and 7; 15.1 and 8)

[97] Note again that although this distribution is naturally suggested by the text, on the basis of the semantics of the two verbs one would rather expect it to be the other way around.

[98] This process is in some ways parallel to the formation of functional organisms from a haphazard meeting of bodily parts in Empedocles' zoogony, from which only the appropriate conjunctions survive.

[99] Cf. LSJ *s.v.* II. Also, Arist. *GC* 322b8–9: Ἔστι δ' ἡ σύγκρισις μίξις· πῶς δὲ μίγνυσθαι λέγομεν, οὐ διώρισται σαφῶς.

(ii) conjunction and coagulation of lumps of matter in order to form larger entities:
συνίστημι (coll. 9.6; 10.13?; 17.2, 8, 15?; 21.3; 25.9)
σύνειμι (col. 25.7)
συνέρχομαι (col. 25.8)
πήγνυμι (col. 15.4)
συμπήγνυμι (col. 9.8)
θόρνυμαι* (col. 21.3, 4 and 6)
ἀφροδισιάζω* (col. 21.6 and 8)
μίγνυμι (coll. 9.9; 21.9)
εἴκω (col. 21.10 and 11)
ἁρμόζω* (col. 21.12)

WHAT'S NEXT?

An intriguing sentence in col. 17.6–9 insinuates that the author has a vision also about the future development of the cosmos. Before this sentence, in the first half of the column, the author argues for the eternity of air, which at a certain moment in the history of the world received the name Zeus. Then he turns to Zeus's epithet 'the last'. I quote the Greek as the text is doubtful.

> καὶ ὕστατον ἔφησεν ἔσεσθαι τοῦτον, ἔπειτ'
> ὠνομάσθη Ζεὺς ̣καὶ τοῦτο αὐτῶι διατελεῖ ὄνομα ὄν,
> μέχρι εἰς τὸ αὐτὸ ε̣ἶ̣δ̣ος τὰ νῦν ἐόντα συνεστάθη
> ἐν ὧιπερ πρόσθε̣ν̣ ἐόντα ἠιωρεῖτο.

It may seem that there is a problem with the tenses of the verbs here. Fortunately, the past of ἠιωρεῖτο is fine with πρόσθε̣ν̣ ἐόντα. On the other hand the Attic future or present of διατελεῖ may appear to be difficult to square with the aorist of συνεστάθη. The former, in conjunction with μέχρι, refers to the future, whereas the latter seems to designate a past action.[100] We have two options. One possibility is to adjust διατελεῖ to the past of the other verb, for instance taking it as an historical present.[101] This would mean that the name Zeus has already ceased to be a valid name for the air. Zeus was the denomination applicable during the period of *diakosmesis*,

[100] Tsantsanoglou retains the tenses as they are in the papyrus and translates 'and it continues to be its name, until present beings were set together etc.' This is a faithful rendering of the Greek, but hard to make sense of.

[101] This is the solution adopted by L&M: 'and this continued to be its name until the things that are now were set together etc.'

but as soon as the world took its present shape the denomination lost its validity. Theoretically this is not impossible; it is, however, rather difficult to imagine that Zeus, the primary subject of the Orphic hymn, was for the Derveni author already *passé*. Moreover, in such a case we would be obliged to take all the other references to Zeus as historical presents; and this is not very likely. Besides, ἔϲεϲθαι in the first half of the sentence also makes it more plausible that we are speaking about the future. It is true on the other hand, that we have aorists both before and after this sentence indicating that the things that are have already been compounded or, in a looser sense, born. The other option is to take the aorist of ϲυνεϲτάθη referring to the future in the sense of 'they will have got set together'.[102] This seems to me clearly preferable. The translation will be:

> And he said that this will be 'the last', because it was named Zeus, and it will continue to be his name until the things that are now have got set together into the same form in which they were floating as they were before.[103] (col. 17.6–9)

Clearly, the author distinguishes two different states of the same matter (νῦν versus πρόϲθεν). Or rather three states, as there is a former state, then an intermediate situation, followed by the replica of the first situation. We are now in the intermediate phase and the name Zeus is applicable until the return of the first arrangement: until the moment when 'the things that are now have got set together into the same form in which they were floating as they were before'.[104] Moreover, the author claims here that the beings were 'floating' or 'suspended' in the previous situation. The verb 'floating' presumably describes the state of lumps of matter as they are suspended, enveloped and dominated by the air. The author will use the same verb for describing the position of the stars hovering at a well-determined position in the air (col. 25.4 and 7).

Does this previous situation of 'floating' refer to the initial mixture we have discussed, when fire was mixed with the other elements? It does not appear so, since in *that* state matter was ignited and probably stirring as it

[102] It is true, however, that, as LSJ indicates, μέχρι takes a subjunctive – usually though not always preceded by ἄν – for future outcomes. This is why I previously suggested emending the verb to ϲυ{νε}ϲταθῆ<ι>. Janko (2002) accepts this emendation, but I am now more tempted to keep the original reading as found in the papyrus, certainly with future sense.

[103] This translation is quite close to the one that has been proposed by Janko: '(Orpheus) said that (Air) would be "last" because he was named "Zeus", and this will continue to be his name so long as the things which now exist have been put together in the same element in which they were suspended when they were pre-existent.' I cannot, however, follow him in taking εἶδοϲ in the sense of 'element'. I think this word needs to refer to the structure or constellation of the beings. But, in the end, our interpretation of the passage, developed independently of each other, comes to more or less the same.

[104] It is not clear whether εἶδοϲ is to be taken distributively or collectively.

was dominated by fire (cf. col. 9).[105] This reasoning leads to the conclusion that the previous state of floating referred to here is the state of affairs preceding the ignited state. If so, we can distinguish the following cosmic states: first 'floating' dominated by the air, then the ignited state, dominated by fire, then the *diakosmesis*, the present state, which leads to the prospective return of the pre-fiery state as a complete domination by the air. The name Zeus applies between the *diakosmesis* and the return of the first state. This could also be the force of ὕcτατον in the lemma: Zeus is the last phase before the world returns to its starting-point.

But no matter what stance one eventually takes on the interpretation of the details, this passage demonstrates that the author's cosmogony is cyclical.

ONTOLOGY AND PHYSICS

Having made an attempt at reconstructing the outlines of the cosmogonical theory, let us now examine the ontology and physics implied in it. This analysis is bound to be even more conjectural than what preceded because it involves extracting from the text general notions such as might never have been thought of by the author. And, as always, trying to make explicit what has perhaps never been so is risky. On the other hand, such an analysis can considerably enrich our picture of the philosophical framework within which the Derveni author interprets the poem, and can even modify some details of our reconstruction of the cosmogonical theory.

THE ELEMENTS

We have seen above that the author distinguishes two ontological levels: the level of entities we encounter in everyday experience on the one hand, and the level of the constituent elements of these entities on the other. The former can come into existence and then perish, whereas the latter have a stable existence not disturbed by the generations and dissolutions going on at the level of transient beings.

What, then, are the things that enjoy their temporary existence now, and what are the stable ingredients of which they have been made? The answer to the first question is quite simple. As the only real alternative in early Greek philosophy to a basically realist position was the Eleatic strict monism, and as the Derveni author allows without any further ado the plural in expressions such as 'the things which exist', it is reasonable to

[105]]ἰ̣δ̣οc would give a good sense but is too short to replace ε̣ἰ̣δ̣οc in line 8 of col. 17.

suppose that – even though the remaining text only mentions the earth, the sun, the moon and probably the stars from the world as we know it[106] – 'the things which exist now' are simply the physical bodies of everyday experience.[107] There is nothing in the text to suggest that we should think of anything more subtle than that.

As regards the level of deeper ontological analysis, that is, the stable ingredients, the interpretation of the text gets more problematic. The theoretical possibilities within the context of late Presocratic philosophy are relatively varied, ranging from the physical monism of Diogenes of Apollonia, to the countless different homoiomereous bodies of Anaxagoras, through the four elements of Empedocles, and the atoms of Democritus and Leucippus. Where on this palette does the Derveni author stand?[108] Let us first make an inventory of the elements mentioned in the surviving part of the text.

Air

From the standard stock, we certainly have the air. As we have seen in the previous chapter, the air gets equated with the cosmic Mind, also called 'the god'. Moreover, this divine air/Mind has taken over the role of at least most, but possibly of all the gods and goddesses mentioned in the poem via the author's allegorical interpretations and complex system of identifications. But what can we say about the ontological status of air?

In the first lines of col. 17 we read the following:

> . . . it existed before it was named. Then it was named. For air existed even before the things that are now were set together and always will exist. For it was not born, but existed. And the reason why air received its name has been made clear above. (col. 17.1–4)

The author tells us here in so many words that the air is distinguished from the other things by the fact that it has always had a stable existence throughout the history of the world, as opposed to the things which were 'set together'. Air was not born even in the restricted sense of composition. Furthermore, the last sentence refers to a probably etymological explanation of a name (Zeus?) Orpheus used for air. It is highly regrettable that this etymology has been lost somewhere in the destroyed bottom part of a previous column.

[106] To these we can perhaps add that col. 24 speaks about those who work the land and those who sail, without implying that all the natural objects and persons involved in these activities are but appearances. Probably all the ritual and other objects mentioned in the first seven columns can be listed here as well.

[107] Rusten (1985) 127–8 seems to restrict the cosmogony to the generation of the heavenly bodies.

[108] Cf. Laks (1997) 128.

Fire

As col. 9 makes clear, fire should certainly be counted among the elements involved in the Derveni author's cosmology:

Now, knowing that fire, [in as much as] it is mixed with the others, agitates the things that are and hinders them from getting set together because of fomenting, he removed it to an adequate distance, so that once it is removed, it does not hinder the things that are from coagulating. For whatever is kindled is dominated (sc. by fire), and when dominated, it mixes with the other things. (col. 9.5–10)

This passage proves that fire, together with air, must have existed before the present cosmic order and the 'things that are now' emerged. Further, the expression 'the brightest and hottest' in the first line of column 14, presumably with reference to the particles constitutive of the sun, might be taken as an indication that fire is not entirely homogeneous but comprises bits showing different degrees of the characteristic features of fire. If this is so, then the sun and the stars are made of the purest form of fire. However, it is just as possible that it only means that the fire, out of which these heavenly bodies are composed, is 'the brightest and hottest' among all the elements.

Lunar matter

As I have argued above, col. 25.1–3 shows that the Derveni author explicitly distinguishes the lunar matter from the solar/stellar matter on the basis of their respective thermic and visual characteristics.[109] As regards the lunar matter, the only positive piece of information we have is that these lumps of matter are [λ]ẹυκότατα μ̣ὲν τῶν ἄλλωγ. I have also tried to show that there is some evidence suggesting that the moon does not have her own light but reflects the sun. If so, the moon is neither hot nor bright, and thus does not evince the basic characteristics of fire. But whether it should be treated as a type of matter in its own right, or is somehow derived either from air or from fire, or is conceived of as a mixture of the two, or even is a form of earth, I am unable to tell.

Water

Apart from a recipe for wineless libations in col. 6.6, water only appears in the allegorical interpretation of the river-god Achelous in the ultimate,

[109] See 245 above.

already badly damaged lines of col. 23. As I have argued above, the remaining fragments do not allow us to decide what the Derveni author wants to say about water there. He might have incorporated it as an element and the physical correlate of the mythological name Achelous, as suggested by Tsantsanoglou;[110] but it is just as possible that the author is only rehearsing there the erroneous interpretation of 'the ignorant', as I am more inclined to think.

On the other hand, we have no reason to suppose that the author denied the existence of water (at least at the phenomenal level).[111] But water does not play any recognisable part in the cosmogony, and there is no indication whether or not we have to count elemental water among the ultimate building stones of reality; and if so, how this element is characterised by the author.

Earth

Next we should examine whether or not elemental earth has any recognisable part in the author's system. The word occurs some nine times in the remaining text. What makes the answer nonetheless problematic is that γῆ or γαῖα can take at least three different meanings. It can be used as the name of the primordial goddess, wife of Ouranos and mother of Kronos. Second, it can refer to the earth as a cosmological entity, and, third, it can mean elemental earth. The presence of the first and second meanings is fairly certain: the poem apparently mentioned the primordial goddess Ge, and the cosmological system of the author no doubt included the earth. It is less clear whether or not the author treats elemental earth, along with air, fire, and possibly water. And if so, whether elemental earth was considered part of the original ontological population (as e.g. in the system of Empedocles) or treated as a later development, something which is composed of an archetypal form of matter. Let us now take the occurrences of γῆ or γαῖα and try to decide what they refer to.

The first reference to the earth is in col. 14:

> He says that this Kronos was born from the sun to the earth because he became the cause through the sun that they were struck against each other. (col. 14.2–4)

This is apparently the physical interpretation of Kronos' mythological genealogy. In the previous column, the author explained Ouranos' severed

[110] Cf. his suggested translation in Laks and Most (1997) 20 n. 59: 'to the water in general he gives the name Achelous.'

[111] Note also the mention of sailors in col. 24.9.

phallus as the sun, so he can now confidently substitute Ouranos with the sun. The maternal side remains γῆ. However, it is difficult to imagine that both father and son are explained in physical terms, but that the mother is still to be understood as a goddess without any physical, cosmological significance; therefore, it is highly probable that in this sentence γῆ is to be rendered not as Ge, the goddess, but as the earth.

There is a longer treatment of Earth/earth in col. 22:

> Earth (Ge), Mother (Meter), Rhea and Hera is the same (*or*: are one and the same). She/it was called Earth (Ge) by convention; Mother, because all things are born from her (*or*: from this one). Ge and Gaia according to each one's dialect. And (she/it) was called Demeter as the Mother Earth (Ge Meter), one name from the two; for it was the same.
>
> And it is said in the *Hymns* too: 'Demeter Rhea Ge Meter Hestia Deio'. For (she/it) is also called Deio because she/it was torn (*or*: ravaged: ἐδηιώθη) in the mixing/sexual intercourse. He will make it clear when, according to the verses, she is born. . . . And (she/it) is called Rhea because many and . . . animals were born . . . from her. Rhea and . . . (col. 22.7–15)

It is not entirely obvious whether or not we have a physical allegory here; and if so, whether the goddesses are identified with the earth or with elemental earth.[112] Our only clues in deciding this question are the etymologies; yet they do not provide an unambiguous answer.

The phrase 'all things are born from her' can, on the face of it, mean that the solid entities of the world are formed out of elemental earth. But it is just as possible that the reference is to the earth as the source of life. Yet again, in view of the parallel in col. 26 (where Rhea/Demeter is explained as Mind 'because the Mind is the mother of the others') it is also possible that there is no such physical explanation here, and that all the goddesses are understood as Mind.[113] The same reasoning applies to the already severely fragmented etymological explanation of Rhea towards the end of the remaining part of the column.

The etymology of Deio, however, makes it more likely that the author explains the goddesses in physical terms. When he says that 'For (she/it) is called Deio because she/it was torn (*or*: ravaged: ἐδηιώθη) in the mixing/sexual intercourse', I am tempted to understand this as referring to the formation of separate entities. On this reading, mixing refers to the compounding of the entities, just as in the previous column in the explanation of Aphrodite.[114] But the emergence of separate entities is concomitant with

[112] Obbink (1997) 42 maintains (without arguments) that the passage is a physical allegory, identifying Demeter, Hera and Rhea with elemental earth.

[113] The feminine in line 7 (ἡ αὐτή) makes it perhaps less likely.

[114] Cf. col. 21.7–10.

the other aspect of the process: the original mass is now torn apart. This is how 'Deio' was 'torn in the mixing'.

At any rate, the identification of goddesses in this passage links to other passages; and this connection creates some tension in the identification. On the one hand, we have seen that in col. 14 the goddess Ge, the mother of Kronos, was most probably understood as the earth. We have also seen that Ge (also as Gaia) is identified with a group of goddesses in col. 22, among whom we find Rhea, the Mother and Demeter. But in col. 26, Rhea/Demeter, Zeus' mother, is explicitly identified with Mind. How can the earth be also Mind? Granting the author some degree of consistency (i.e. assuming that he cannot equate a certain element from the poem with *a* in one passage and with *b* in another without believing that *a* and *b* are related to each other), I see one way out. The air must be the common denominator between the earth and Mind; presumably in such a way that the earth was formed out of air.

This, after all, is not uncommon in early cosmogonies. Excluding now the aerial monists Anaximenes and Diogenes of Apollonia in whose systems it is obviously so, earth comes from air also in most of those cosmogonies which work with the model of a gradual separation or complexification. Especially so where the fundamental opposition is between a bright, hot and light stuff on the one hand, and a dark, damp and heavy stuff on the other; and where, in accordance with a more archaic understanding of the word, ἀήρ is the emblematic representation of the darker, damper and heavier. A clear example is Anaxagoras, but one can think of 'night' in Parmenides' *doxa*.[115]

To these passages we may add the first lines of col. 18, in which the earth appears once again, if we accept a highly likely supplement proposed by Janko:

> . . . speaking about those moving downwards. But he means that the [?earth] and the other things are all in the air, there being breath. (col. 18.1–2)

The supplement is supported not only by the space in the lacuna and the feminine article, but also by the phrase 'those moving downwards'. The context here is clearly physical or cosmological. If the supplement is correct, then the interest of the passage would be that it distinguishes the earth from the air, and places it on the side of 'the others': both the earth

[115] Cf. Anaxagoras B1, B15 and B16. For Parmenides, see esp. A37.

and the other beings are in the air, and, as I shall argue below, are in this way dominated by the air.[116]

What can we say, then, about the ontology of the Derveni author? I think the evidence is not sufficient to give a definite answer to this question. It is not impossible that the air is not only the divine element in the cosmos but also the archetypal form of matter from which other elements were separated out.[117] Indeed, as I tentatively argued above, there are some indications that earth, and the earth, is derivative of air. On the other hand, the highly problematic sentence in col. 17.6–9 suggests that even in the first cosmological phase there were forms of matter other than air, which could 'float' in the air.[118] Moreover, it seems to me that, together with air, fire is also an original component of the universe, for some form of fire is there and is active already in the pre-cosmic mixture. True, this does not rule out the possibility that fire had been formed of air at an even earlier stage of cosmic development. On the other hand, fire is apparently an independent force, and cannot be turned (back) into air; Mind/air can dispose of the unwanted quantity of fire only by changing its location. And, as I shall try to demonstrate below, the entire dynamics of the author's cosmogony is based on the air–fire duality.

On the basis of these considerations, my suggestion is that the author was not a monist but a pluralist. In the primary ontological population we have to count at least the air, which is divine and intelligent, as well as fire, and probably also other forms of matter that these two can 'dominate'. This distribution raises the interesting possibility that the author considers air and fire active elements, while the other elements, water and earth, are passive elements: a scheme somewhat reminiscent of the Stoic theory. It is also worth noting that air is unambiguously identified with the divine characters of the poem; it may be that this is also the case for earth, and (less likely) for water. Fire, on the other hand, is not identified with a divinity as such, but with the phallus, i.e. the generative organ of Ouranos, which gets severed by Kronos and then swallowed by Zeus. This is what the author explains as 'separation' and 'encircling'. Moreover, the superlatives used in describing the characteristics of the components of the heavenly bodies (coll. 14.1 and 25.1) suggest that fire is not homogeneous but that different parts of it show different degrees of purity.

[116] Instead of his original suggestions τὴν δ[ὲ γῆν], Janko (2002) opts for Burkert's suggestion τὴν δ[ίνην]. I briefly discuss this supplement in the Appendix 377–8 below.

[117] E.g. Boyancé (1974) 103 is of this view: 'en dernière analyse tout est composé d'air.'

[118] See 258–9 above.

THE CONSTITUTION OF MATTER

There is very little we can know for certain on this point. For one, it is clear that matter can be 'split up into small lumps'. These can collide, set together and behave as we have seen above;[119] they have qualities, and even show grades of physical characteristics: there are some which can be characterised as 'the brightest and hottest', others as 'the whitest'. On the whole, it is on the basis of such qualities that they can find their 'like'.

On this basis we can ascribe some kind of corpuscularism to the author. But this is a very general characterisation and leaves room for much variation.[120] It is not clear, for example, whether these lumps of matter are homogeneous on the inside or whether they show some sort of more complex internal structure. Or when the original mixture broke down into these smaller lumps of matter, were the borderlines dictated by any underlying structure of matter or was it just the play of hazard? All these questions – and they could certainly be multiplied – must remain open. It is not even unlikely that these problems were not of much concern to the author.

MECHANICS AND DYNAMICS

The author's cosmogonical account clearly involves a physical theory in the sense that it explains the formation of the current cosmic arrangement in terms of mechanical forces exercised on matter. The keyword in this dynamics is 'domination' (ἐπικρατεῖ). In the extant part of the papyrus there are two kinds of matter which are said to dominate other kinds of matter: air and fire. In what follows I shall try to examine more closely what domination can mean in the Derveni author's theory.

Domination by air

The air is explicitly said to dominate all things:

> . . . existing things have been called each single name by reason of what dominates (them); all things were called Zeus according to the same principle. For the air dominates all as far as it wishes.[121] (col. 19.1–4)

[119] See 152–7 above.

[120] This meagre information has prompted comparisons with the atoms of Leucippus and Democritus, the 'seeds' of Anaxagoras, the corpuscularism of Empedocles, and the Pythagorean ξύσματα.

[121] The *ZPE* edition has πάντᾳς in line 3. This is problematic for two reasons. First the masculine is hard to make sense of in this context: it would certainly need some explanation why men (or gods) are dominated by air. Apart from this question, the genitive is far better attested with ἐπικρατεῖ

Again, in col. 23, in the passage establishing the equivalence between Zeus, Okeanos, and the air, it is the 'great strength' of the air which constitutes the basis for the identification.[122] It is reasonable to suppose that this 'great strength' is related to air's outstanding capacity to dominate other things.

But in what sense or in what way does air dominate the other things? Interestingly, in three of the four remaining passages where air is mentioned, air is said to contain other things, or more precisely, the other things are said to be 'in the air'. In col. 18.2, we are told that [?earth] and all the other things are in the air.[123] This passage is all the more notable as it is part of the interpretation of Moira. Moira is interpreted in the later part of column 18 as 'the wisdom of the god always and through everything'; and the author urges that she was the wisdom of god even before air received the name of Zeus, i.e. even before the current cosmic arrangement emerged.[124] Then, in the next column, we have the assertion, just mentioned, about air's dominating everything else, immediately followed by the explication of Moira's power of determining the fate of the other beings:

> For the air dominates all as far as it wishes. And when they say that the Moira spun they say that the wisdom of Zeus ordains how the things that are and the things that come to be and the things that are going to be must come to be and be and cease. (col. 19.3–7)

So everything is in the air; air dominates all things; and Moira, the wisdom of Zeus, i.e., the active part of air, governs the generation and destruction of all things.

In col. 21, the Derveni author explains the identity of Aphrodite, Zeus, Harmonia and Peitho. This offers him the occasion to explain an important aspect of the process through which the independent, separated entities of the present cosmic order were formed. The first stage of this process is described as follows:

than the accusative. Thus Tsantsanoglou's new reading πάντῳν̣, as reported in a note to the L&M translation, is certainly welcome.

122 Col. 23.1–7: τ̣ο̣ῦ̣τ̣ο τὸ ἔπος πα̣[ρα]γωγὸμ πεπόηται καὶ το[ῖc μ]ὲν | π̣ο̣λ̣λοῖc ἄδηλόν ἐcτι τοῖc δ̣ὲ ὀρθῶc γινώcκο̣υcι | εὔδηλον ὅτι Ὠκεανόc ἐcτιν ὁ ἀήρ, ἀὴρ δὲ Ζεύc· | οὔκουν ἐμήcατο τὸν̣ Ζᾶνα ἕτεροc Ζ̣εὺc ἀλλ' αὐτὸc | αὑτῶι cθένοc μέγα. οἱ δ' οὐ γινώcκοντεc τ̣ὸν | Ὠκεανὸν ποταμὸν δοκοῦcιν εἶναι ὅτι εὐρὺ ῥέοντα | προcέθηκεν.

123 The construction of the badly fragmented first sentence is not, however, unequivocal. Apart from the supplement δ[ὲ γὴν], it is not clear where exactly the words]α φάμενοc [. . . .]ν, in superscript above line 1, belong. Furthermore, ἐόν can easily be ἐόν, as proposed also in the *ZPE* transcript. And more to our present concern, ἐν τῶι ἀέρ̣ι̣ might refer to π̣[νε]ῦμα ἐόν/ἐόν, and not to καὶ τἆλλα πάν[τ]α εἶναι.

124 Col. 18.9–10: πρὶμ μὲγ γὰρ κληθῆ̣ναι Ζῆνα, ἦμ Μοῖρα | φρόνηcιc τοῦ θεοῦ ἀεί τε κα̣ὶ̣ [δ]ιὰ παντόc.

By saying 'by mating', he (sc. Orpheus) makes clear that divided up into small (pieces) they were moving and mating in the air, and as they were mating, they all got put together with each other. They were mating until each came to its like. (col. 21.1–5)

Thus once again, we find the things 'in the air'. The passage is in accordance with the assertion from col. 17, according to which 'air existed even before the things that are now were set together' – for, obviously, otherwise they could not have been in the air during the process of composition. It is the wisdom of the god, the active part of air, which 'ordains how the things that are and the things that come to be and the things that going to be must come to be and be and cease'. The air, in which the things move around, get compounded and so forth, governs the whole process of generation of new entities. The air dominates the others which are 'in the air', and the fate of these things is decided by the air as they are enveloped by it.

This picture is borne out by the last passage mentioning the air. In col. 25, the Derveni author is discussing the constitution of the stars:

And there are other things now in the air, floating far away from each other. But during the day they are invisible being dominated by the sun, whereas during the night it is visible that they exist. They are dominated because of their smallness. Each of them floats in necessity, so that they do not come together with one another; for otherwise all those which have the same characteristics as those out of which the sun was set together would come together in one mass. (col. 25.3–9)

As these fiery particles have a tendency to come together driven by the force of 'like to like', a counter-force must be exercised: each of the particles floats 'in necessity'. The wording ἐν ἀνάγκηι – although it well expresses the idea that these particles are detained under the power of a strong external force[125] – does not specify exactly what this counter-force is. As the stars are expressly said to float in the air, we can, for example, exclude the view attested in some other early cosmologies, that the (fixed) stars are, as it were, glued or nailed to a solid firmament.[126] Barring the existence of a more complex physical theory to explain how the stars are kept apart (beyond the expression ἐν ἀνάγκηι the author maintains complete silence on this), the likeliest interpretation is that they are controlled by the air in which they float (ἐ̣ν τῶι ἀέρι ἑκὰς ἀλλήλων α[ἰ]ωρούμενα). Indeed, it is

[125] The expression ἐν ἀνάγκηι is not very common in pre-Hellenistic times. Normally it expresses external limitations, need and constraints, *force majeur*. Cf. Xen. *Hell.* 4.4.23; *An.* 2.5.21; Democritus B239 DK (ἐν ἀνάγκηισιν). The expression is used in the Septuagint and becomes more frequent in patristic literature.

[126] This is, for example, the view ascribed to Anaximenes by Aëtius (A14 DK). For a recent discussion see Wöhrle (1993) 27–8 and 72–3. Cf. also Empedocles A54 DK.

possible that ἀνάγκη was identified with air: as they float in the air, they float in necessity (αἰωρεῖται δ' αὐτῶν ἕκαστα ἐν ἀνάγκηι).

Another occurrence of the otherwise relatively rare word αἰωρεῖται may give further support to this interpretation. The word turns up in col. 17:

> . . . it existed before it was named. Then it was named. For air existed even before the things that are now were set together and always will exist. For it was not born, but existed. And the reason why air received its name has been made clear above. But it was thought that it was born, because it got the name Zeus, just as if it did not exist previously. And he said that this will be 'the last', because it was named Zeus, and it will continue to be his name until the things that are now have got set together into the same form in which they were floating (ἠιωρεῖτο) as they were before. And it is clear that they became such because of this, and having come to be, they are all in it. (col. 17.1–11)

Thus in more than one of the cosmic arrangements, partly in the past and partly in the future, beings are said to be 'floating' – and it is hard to imagine that they are floating in anything else than in the air. Especially so as the main focus of the whole gloss is on air/Zeus, and in this context the claims about the things floating serve as an illustration for the thesis of the supremacy and causal role of the air in the generation and cosmic arrangement of entities. To sum up, αἰωρεῖται must signify the physical contact through which the entities are enveloped in and dominated by the air, and through which air can attain the stability of separate entities. Indeed, it seems to me quite plausible that the author's physical theory about the things being in and hence dominated by the air was connected to the memorable image of all the things being in Zeus's belly in the Orphic poem – especially so as Zeus is after all but one of the names of the divine air.

Incidentally, the recurrent use of the verb in connection with the way entities behave when surrounded by air makes one wonder if the author had not employed the phonetic proximity of ἀήρ and αἰωρέω (etymologically and semantically connected also to ἀείρω) in developing this theory.[127]

We should add to these passages the one speaking about the formation of the sun in col. 15. Here it is stated that the fiery particles were not only separated (χωρ[ι]ζομένου) from the rest, but also encircled (ἀπολαμβανομένου ἐμ μέσωι). This 'encircling' must refer to the means by which air delimited, confined, and bridled the newly formed aggregate of fiery particles.

[127] Tentatively suggested by Laks (1997) 130.

The above account of the way the air 'dominates' other things has some further implications. For it makes unlikely an interpretation that has been repeatedly suggested, and which also influences the translations of the first lines of col. 19 as proposed by both Laks and Most, and by Janko. On the basis of an alleged parallel with Anaxagoras B12 ('in whatever there are things predominating, these things are, and were, most manifestly each object'),[128] taken in conjunction with the Theophrastean doxography on this point (to the effect that the physical entities of Anaxagoras are always composite 'but characterised according to what predominates'),[129] scholars have maintained that when the Derveni author speaks about domination as determining naming, his claim should be viewed in the light of the Anaxagorean analogy. We have composite entities, and an individual entity *is that*, in the version of Anaxagoras, and *is called that* in the version of the Derveni author, which is predominant in the entity in question. Gold is gold, or can be called gold, because, although it contains all kinds of other stuff, gold dominates in it quantitatively. This must be the reasoning on the basis of which Laks and Most render εκ[. . . .] τ̣ὰ̣ ἐό̣ντα ἓν [ἕκ]αϲτογ κέκ[λητ]αι ἀπὸ τοῦ ἐπ̣ικρατοῦντοϲ as 'each single thing had received its name from what dominates *in* it' while Janko translates it as 'each individual thing has been called after the dominant (element) *in* it' (italics are mine in both cases).[130] Yet the other passages show clearly that when the author claims that 'all things' are dominated by the air, it does not mean that all the things have a dominant proportion of air in them. The air dominates the other things not because the air is in the other things, but because all the other things are in the air.[131]

Moreover, domination does not appear primarily a question of quantitative superiority. It does not exclude it, but the accent is elsewhere. The concept as used by the author rather lays stress on the point that, without explicit reference to quantitative determinations, there is a preponderant force under the effect of which certain characteristics or tendencies of the

[128] Text and translation by Sider (1981). [129] Theophrastus' *Physikai doxai* fr. 4.

[130] Laks and Most (1997) 18 and Janko (2001) 27. See also Janko (1997) 64: 'Things are named according to which of the elements dominates (ἐπικρατεῖν) in them.'

[131] Discussing these passages, Laks (1997) 130 writes that 'air becomes the place where things evolve, rather than [as opposed to Diogenes of Apollonia] their "substance".' I think 'place' is an unfortunate expression in this context. Apart from the question of what concept of 'place' we are supposed to use here, before Plato and Aristotle, 'place' implies some kind of neutrality and passivity, whereas, as I have tried to show, the air's role is active, actually determining the course of the whole cosmic development. Burkert (1997) 170 n. 12, on the other hand, has to assume that ἐπικρατεῖται figures here in a different sense than in col. 25.

dominated are suppressed or changed.[132] It seems to me that although proportionately air might well be of the largest mass in the author's cosmos, it is not the quantitative aspect that is underlined in air's superiority. One has to remember that ἐπικρατέω is primarily a political (and military) term, and the author does not need to think in terms of majority rule in a democracy.[133]

The force exercised by air in order to avoid the stars' gathering in one mass requires further attention because it helps us to identify the means and purposes of the author's physical explanation. For it does not seem that the fact that these bits of solar matter 'are in the air' is sufficient in itself to account for the blocking of mutual attraction between them. When the general principle of 'like to like' *is* at work, and smaller parts of matter are drawn together, these lumps are in the air just as much. This was the situation, for instance, during the *diakosmesis* when matter was divided into small pieces and these pieces were in the air, and yet they could move around, collide and unite with each other under the auspices of Aphrodite/Zeus/Peitho/Harmonia. Indeed, the author tells us explicitly that the lumps of matter were moving and mating 'in the air' (col. 21.1–5).

Hence the fact that the stellar matter is enveloped in air is not enough to explain why the stars are still apart. If we left the stars alone (εἰ γὰρ μή), the general force of 'like to like' would take effect and 'all those which have the same characteristics as those out of which the sun was set together would come together in one mass' (col. 25.8–9). So we still have to explain why the stellar matter does not gather into one mass. Therefore, we have to add an ad hoc auxiliary hypothesis to our primary theory implying an extra force which is capable of countering the primary physical force of 'like to like', so that the resulting more complex model is able to explain the observational data that the stars remain apart. This can be done, for example, by assuming that the intelligent and divine air purposefully exercises a special power, specifically in order to keep the stars apart.[134]

Although one could take the addition of this seemingly arbitrary extra force at once as a serious limitation on the author's theory, this is not necessarily the case. It often happens that an ad hoc hypothetical extra factor has to be introduced into a scientific theory, without any independent

132 The point is well illustrated in Aristotle's *GC* 321a35-b2: in the mixture of wine and water, wine is the ἐπικρατοῦν not because there is more wine in the mixture than water, but because the characteristics of wine prevail.

133 The same probably holds true of Anaxagoras: it is not its quantity which makes Mind the 'strongest' and able 'to control' (κρατεῖ) not just the living beings but also the vortex (cf. B12 DK).

134 If indeed there was such an identification, this is the air/Mind *qua* Ananke.

justification, just in order to bridge the gap between the original model and the observational data. As a matter of fact this is a normal way to amend and develop scientific models. Without unduly pressing the point, the extra force keeping apart the stars in the Derveni author's cosmology is not incomparable with Einstein's 'cosmological constant'. Einstein felt obliged to introduce this hypothetical repulsive force of unknown origin into his cosmological equations just in order to balance the effects of gravitation, i.e., to account for the observed fact that the cosmic masses do not collide with each other under the gravitational effects of a finite mass-energy density. The introduction of the cosmological constant was a methodologically valid move, irrespective of the fact that later Einstein, having learnt and accepted Hubble's more explanatory theory of an expanding Universe, called it 'the biggest blunder of his life'.

Yet there is a salient difference – and this is where we can pinpoint the difference between *pure et dure* scientific explanation on the one hand, and natural theology on the other. To square the original physical model with the data provided by observation, the scientist introduces a further hypothesis. Einstein introduces the cosmological constant, a further physical force, or Callippus amends Eudoxus' model by the addition of a fourth homocentric sphere. However, as far as we can judge from the remaining part of the papyrus, the Derveni author does not give a further mechanistic, physical explanation for the extra force to be introduced. It looks rather as if this force, in contrast to the force of 'like to like', is not just a brute mechanical force but the manifestation of the supreme power and will of the divinity. This feeling is definitely reinforced by the closing sentences of the passage. Having sketched out the counterfactual scenario of the cosmic catastrophe which would occur if this counterforce were not in effect, the author rounds off the picture by making it clear that: 'If the god had not wished that the things which are now should exist, he would not have made the sun. But he made it of such a sort and of such a size as is explained in the beginning of the account.' For the Derveni author, the disagreement between the outcome of the primary physical explanation, on the one hand, and our experience of the cosmos, on the other, is not a call for a more complex physical model, but the proof that the maintenance of the cosmic order requires the working of the divine intelligence.

Domination by fire

In col. 9 fire is said to dominate things through ignition, while domination by fire leads to *mixis*. This is relatively straightforward. The only notable aspect is the place of 'domination' in the logical order of events: 'For whatever is kindled is dominated (sc. by fire), and when dominated, it mixes with the other things' (col. 9.9–10). This formulation gives the impression that 'domination', and not ignition, is the immediate cause of mixing.

Incidentally, the point I have made above about 'domination' not primarily being a question of quantitative determinations seems to be borne out by the sun's domination over the stars:

> But during the day they are invisible being dominated by the sun, whereas during the night it is visible that they exist. They are dominated because of their smallness. (col. 25.4–6)

The last, explanatory phrase strongly suggests that it is not self-evident that domination is based on quantitative differences: it is just one of the possibilities. It also has to be noted that this is a somewhat special case of domination, for the sun and the stars are made of the same stuff showing the same power or characteristics. The point is clear, yet it is interesting to note that something can dominate smaller entities composed of the same kind of matter.[135] However it is still in accordance with our previous understanding of domination: a preponderant force makes the subordinate incapable of realising its own characteristics or power.

Moreover, although the word ἐπικρατέω does not appear in that context, we have seen that the major influence fire exercises on others is to communicate some kind of motive energy to them.[136] In the pre-cosmic era this effect was described by the verb ταράccω (col. 9.6), whereas when fire took the shape of the sun, it made things collide with each other, κρούεcθαι πρὸc ἄλληλα (e.g. col. 14.2–4).

THE DIVINE AIR AND THE FIRE

Clearly, the successive phases of cosmic history are explained as the continued interplay of two active forces: the air, the physical aspect of the divine Mind on the one hand, and the fire on the other. Yet the relationship between these two active elements is far from symmetrical. To begin with, air, as is befitting for a cosmic divinity, is able to control fire. It can change fire's location (col. 9.7–8), gather the appropriate amount of it in one mass, confine it to a particular location (col. 15.1–5), disperse the remainder all

[135] Janko, however, thinks that the last sentence also speaks about the nocturnal conditions, and that the stars are dominated by the moon.

[136] The motive force of fire is first attested in the medical literature. Cf. *De victu* 1.3.9: τὸ μὲν γὰρ πῦρ δύναται πάντα διὰ παντὸc κινῆcται. Hippon 38 A3 DK; περὶ cαρκῶν esp. 1–4; *De hebdom.* 1.

around, and fix the position of these smaller bits (col. 25.3–9). As I have argued above, the text does not contain any strictly physical, mechanical description for these activities of the air/Mind. It is possible that the *pneuma*/Moira, i.e., the active, moving part of air, is somehow responsible for these actions. It is, however, also possible that the author did not give any specific physical account of the way air can bring about all these changes. Conceivably, it is simply in the power of the divine air to do all these things.

More important is the fact that the air *qua* Mind is endowed with mental and other psychological faculties. Due to its wisdom (equated with *pneuma* and Moira), it is capable of deliberation and takes the major decisions on the fate of all things by its own volition (cf. col. 17.4–7). Furthermore, it obviously had a clear image of the cosmic order to be attained right from the start. Accordingly, it analysed the pre-cosmic situation, acknowledged the reason for the incapacity of the beings to compound, and then took the necessary steps, created the sun and the stars and so forth. The crucial sentences in col. 25 make it clear that the divine air took these measures with the intention of realising a well designed plan:

> If the god had not wished that the things which are now should exist, he would not have made the sun. But he made it of such a sort and of such a size as is explained in the beginning of the account. (col. 25.9–12)

Yet this claim directs attention to a further remarkable aspect of the Derveni author's cosmogonical model. Although the cosmic god is said to be able to dominate all things 'as far as it wishes' (col. 19.3–4), it did not mould the beings in an immediate way. Instead, it operated by means of fire. In the initial situation fire prevents beings from taking shape. However, once fire is turned into the sun, it becomes the major generative principle (cf. col. 13.7–11). The effect is different, but not because fire's activity or power has been changed. All it does through these cosmic processes is blindly emit its heat, without any purpose or intentions, and with no consideration about the effects. So, contrary to the intelligent and purposeful actions of the divine air, fire's is a brute, mechanical force only capable of heating and initiating mechanical reactions. The most important characteristic of Mind/air on the other hand is that it is able to put at its own service the mechanical causation triggered off by fire, and thus to use this brute force in order to actualize its cosmic vision. In the next chapter I will try to show that the purposeful application of mechanical causation is the trait which makes the Derveni author's νοῦς a true relative of the νοῦς of Anaxagoras, and even of that of Timaeus.

ORPHEUS AND THE DERVENI AUTHOR

Before turning to the Derveni author's affiliation to other cosmologists in the next chapter, I shall round off the above survey by considering the relationship between the theogony implied in the poem and the cosmogony advanced by the author. On the face of it, there is hardly anything connecting the two. This is why it is customarily held that the author – as indeed all who engage in the business of allegoresis – loses sight of and violates the apparent meaning of the text. I would like to maintain, however, that at a deeper level there are some important convergences between the two accounts.

(i) As I have argued above,[137] one distinguishing feature of the Derveni theogony lies in the fact that, in contrast to the Hesiodic *Theogony*, this genealogy starts with the Night-Aither pair, a pair that evinces a set of basic polarities. We have also seen that Night has a far more important role than Aither. This structure, I believe, is finely echoed in the Derveni author's cosmogonical theory. There are two main actors in the story, air and fire, and the dynamics of the cosmic development are explained in terms of the interplay between the air–fire pair. Aither is easily identifiable with fire, and the author probably did identify the two.[138] On the other hand, Night shows strong affinities with air and probably was actually identified with it by the author. Further, Night's superior importance in the poem is reflected by air's preponderance over fire in the author's theory. We have also seen that the author's cosmogony may have started with air and fire, and the other forms of matter developed later from these two.

On this interpretation, we get remarkably close to the scheme that Diodorus of Eritrea and Aristoxenus (*apud* Hippolytus *Ref.* 1.2.12) say Pythagoras received from Zaratas (Zoroaster). According to this dualistic scheme, quoted on page 167 above, the sexual distinction (father-mother) is further expressed in terms of the light-dark opposition, and thus corresponds to the Aither-Night pair, but represents also the fundamental physical pairs of opposites (hot–cold, dry–moist, light–heavy, swift–slow). It is tempting to speculate about the Persian background of the dualism of the first divinities I proposed for the Orphic theogony, while this dualism can also easily be interpreted in terms of pairs of opposites in a physical, cosmological context. References in Aristotle and Eudemus of Rhodes

137 167–8 above.

138 Aither was probably in fact explained as fire by the author. Cf. the first line of col. 14, where the expression ἐ]{ḳ}χθόρηι τὸ{ν} λαμπροτάτον τε [καὶ θε]ρμό[τ]ατον, referring to the solar matter, answers αἰθέρα ἔχθορε from the lemma.

show that interested Greeks of the age could know about the dualistic theo-cosmogony of the magoi, and Eudemus also refers to the light-dark opposition.[139] The mention of the magi in col. 6 of the papyrus, and their treatment as experts on religious matters, might show that the Derveni author was acquainted with this material.[140] An obvious difference, however, is that the good–bad opposition, very prominent in the Persian system, is, as far as I can see, absent from this physical scheme of opposites.

(ii) The castration of Ouranos, also in Hesiod, is generally interpreted by scholars as the mythological expression of the separation of earth and sky, the most essential moment in the birth of the cosmos. The Derveni author interprets this episode as the separation of the fiery particles leading to the formation of the heavenly bodies. In this case also, the separation is the initial and most important moment of the cosmogony. Moreover, this primal separation is generally followed in mythological narratives by the emergence of the principle of (sexual) generation. In Hesiod, Eros' birth immediately follows the birth of Earth, and probably both in Hesiod and in the Derveni theogony Aphrodite's birth is directly caused by the castration of Ouranos. In the Derveni author's cosmogony the formation of the sun is a direct consequence of the primary act of separation, and the sun is explicitly treated as the principle of generation (col. 13.7–12).

(iii) I have also argued that, in contrast to the Hesiodic narrative, Kronos' relationship to Zeus is mitigated in the Derveni theogony. By castrating Ouranos, Kronos provides the means by which Zeus can eventually create the world anew. Moreover, Kronos appears actively to help Zeus with oracular advice. In the Derveni author's interpretation Kronos is not just identified with Mind, and thus with Zeus (the same could probably also be said of Ouranos), but the physical action described through the etymology of Kronos clearly helps the *diakosmesis* marked by the name of Zeus; indeed, it is perhaps the most important phase in the formation of separate entities.

(iv) When considering the Derveni author's application of the Parmenidean principle, we have seen that the author rightly stresses the point that the central event of the theogony as described in the poem, the new creation effected by Zeus, is the rearrangement of the already existing entities. This is what physical cosmogony is all about.

139 Arist. *Met.* 1091b10; Eudemus fr. 150 Wehrli (= Damascius *De princ.* 1.322 Ruelle). On the magi in the Greek context, see now Burkert (1999a) ch. 4 with bibliography.

140 I have tried to argue however that the reference is not necessarily to Persian magi. See 181–2 above.

(v) I have also raised the possibility that the author's insistence on air's containing and thus dominating the other beings is the physical explanation of Zeus containing in himself all the beings of the world.

(vi) We have also seen that a further salient feature of the Orphic theogony is that with its idiosyncratic story about Zeus, it approximates a creationist model.[141] The Derveni author's cosmogony employs the creationist model probably more extensively than any known Presocratic theory. Moreover, we have noted that the verb μήϲατο is used in the poem to describe the activity by way of which Zeus re-created the gods. This term stresses the intellectual nature of Zeus' action. The Derveni author, in turn, describes the way a divine *Mind* contrives a new cosmic arrangement, and lays special emphasis on the intellectual aspect of the *diakosmesis*.

(vii) Finally, we have seen that Mind/air uses fire as his main device in the realisation of the great cosmic project. It is worth recalling at this point that Zeus's traditional attribute and the tool by means of which he keeps order in the world is the thunderbolt. As the lemma in col. 19.10 says:

> Ζεῢϲ βαϲιλεὺϲ, Ζεὺϲ δ' ἀρχὸϲ ἁπάντων ἀργικέραυνοϲ.
> *Zeus the king, Zeus who rules all with the bright bolt.*

[141] 179–81 above.

CHAPTER 7

Anaxagoras

Having made an attempt to reconstruct the Derveni author's views on the supreme divine being and the formation of the present cosmic order on the basis of a close reading of the text, I shall now try to examine the place of this set of tenets in the context of Presocratic philosophy. The following survey – which is meant to further the project announced by Walter Burkert in his seminal paper 'Orpheus und die Vorsokratiker' published in 1968, so already thirty-five years ago – has two obvious objectives. On the one hand, it hopes to elicit the major doctrinal influences on the author. On the other hand, these systematic comparisons exhibiting both the similarities and the divergences will, I hope, enhance our understanding of the internal dynamics of the author's world-view.

This survey is certainly not exhaustive; I shall consider only those authors whose work I find particularly illuminating for the Derveni text. Thus, I shall not discuss the Atomists for example, since apart from a very few, and possibly accidental, verbal resemblances (of which the most notable is the use of κρούω) I do not think that they have much to do with the author's cosmology and physics.[1] One might have expected a comparison also with Empedocles. I was, however, unable to establish doctrinal correspondences with Empedocles that could illuminate aspects of the Derveni author's teaching. I shall nevertheless come back to Empedocles in the last chapter where I discuss the Derveni author's overall intellectual outlook.

A word of caution is in order at this point. The following analyses necessarily involve dealing with Presocratic fragments and theories the understanding of which is highly controversial. I certainly cannot in the present context rehearse and assess all the interpretations available. Nevertheless, I shall try to avoid *ignotum per ignotius* explications and the pitfall of giving

[1] Burkert (1968) 98–9 tries to assimilate the expression κατὰ μικρὰ μεμερισμένα (col. 21.2) to the atoms of Leucippus and Democritus. I see no reason to think that the lumps of matter from which the presently existing entities of the world have been built up share the basic features of atoms.

preference to certain readings just because they fit well with my view of the Derveni author.

Soon after Kapsomenos published the first, provisional and very limited transcript of the papyrus in 1964, the pioneering studies of Merkelbach (1967) and Burkert (1968) established that Anaxagoras must have had a predominant place in the Derveni author's intellectual background.[2] Burkert's subsequent paper already bears the title: 'La genèse des choses et des mots. Le papyrus de Derveni entre Anaxagore et Cratyle'.[3] Since then, Anaxagoras, together with Diogenes of Apollonia, has been treated as the main source of the author's ontological and physical views,[4] to such a degree that Burkert has recently attempted to decide on a disputed point in the Anaxagorean cosmology – whether or not there is a plurality of worlds for Anaxagoras – by the help of the Derveni text.[5]

To be sure, there is nothing surprising in finding the author being influenced by Anaxagoras, as Anaxagoras' important doctrines reverberate from many different corners from the middle of the fifth century onwards. He might have been the one who brought Ionian-type physical speculation to Athens.[6] His direct influence is manifest not only on his immediate disciples such as Archelaus of Athens and Metrodorus of Lampsacus, but also on thinkers like Diogenes of Apollonia. Furthermore, Plato's account of his master Socrates (who had apparently been associated with Archelaus)[7] reading Anaxagoras with great interest is also confirmed by Xenophon.[8] And Anaxagoras' sway was not confined to circles of professional philosophers. His theories about the annual flooding of the Nile and the role of the mother in procreation were, for instance, echoed by the tragic writers.[9] Some ancient sources even call Euripides his 'disciple'[10] and, indeed, numerous and varied Anaxagorean ideas are detectable in Euripides' verses.[11]

Moreover, there are indications that Anaxagoras' fame and influence were of an even wider scope, at least in Athens. A remark in the *Apology* about the availability of Anaxagoras' treatise suggests that it was copied,

[2] Merkelbach (1967) 10; Burkert (1968) 98. [3] Burkert (1970).
[4] The description given by KRS 30 is typical: 'an allegorical commentary by someone versed in Anaxagoras and Diogenes of Apollonia'.
[5] Burkert (1997). [6] KRS 385. [7] Cf. e.g. Simpl. *in Phys.* 27.23.
[8] Plato *Phd.* 97b8–99d2; Xen. *Mem.* 1.4.
[9] Aesch. *Suppl.* 559; fr. 300; Soph. fr. 797; Eur. *Hel.* 3; fr. 228 ἐν τῇ Αἰθιοπίᾳ, as printed in 59 A91 DK, is preferable to the manuscript tradition of Hippolytus 1.8.5. ἐν τοῖϲ ἄρκτοιϲ.
[10] Cf. e.g., Strabo 14.645 Cas. [11] Cf. 59 A20 a–c; A30; A33; A48; A112 DK. Cf. KRS 355 n. 2.

circulated, and probably read in relatively large numbers.[12] The sad story of his trial – the real target of which was perhaps Pericles himself[13] – also affirms Anaxagoras' importance in Athenian public life.

On the other hand, it is not clear how much influence he had outside Athens. He was certainly an important public figure in Lampsacus – or at least this is what the amusing anecdote about Anaxagoras being granted a wish towards the end of his life and asking for a holiday for school children, and the information about his public burial suggest.[14] It is not improbable that he even founded a school of his own in Lampsacus.[15] However, even though echoes in Metrodorus, Diogenes, and perhaps also in Melissus and the Atomists show that his ideas were influential in Ionia, very little can be known about his reception in the west.

Anaxagoras' impact was thus wide-ranging and complex. Let us see how he may have influenced the Derveni author. Funghi aptly summarises the common elements as ascertained by previous scholarship: (i) the presence of a cosmic Mind and its definition; (ii) the theory of the pre-existence of matter, (iii) and the corpuscular constitution of reality, (iv) which is produced by a process of differentiation, and (v) acquires its name on the basis of the predominant constitutive element.[16] In what follows, I shall systematically compare the respective doctrines along these lines. Eventually, I shall try to show that, even though it has received relatively little attention, (i) is the most important, whereas (v) – which by contrast has been stressed repeatedly – is hardly a point of contact at all.

MIND

Anaxagoras' physical, cosmological doctrines were no doubt of great importance, yet one is inclined to agree with the Socrates of Plato's *Phaedo* that Anaxagoras' greatest innovation was the introduction of Mind as the ultimate cause of cosmic order.[17] It remains so even if both Plato and Aristotle hasten to point out the limitations of the role Anaxagoras ascribes to Mind. Nonetheless, when it comes to characterising the Anaxagorean Mind and defining the part it plays in the history of the cosmos, one has to face intimidating difficulties, as is well testified by, among others, the recent studies of A. Laks, J. G. DeFilippo and J. H. Lesher.[18]

[12] Plato, *Apol.* 26d. There are, however, some doubts about the interpretation of this evidence.
[13] Plutarch *Pericles* 32 = 59 A17 DK; cf. Diodorus 12.39.
[14] D.L. 2.14–15; Alcidamas *apud* Arist. *Rhet.* 1398b15 = 59 A23 DK.
[15] See *HGP* II 268–9.
[16] Funghi (1997) 34, slightly paraphrased.
[17] *Phd.* 97b8-d4 = 59 A47 DK; cf. Arist. *Metaph.* 984b15–20 = 59 A58 DK.
[18] Laks (1993); DeFilippo (1993); and Lesher (1995).

It has to be noted first that Anaxagoras' choice of the term νοῦς was something of a novelty. No doubt, some of the earlier Presocratics also incorporated supreme intellectual powers into their respective systems,[19] yet, as far as we can see, none of them used the word νοῦς for these entities. Moreover, the Anaxagorean usage considerably strained the semantics of the word. Accordingly, Kurt von Fritz has justified the restricted scope of his 'Nous, noein, and their derivatives in Pre-Socratic philosophy (*excluding Anaxagoras*)', by maintaining that one has to write a separate study on the Nous of Anaxagoras because it does not readily fit into the history of the semantic development of the word.[20] This point is not mitigated by the fact that, paradoxically, we have to rely to a great extent on the ordinary meaning of the word in trying to define Mind's function and characteristics.[21] At all events, the term was so much of a trademark of Anaxagoras that he himself was sometimes mockingly called the 'Mind'.[22]

As regards the Derveni author, we have seen that among all the denominations which are applicable to the cosmic god, the name 'Mind' was most probably the one which did not come from the poem but which the Derveni author himself introduced. This terminological point in itself should be a clear indication of a – direct or indirect – Anaxagorean influence.[23] The fact that the Derveni author retained the Anaxagorean denomination for his own supreme being becomes even more significant when one considers that the author was apparently also profoundly influenced by Diogenes of Apollonia who, for his part, preferred the term νόησις.[24] Thus we have every reason to assume that the author was in some way indebted to the Anaxagorean theory of Mind. It remains to see how far this indebtedness goes.

In comparing the two homonymous ultimate cosmic intellects – that of the Derveni author on the one hand, and that of Anaxagoras on the other – we should eventually describe what, or perhaps who, the respective Minds are; and what the major cosmic acts of the respective Minds consist in.

[19] For an overview of the material, which as a matter of fact is not very rich, see Lesher (1995) 133–5. The most important loci are Xenophanes B25 DK and Heraclitus B41, B32 and B108 DK.

[20] von Fritz (1945) 23. See also von Fritz (1971) 576: '... habe ich den ΝΟΥΣ des Anaxagoras ausgelassen, weil er in der Geschichte dieser Bedeutungsentwicklung sich nicht wollte einordnen lassen. Er paßt auch, gerade mit seinem wichtigsten Aspect, in dieses Entwicklungsgeschichte durchaus nicht hinein.'

[21] Cf. Lesher (1995).

[22] 59 A1 DK with reference to Timon's *Silloi*, see also Harpocrates in 59 A2 DK.

[23] It is probably ultimately due to Anaxagoras' influence that the Demiurge of the *Timaeus* is sometimes called Mind, and that Aristotle's supreme being can also bear this name.

[24] On the point that this terminological difference indicates a critique of doctrine, see Laks (1983) xxxix.

However, these two questions are so intimately entwined that it is hardly possible to analyse them separately. Hence I shall approach the problem by asking the following questions, which, in a way, intersect with the above two points: (i) what are the more immediately theological implications of the respective concepts of Mind in the two systems? (ii) how far are they characterised in anthropomorphic terms? (iii) in what sense can they be treated as causes? and (iv) how are they related to matter?

Mind and the divine

Anaxagoras' place in the early history of natural theology is remarkably controversial. Jaeger – drawing on Deichgräber's analyses of the paratactic, hymnic style of the description of Mind at the beginning of B12 – has maintained that Anaxagoras' concept of Mind has serious theological implications.[25] Cleve and (more recently and more forcefully) Sider have argued against such a view, also denying the relevance of Deichgräber's contribution.[26]

As a matter of fact, we can observe the same contrasting assessment of Anaxagoras already in the ancient sources. Aëtius, for instance, states without further ado that Mind is a god, or the god, and Philodemus may have treated the Anaxagorean Mind on a similar basis in the *De pietate*.[27] On the other hand, we cannot forget that Anaxagoras was expelled from Athens on charges of impiety, and he was apparently severely opposed to traditional religious beliefs and praxis.[28] Moreover, according to some later sources, Anaxagoras was even named 'the Atheist'.[29] One obvious way of doing away with this (probably) seeming contradiction is simply to refer to the feud between traditional religiosity and the new type of natural theology. Yet, as some scholars have maintained, the contrast can go deeper than that, and is perhaps internal to Anaxagoras' treatment of Mind.

It has to be noted first that Mind is never called a god or even divine in the extant fragments of Anaxagoras.[30] Yet, irrespective of the question whether or not Anaxagoras himself thought of Mind as a god, there is no doubt that in the sway of his concept and characterisation of Mind, people subsequently spoke about their respective amended versions of Mind

[25] Jaeger (1947) 160–71; Deichgräber (1933) 347–53. So also Gerson (1990) 28–32.
[26] Cleve (1943); Sider (1981) 96. Cleve's position was criticised by Vlastos (1952) 114 n. 76.
[27] Both in 59 A48 DK. [28] Cf. the anecdote transmitted by Plutarch in *Pericl.* 6.
[29] Irenaeus in 59 A113 DK: 'autem, qui et atheus cognominatus est'.
[30] See however Guthrie *HGP* II 279: 'It is nowhere in the extant fragments called God, but this may be accidental and it is impossible that Anaxagoras should not have thought of it as divine (θεῖον).'

with overtly theological implications. Diogenes said that Intelligence/air 'seemed' to him to be a god,[31] just as Timaeus' Demiurge/Mind is also freely called a god in the Platonic text. These, however, are all philosophical approaches to the divinity, and it is highly unlikely that any of these authors considered their cosmic intellects objects of ritual worship and cult.[32]

Hence the special interest of the Derveni author's case. To begin with, the author surely uses the denominations 'Mind' and 'the god' synonymously.[33] More importantly, if we accept that the Derveni author was primarily a religious figure, a *mantis*, an Orphic initiator, so that his interpretation of the poem via physical and cosmological speculations needs to be viewed against the background of the eschatological, ritual, religious theme of the first six columns – as has been convincingly argued for by Laks,[34] and as I have also tried to show in chapter two above – then this case would prove that the Anaxagorean conception of the cosmic Mind not only fostered philosophical discussions on the nature of the divine, but could also infiltrate into priestly circles directly involved with rites and other forms of religious practices. On this reading, the theoretical contribution of Anaxagoras could become a major impetus both for the Platonic Socrates in pursuing a more thorough and radical *philosophical* analysis of the teleological structure of the world, and for an intellectually active Orphic initiator to reinforce some form of *traditional religiosity* by elaborating a more up-to-date and appealing theoretical framework to prop up his priestly lore and practice.

The story can get a further twist, showing the other side of the coin, with the nebulous and intriguing figure of Hermotimus of Clazomenae. As Aristotle maintains

> When one man said, then, that Mind was present – as in animals, so throughout nature – as the cause of the world and of all its order, he seemed like a sober man in contrast with the random talk of his predecessors. We know that Anaxagoras certainly adopted these views, but Hermotimus of Clazomenae is credited with expressing them earlier. Those who thought thus stated that there is a principle of things which is at the same time the cause of beauty, and that sort of cause from which things acquire movement.[35]

[31] Diog. Ap. B 5: καὶ μοι δοκεῖ τὸ τὴν νόηcιν ἔχον εἶναι ὁ ἀὴρ καλούμενοc ὑπὸ τῶν ἀνθρώπων, καὶ ὑπὸ τούτου πάνταc καὶ κυβερνᾶcθαι καὶ πάντων κρατεῖν· *αὐτὸ γάρ μοι τοῦτο θεὸc δοκεῖ εἶναι* κτλ.

[32] Hecuba's prayer in Eur. *Tr.* 884–7 is an interesting attempt to recycle philosophical speculations into religious practice by raising the possibility that the philosophers' god can become the addressee of a prayer: ὦ γῆc ὄχημα κἀπὶ γῆc ἔχων ἕδραν, | ὅcτιc ποτ' εἶ cύ, δυcτόπαcτοc εἰδέναι, | Ζεύc, εἴτ' ἀνάγκη φύcεοc εἴτε νοῦc βροτῶν, | προcηυξάμην cε.

[33] Cf. coll. 18.11 and 25.10. [34] Laks (1997). [35] Arist. *Metaph.* 984b15–22 (trans. Ross, modified).

It is not without irony that Hermotimus of Clazomenae is apparently a candidate for Aristotle's laudatory title 'a sober man'. For other sources tell us that this person was something of a mystic ecstatic. A *mantis* and *theoclytos*, whose soul was credited with being able to travel freely to remote places detached from his body, until his enemies, led by his treacherous wife, burnt his body as it was lying dormant unattended by his soul.[36] On the basis of this, Hermotimus is treated by some modern interpreters in a group with Abaris, Epimenides, and others, as one instance of 'Greek shamanism'. Besides, Heraclides Ponticus, as reported by Diogenes Laertius, informs us that Hermotimus was one of the incarnations of Pythagoras' soul.[37]

The relationship between these two seemingly unconnected sets of information about this fellow-townsman of Anaxagoras might be enlightened by another testimony of Aristotle, this time from the *Protrepticus*:

For mind is the god in us – whether it was Hermotimus or Anaxagoras who said so – and mortal life contains a portion of some god.[38]

If Hermotimus actually said something in this vein (which I am more inclined to accept than that he was speaking about a diacosmic Mind),[39] then the connection, however removed, with the Derveni author would give some support to Guthrie's intuition that the core of the Orphic theological speculation was about the mystic union of the divine and the individual soul. And if there is a grain of truth in Aristotle's clearly second- or third-hand reports about Hermotimus, then we would have some evidence for Guthrie's further claim that mystical thinking about the underlying unity of the microcosmos and the macrocosmos on the one hand, and Ionian physical, cosmological speculation on the other are different realisations of a shared intuition.[40] Also, if Aristotle's remark in the *Protrepticus* has something to do with a historical Hermotimus with some kind of Pythagorean connections, then, at the other end of the line, we might have some indication about the Pythagorean background of the *Timaeus*' conception about the identity, or at least isomorphism, between cosmic and personal intellect. Admittedly, this is all extremely conjectural and hypothetical.

[36] Pliny, *Nat. hist.* 7.174; Tert. *An.* 44; Plutarch, *De gen. Socr.* 592c–e (where he is mistakenly called Hermodorus of Clazomenae).

[37] D. L. 7.5.

[38] Arist. *Protr.* fr. 61 Rose (= Iamblichus *Protr.* 48.16 Pistelli): "ὁ νοῦϲ γὰρ ἡμῶν ὁ θεόϲ", εἴτε Ἑρμότιμοϲ εἴτε Ἀναξαγόραϲ εἶπε τοῦτο, καὶ ὅτι "ὁ θνητὸϲ αἰὼν μέροϲ ἔχει θεοῦ τινοϲ".

[39] The contention of Aristotelian commentators that Hermotimus, together with Anaxagoras, was the first to posit an efficient cause is evidently an extrapolation from Aristotle's own words and has no independent value. Cf. Alex. Aphr. *in Met.* 32.15; 33.10; Asclep. *in Met.* p. 43.

[40] *HGP* I, 132; Guthrie (1952a) and (1955) 316.

At any rate, it is already interesting that Aristotle himself refers to a frantic eccentric as the possible ultimate origin of an important doctrine of his own. And as we go back to our more immediate concern, it is interesting to speculate that Anaxagoras' doctrine of the Mind might have been in some way influenced by a mystic, religious mode of thought.[41] It is a consideration which could complement our previous point that the Derveni author shows the re-appropriation of the theoretical elaboration effected by Anaxagoras into a more immediately religious context, by showing that the Anaxagorean concept in itself had been in some way influenced by religious thinking. It could thus reinforce the feeling that the concept of a cosmic divine intellect was an interface and a field of continued interplay between philosophy and a certain religious mentality with higher intellectual demands.

Undoubtedly, much more could be said about the philosophical and religious ramifications of the idea of a cosmic Mind. Yet this very phenomenon warns us that it is not enough to trace the itinerary of such a pervasive theme. For it is just as important to see that what might appear to be the very same idea can in fact receive different meaning and significance when applied in different contexts with different objectives.[42] This is the theme I pursue in the next section by pointing out some shifts of emphasis between Anaxagoras and the Derveni author.

The characterisation of Mind

Anaxagoras' intellectual effort to posit an ultimate and intelligent cause of cosmic order can be treated as part of that tradition which, from Xenophanes onwards, tried to work out a more abstract, philosophical conception of the divine as opposed to the gods of the traditional lore. An important part of this task was to try to expel or modify a great number of the commonly accepted characteristics of traditional gods: to argue that they are not only deathless but also not generated, that they are not to be imagined as likenesses of men and women, not likely to be involved

[41] This question has been elaborated in very general terms in Detienne (1964). It has to be noted that the Hermotimus connection remains relevant even on such a cautious interpretation of the material as the one conjectured by Diels and Dodds: 'Aristotle, indeed, thought there were grounds for believing that Hermotimus anticipated his more famous townsman Anaxagoras in his doctrine of *nous*; but this may mean only, as Diels suggested [on Anaxagoras A58], that for evidence of the separability of *nous* Anaxagoras appealed to the experiences of the old local shaman' (Dodds (1951) 143).

[42] This can be a major objection to the treatment of the theme of the cosmological and eschatological role of fire in Kingsley (1995). For he convincingly shows the pervasiveness of the theme, but, it seems to me, fails to make it clear that, despite the underlying thematic unity, the idea does not remain the same when employed in different contexts.

in marriages, intrigues and so on. And just as importantly, to reduce the number of superhuman agents, so as to retain only one or two ultimate causes instead of the flock of traditional gods. This meant a radical challenge, as all the traditional textual and visual representations of the gods, also employed in religious praxis, often treated the gods to a large extent as anthropomorphic both physically and psychologically.[43] It is in this context that Kurt von Fritz, among others, speaks about disanthropomorphisation as one of the major features of Anaxagoras' conception of Mind.[44]

For our present purposes it is important to note that it is highly possible that it is in the same context – i.e. disanthropomorphisation of the traditional gods – that the disciples of Anaxagoras undertook the allegorical interpretation of traditional epic poetry. As Syncellus informs us, 'the Anaxagoreans' interpreted the Zeus of epic texts as Mind, and the Athena as *techne* etc.[45] Furthermore, Favorinus, as reported by Diogenes Laertius, tells us that Anaxagoras

> seems to be the first to declare that Homer's poetry was about *arete* and *dikaiosune*; this argument received a more elaborate defence from Metrodorus of Lampsacus, an acquaintance of his, who was the first to devote himself to the study of the poet's physical doctrine.[46]

It does not necessarily mean that Anaxagoras himself was in the business of allegoresis,[47] but it is nevertheless remarkable that some ancient sources connect the very origins of moral and physical allegory with the name of Anaxagoras and his immediate followers. There is no need to stress the importance of these points with regard to the Derveni author's treatment of the gods of the Orphic poem. Indeed, David Sider has recently maintained that the connection is so overwhelming that 'it strongly points to Lampsacus as the source for the writing of [the] text of the Derveni papyrus.'[48]

However, 'disanthropomorphisation' and the corresponding critique of traditional representations of gods is just one side of the project. The positive

43 This is just a very schematic formulation as the question is obviously much more complex. In particular, one should not ignore the important qualifications stressed e.g., by Vögelin (see especially Vögelin (1957) 174–8) and Snell (see Snell (1986) ch. 9): it was not the case that there was a well-defined concept of man which was then projected onto other, non-human entities of the world, but rather that the border-lines between human and non-human were not fixed yet.

44 von Fritz (1971) 588.

45 Syncellus *Chron.* 140 c i p. 282, 19 Dind. = 61 [Metrodorus of Lampsacus] 6 DK. ἑρμηνεύουϲι δὲ οἱ Ἀναξαγόρειοι τοὺϲ μυθώδειϲ θεοὺϲ νοῦν μὲν τὸν Δία, τὴν δὲ Ἀθηνᾶν τέχνην, ὅθεν καὶ τὸ "χειρῶν ὀλλυμένων ἔρρει πολύμητιϲ Ἀθήνην". Incidentally, the verse quotation is supposed to be an Orphic fragment (*OF* 347).

46 D.L. 2.11 = 59 A1.11 DK.

47 So Pfeiffer (1968) 35 n. 3. Cf. also Richardson (1975) 69–70.

48 Sider (1997) 138.

side of the endeavour consists in positing an ultimate cosmic agent which is assumed to be responsible for changes and order in the cosmos. The two sides of the project – disanthropomorphisation as demanded by a more abstract way of thinking on the one hand, and speaking about the workings of an intelligent agent on the other – means that the philosopher has to balance on a very narrow edge. The express recognition that this is actually an *impossibly* narrow edge, so that one is bound to fall back on anthropomorphic imagery and metaphor when speaking about a cosmic cause, will require conscious reflection on the possibilities and limitations of the philosophical idiom and the admirable intellectual frankness of Plato.

Not surprisingly, the characterisation of Anaxagoras' Mind, although a great step towards the elaboration of an abstract cosmic cause, remains heavily loaded with anthropomorphic language. In the remaining fragments,[49] and above all in fragment B12, it is said to be endowed with exceptional intellectual capacities (γνώμην γε περὶ παντὸϲ ἴϲχει B12.10; πάντα ἔγνω B12.16), possessing a vast power (αὐτοκρατέϲ B12.2; ἰϲχύει μέγιϲτον B12.10–11); owing to this power, it governs the rest (κρατεῖν B12.8; ὅϲα γε ψυχὴν ἔχει . . . πάντων νοῦϲ κρατεῖ B12.11–12; ὁ δὲ νοῦϲ ὅϲα ἐϲτί τ' ἐκράτηϲε[50] B14.1), and actively changes the course of events (τῆϲ περιχωρήϲιοϲ τῆϲ ϲυμπάϲηϲ νοῦϲ ἐκράτηϲεν, ὥϲτε περιχωρῆϲαι τὴν ἀρχήν B12.12–13; πάντα διεκόϲμηϲε, καὶ τὴν περιχώρηϲιϲ αὕτη B12.18–19; ἤρξατο ὁ νοῦϲ κινεῖν B13.1) – and perhaps, at least on the reading of some scholars,[51] finally takes its leave (ἀπεκρίνετο B13.2).

The Mind/air of the Derveni text is described in remarkably similar terms. It has intellectual abilities (γιγνώϲκ[ων] οὖν col. 9.5), has notable strength (cf. the reasoning about Okeanos in col. 23, with my interpretation on 199–200 above), has the power to control everything (πάντῳγ γὰρ ὁ ἀὴρ ἐπικρατεῖ col. 19.3), and is even said to equal everything else ([αὐ]ṭὸν Νοῦμ πάντων ἄξιον εἶṿαι μόν[ο]ν ἐόντα col. 16.10). The Mind is actively involved in the cosmogonic process (ἐξήλλαξ[ε col. 9.7; κρούντα τὸν Νοῦμ πρὸϲ ἄλληλ[α] col. 14.7; ἐπόηϲεν ἥλιον col. 25.10), and owing to its role in the *diakosmesis*, can be described as the parent of everything else (μήτηρ ὁ Νọῦϲ ἔϲτιν τῶν ἄλλων col. 26.1).

Apart from the two Minds' respective relations to matter – to which I shall return later – there are two significant contrasting points, and these, I

[49] References are to the text as printed in Sider (1981).

[50] Emendation by Sider. For arguments, see Sider (1981) ad loc.

[51] So, e.g., Guthrie in *HGP* II 274 (with bibliography for former propounders of the same view). Contra Sider (1981) with bibliography of authors of the same view, to which add notably Lanza (1966) 234.

submit, are closely connected. As noted above, Anaxagoras does not qualify his Mind as god, whereas the Derveni author naturally applies this term to his supreme cosmic agent. More important, no term expressing volition is used with regard to Anaxagoras' Mind, whereas the Derveni author repeatedly refers to the intentions of the divine Mind/air. It is declared to be able to dominate everything else *as far as it wishes* (col. 19.3–4), the present cosmic order is said to have come about in accordance with the godhead's *volition* (col. 25.9–10) as indeed, in general, its wisdom *ordains* the generation and passing away of beings (col. 19.5–7).

One could object that the comparison based on the absence of expressions attributing volition to the Mind in Anaxagoras is just an *ex silentio* argument depending on the transmission of Anaxagoras' text. It is notable, however, that such formulations are carefully avoided in the long description of Mind and its activities in B12.[52] Hence it is much more likely that this is a genuine difference between the respective conceptions of Mind, and indicative of the contrasting intentions of the two authors. It is worth recalling at this juncture that in pointing out those features which distinguish the new concept of a cosmic cause developed by the Presocratic philosophers from the traditional representations of gods, interpreters generally stress that philosophers consciously tried to avoid the volitional motifs. And this is exactly what distinguishes the two Minds: Anaxagoras is striving to develop a more abstract concept of a cosmic cause, whereas the author of the papyrus is ready to retain more of the anthropomorphic traits of the traditional divinities, and notably the idea that the supreme being has intentions. This is a clear indication that the Derveni author's supreme cosmic agent remains a personal god with immediate religious significance.[53]

Mind as cosmic cause

We can pursue the same line of thought by pointing out the similarities and differences in the two Minds' respective positions in cosmic causation. It is clear that Mind's functions in Anaxagoras, its overwhelming power

[52] Lesher's thesis that in B12.10 DK γνώμη means 'decision' and hence the translation of the sentence should be '[Mind] holds every decision concerning everything' (Lesher (1995) 138–9) certainly weakens but does not rule out this interpretation. Apart from this point, Parmenides' question concerning the privileged moment of commencement (B8.9–10 DK) remains pertinent. Why did the Mind 'choose' this moment? Furthermore, what had it been doing before?

[53] The problem emerges with a new complexity in the full-grown teleological system of the *Timaeus*. As by force of his essential characteristic of being good he *must* act in such a way that the created world become 'the best possible', how far does the Demiurge act on his own volition? Cf. Scolnicov (1997) 365.

notwithstanding, are severely limited. To begin with, Mind can only exercise its power on the pre-existent matter. What is more, this pre-existent matter – at least on one possible interpretation – already contains in it all the different possible types of matter (what were to be called homoiomerous bodies by later generations of philosophers), although in a latent form, with their respective distinctive features.[54] Even though gold or bone as we know them will only emerge as a result of the process of separation, the constituent elements whose quantitative dominance will cause a separate entity to be gold or bone are already there from the beginning.[55]

The presence of the famous 'seeds' in the original mixture probably means a further restriction,[56] especially if one aligns oneself with those commentators who take this term to mean something like 'matrix'. On this reading, a 'seed' comprises all the ingredients and *dynameis* of a future composite stuff in the appropriate proportions and even with the prefiguration of the structure of the future entity.[57] We go even further if, as David Furley has suggested,[58] Anaxagoras was using the term with its everyday meaning – i.e., the seeds of animals and plants – for in that case even the most complex living organisms are already there in germ from the beginning, only waiting for further stuff to be added and 'separated out' through nutrition. If this is the right interpretation, Anaxagoras takes the Parmenidean warning very seriously: there is nothing in the present-day cosmos which was not there, although in a concealed form, already from the start.

[54] Lucretius 1.875; cf. Furley (1989) 53. On the interpretation of Cornford, Vlastos and Schofield, 'everything in everything' should mean 'in every object or stuff, there is every opposite'. Contra KRS 366.

[55] On the more restrictive interpretation, the homoiomerous stuffs as such are not there from the beginning, but only the opposites. As distinct entities appear, certain qualities begin to dominate in them, and as a result of the simultaneous domination of certain qualities, a particular entity can be seen as the manifestation of a particular homoiomerous stuff, the one which is characterised by the predominance of these qualities. This lump of matter is a shining piece of gold not because gold as such predominates in it over the bone, flesh and wood in it, but because heaviness, hardness, brightness, and other characteristics of gold predominate in it. As David Sedley has pointed out to me, one of the great advantages of this interpretation is that it is much more in line with Anaxagoras' theory of perception. It seems to me, however, that B11 DK constitutes a difficulty for this reading. In this fragment, on the one hand, we have a statement that in everything there is a portion of everything, but, on the other hand, Anaxagoras says that there are things (from the set of 'everything') in which there is also mind. Hence, it seems unlikely that 'everything' refers to qualities and not to substances.

[56] Cf. B4b. David Sedley has suggested to me, however, that, contrary to the standard interpretation, B4b should refer not to the original mixture, but to the state of things just before life first appeared. Naturally, on this reading, the world is less 'ready' from the start.

[57] Along these lines, although with considerable differences with regard to the details and wording, Vlastos (1950) 461–70; Sider (1981) 131–2; Teodorsson (1982), who even uses the term 'programmes' (p. 85).

[58] Furley (1989) 55–8.

Moreover, the separation and conjunction of different types of stuff, which eventually make the latent apparent, is directly caused by the vortex and the mechanical factors associated with it (cf., e.g., B9). Also, the area affected by the rotation seems to grow mechanically as the edge of the whirling matter drags with itself the contiguous portion of the surrounding mass. Hence, as far as the cosmogony goes, the only action actually attributable to Mind is that it gave the original impetus, pushed, as it were, the button of the centrifuge. The rest was due to blind mechanical forces. This is exactly what makes Anaxagoras' doctrine of Mind so frustrating for Plato and Aristotle.[59]

It would follow from this reasoning that Mind is no more than a pure function, the act of setting the whirl in movement.[60] Yet there is another side to Anaxagoras' theory. To begin with, if the above account were the whole of the story, Mind's name would be completely inappropriate. And indeed, Anaxagoras himself says

> And Mind knew (ἔγνω) all the things that are mixed with each other, and separated out and divided off; and what kinds of things were to be – those that were, and are no longer, what are, and what will be – Mind arranged all in order, also this rotation through which now revolve the stars, the sun, and the moon, the air and the aither which separate out.[61]

The accent here is on the intellectual capacities of Mind. The full force of ἔγνω is difficult to reproduce in translation. As Sider notes, it is more than simply 'to know' (although I do not find his 'gave heed' satisfactory either). It also comprises the senses 'to understand' and 'to discern' with the connotation of distinguishing.[62] Anaxagoras emphasises here that Mind is and has always been aware of everything, and it foresaw everything from the start. Thus, although the only physical act it actually did was to give the primal nudge to the vortex, it was nevertheless entirely alive to all the consequences of its initiative. It understood the nature and effect of the mechanical reactions triggered off by the vortex, and could thus predict all the results. It did not create the heavenly bodies or this piece of gold or bone directly, but the future existence of all these entities was a necessary

[59] Cf. Plato *Phd.* 97b8–99d2 and Arist. *Metaph.* 985a18, both in A47 DK.

[60] For an exposition of this view, see e.g. Silvestre (1988).

[61] B12.15–21 DK: καὶ τὰ cυμμιcγόμενά τε καὶ ἀποκρινόμενα καὶ διακρινόμενα πάντα ἔγνω νοῦc· καὶ ὁποῖα ἔμελλεν ἔcεcθαι – καὶ ὁποῖα ἦν ἅccα νῦν μὴ ἔcτι, καὶ ὁποῖα νῦν ἔcτι, καὶ ὁποῖα ἔcται – πάντα διεκόcμηcε νοῦc, καὶ τὴν περιχώρηcιν ταύτην, ἣν νῦν περιχωρέει τά τε ἄcτρα καὶ ὁ ἥλιοc καὶ ἡ cελήνη καὶ ὁ ἀὴρ καὶ ὁ αἰθὴρ οἱ ἀποκρινόμενοι.

[62] Cf. LSJ *s.v.* Note also the aorist, which can, however, be understood as gnomic (so Lesher (1995) 139 n. 14). Laks (1993) develops the idea that the *diakosmesis* is above all the physical counterpart of the cognitive act of discerning, distinguishing.

consequence of the initial act – and Mind was aware of this.[63] Besides, this account accords well with one of the important semantic ingredients of the word νοῦς: 'plan', 'project', or 'design'.[64]

Following this line of thought, one can even ask whether Mind initiated the vortex *because* it wanted the cosmos to be as it is now. Retrospectively, it is obvious that it started a vortex, for this movement accounts for phenomena like the revolution of the stars, so it can take the role of a fairly good explanatory hypothesis.[65] But given the original mixture, was it in Mind's discretion to choose between different types of motion to impart? Could it have chosen, say, the winnowing basket instead of the centrifuge? And if so, what would the world be like now?[66]

On the basis of the above considerations, there would be two equally valid answers to a question asking about the cause of an event or state of affairs in Anaxagoras' cosmos. On the one hand, for any event *b*, there is an event *a* from which *b* is derivable with reference to a mechanical factor φ.[67] For instance, the formation of the earth (*b*) can be derived from the cosmic rotation (*a*), a contingent historical event, plus the general principle – if one wants to avoid the expression natural law – that a rotation drives the heavy elements towards the centre of the rotation (φ). Similarly, the celestial bodies are ignited (*b*) because they travel at a great speed (*a*) and the friction, in general, heats things up (φ).[68] On the other hand, all these events are elements in Mind's grand project, and got their *raison d'être* from it. So the other answer to a why-question would be that this or that is thus and so, because it was designed or determined to be so by Mind.[69]

[63] So, e.g., Jaeger (1947) 163. [64] 'Absicht', von Fritz (1971) 586.

[65] Some problems remain, however. I would imagine that if you started a vortex, you would end up with a disk, a cylinder, a cone perhaps; but how could a vortex result in a *spherical* cosmos?

[66] To put it differently: what is the relationship between the observable revolution of the heavens and the view that the intellectual motion *par excellence* is circular motion? Moreover, how far is the *Timaeus* indebted to Anaxagoras in connecting cosmic circular motion and mental circular motion? On the circular motion of the rational individual soul in the *Timaeus*, see Sedley (1999) 316–24.

[67] I do not find convincing Sider's suggestion that 'Nous represents the ordered workings of what we would call the laws of nature, which allows some things to happen and forbids others' (Sider (1981) 104). I see no reason to assume that the mechanical factors are not independent from Mind.

[68] Incidentally, the fact that there is friction between the celestial bodies and their immediate environment suggests that the speed of the observable rotation of the stars does not correspond to the speed of the vortex. Admittedly, there are other interpretations to account for the fiery nature of the heavenly bodies, but the same kind of physical causal explanation would apply to those scenarios just as well.

[69] Lesher (1995), if I have understood him right, is very close to this interpretation. On the other hand, this account differs from that of Barnes (1982) 410–12 in assuming that we have in Anaxagoras *both* the 'Humean' and the 'Berkeleian' type of causation.

If we now turn back to the Derveni author's theory of Mind, we get a strikingly similar picture, although, once again, with some significant differences. As I have argued in the previous chapter,[70] the Derveni author's Mind does not order the cosmos in an immediate way. His direct cosmogonical activity is restricted to the manipulation of the fiery stuff: the creation of the sun and the scattering of the stars. The rest, the separation and conjunction of matter, and hence the formation of independent entities, is due to the mechanical effects of the heat emitted by the sun. Note also, that although both the respective initial acts (the creation of the sun on the one hand, and starting the vortex on the other), and the immediate consequences (collisions on the one hand, and separation of heavy and light on the other) are different, the regulation of motion is a crucial element in both accounts. Motion is imparted directly by Anaxagoras' Mind, whereas its counterpart in the papyrus sets the right level of motion by placing the fiery particles – the immediate sources of motive force – at the appropriate location.

Anaxagoras' model is certainly more economical. The Derveni author's Mind directly intervenes in the spatial rearrangement of matter – the fundamental process of all cosmogonies – whereas the activity of Anaxagoras' Mind is more simple: it imparts the rotation, and then it is the vortex which does the job of redistributing matter in space. Nonetheless, the way the Anaxagorean Mind initiated the rotation is by no means less mysterious than the way the Derveni author's Mind/air removed the fiery particles.[71] At any rate, the *champs d'action* of both Minds are relatively restricted, and the rest of the *diakosmesis* is explained in terms of causal reactions and mechanistic processes.

The parallelism continues on the cognitive side of Mind's activity. As we have seen above, the Derveni author makes it explicit – more explicit than Anaxagoras – that Mind is aware of the mechanical factors working in the cosmos,[72] and that its initial act was executed purposefully with a view to achieving the expected results.[73] Mind had a plan, and also a clear understanding of the mechanical processes, so it took the appropriate measures which, through the workings of mechanical reactions, would eventually lead to the realisation of the plan. Thus, again, we have a double explanation: an event or state of affairs can be explained either in terms of mechanical

[70] Cf. 274 above.

[71] So, e.g., KRS 364: 'How Mind imparted the first rotatory movement is by no means obvious; it may be that even Anaxagoras himself had no clear mental picture of the process.'

[72] Cf. col. 9.5–8. [73] Cf. col. 25.10–12.

factors or with reference to Mind's project. We find a clear statement of this double explanatory structure with reference to the collisions:

He says that this Kronos was born from the sun to the earth because he became the cause through the sun that they were struck against each other. (col. 14.2–4)

As we know well, Kronos is a name for Mind: Kronos is Mind *qua* the cause of the collisions.[74] Thus Mind is the cause of, or responsible for (αἰτίαν ἔσχε), the collision of the lumps of matter; nevertheless it is through (διά) the sun, i.e., the motive force communicated by the sun, that Mind can achieve all this.

This passage about Kronos sheds new light on the other etymologies and identifications as well. As noted in the reconstruction of the cosmogony, most of the allegorical explanations of the divine characters of the poem refer to different phases and aspects of the cosmogonical process. Apart from Kronos, this is certainly the case with Aphrodite (the *mixis* of bits), Harmonia (lumps of matter get fitted to each other), Peitho (parts of matter yield to each other), Deio (the original mixture is torn apart as the independent entities take shape), Rhea (the birth of animals) and probably also with Night (cooling of the original mixture). This is highly significant not just because we learn about the development of the cosmogony through these explications, but also, and more importantly, because these divinities get identified in the course of the argument with each other and ultimately with the supreme cosmic cause. This means that the author is not content with a simple physical allegory: he wants to emphasise that in all these processes we have to recognise the workings of the divine Mind. At one level of the explanation, the author explains the names of the divine characters of the poem by reference to the different phases of the physical process which leads to the formation of the present cosmic order, but at another level he always emphasises that it is Mind who is responsible for all these cosmic events. Ultimately, all these physical processes are the manifestations of Mind.

Commentators – both ancient and modern – disagree on the question of how far, after having started the vortex, Anaxagoras' Mind remains involved in the subsequent cosmic processes apart from maintaining control over animate beings.[75] This long-standing debate in itself is an indication that Anaxagoras was not explicit on the point. This is far from the case with the Derveni author because his Mind is certainly not satisfied with the role of the designer and initiator, but takes immediate responsibility also

[74] Col. 14.7. [75] Cf., e.g., Lesher (1995) 135 n. 24.

for the subsequent events. The author presents the successive stages of his cosmogonical theory in terms of the different names of the cosmic divinity, emphasising in this way that the supreme being is the cause of each and every step of this development.[76] Mechanical factors are involved, but always and only in so far as they are the means through which the Mind can mould the cosmos to its volition. The underlying structure remains the same as in Anaxagoras, but with a clear and significant shift of emphasis.

Mind and matter

There is some scholarly controversy as to how far Anaxagoras went in separating Mind from the rest of the world. To begin with, Mind is said not to be mixed with other forms of matter,[77] whereas it is a general and much stressed characteristic of the rest of the world that, even in the post-cosmogonical state, different types of matter remain inherently mingled.[78] This is already an important dividing line. It is not entirely clear, however, whether or not Anaxagoras also pioneered the positing of an incorporeal supreme cause. Guthrie, for example, has maintained that the separation of Mind and matter is complete in Anaxagoras, as the philosopher from Clazomenae 'finally grasped the idea of non-material existence'.[79] Nevertheless, most commentators resist this view and urge that Anaxagoras did not go so far, and still conceived Mind in the old corporeal fashion. As they take it, the description λεπτότατόν τε πάντων χρημάτων καὶ καθαρώτατον (B12.9–10) means in fact that the difference between Mind and other forms of matter is a question of showing physical qualities at different degrees and is not an absolute divide.[80] Almost everybody allows, however, that Anaxagoras is manifestly striving to imagine incorporeal existence in his endeavour to make his dualistic scheme complete.

[76] Contrast also with Plato *Tim.* 42e5–6: καὶ ὁ μὲν δὴ ἅπαντα ταῦτα διατάξας ἔμενεν ἐν τῷ ἑαυτοῦ κατὰ τρόπον ἤθει (without ignoring that the precise meaning of this sentence is vexed as well).

[77] 59 B11 DK and B12.2–8 DK. [78] 59 B6 DK, cf. B12.1 DK. [79] *HGP* II 276–9.

[80] Cf., e.g., Sider (1981) 97–8. It has to be noted that one of the major arguments for the corporealist view is that Mind has spatial extension, as it is expressed in 59 B14 DK: ὁ δὲ νοῦϲ ὅϲα ἐϲτὶ τ' ἐκράτηϲε καὶ νῦν ἐϲτιν ἵνα καὶ τὰ ἄλλα πάντα ἐν τῷ πολλὰ περιέχοντι καὶ ἐν τοῖϲ προϲκριθεῖϲι καὶ ἐν τοῖϲ ἀποκεκριμένοιϲ (so, e.g., KRS 364). This, I think, is problematic. On the basis of the 'everything in everything' principle, wherever there is any kind of stuff, no matter whether it is in a pre-cosmic state or has already undergone *diakosmesis*, all the opposites and all the homoiomerous bodies will be present. But can two bodies be at the same place without being mixed with each other? If not, then mind, explicitly said to be unmixed, cannot be at the same place where there is some other body. Or else, it is there but not as a body. Moreover, spatial extension does not necessarily imply corporeality. In the *Timaeus*, the world-soul and its copies in individuals have a well-defined spatial extension without being corporeal (Cf. Sedley (1999) 317–19).

This evidently does not hold for the Derveni author, as his Mind is also air. The non-acceptance of the more radical dualism of Anaxagoras clearly ranges the author with Diogenes of Apollonia, so I shall discuss this question in that context.[81]

We can note right now, however, that the divergence as compared to Anaxagoras is not merely a question of degree. In Anaxagoras' cosmology, the major opposition in the original mass was between air and aither. Whether or not we identify this aither with fire, as Aristotle suggested,[82] it is clear that fire is the more λεπτός of the two, so when Mind is said to be λεπτότατόν τε πάντων χρημάτων, we have an extension in this direction. As David Sider has put it, Mind is 'more ethereal than the aither'.[83] We have a similar opposition in the Derveni author's system between air and fire, and, significantly, Mind is not identified with fire but with air.

COSMOLOGY

Let us now consider the complementary aspect of the two stories and see what happens to the physical world under the action of Mind.

The cosmogonic process

In the respective descriptions of the emergence of the present cosmic order we find, once again, some highly interesting resemblances. True, this is partly due to the fact that both accounts contain elements which are from the standard stock of Greek cosmogonical speculation – yet the correspondences, as I shall argue, go considerably further.[84]

The majority of commentators assume that Mind is the ultimate source of movement in Anaxagoras, and thus matter must have been static in the initial situation.[85] Furthermore, according to the famed dictum of B1, 'all things were together', that is, the mixture was complete, in such a way that an imaginary observer would have been unable to discern strictly speaking any distinct contours in the mass of matter (B4b). On the other hand,

81 Cf. 311–13 below. 82 Arist. *De caelo* 270b24 83 Sider (1981) 98.

84 I shall be able to refer only to the most important moments in the Anaxagorean cosmogony. For a more detailed discussion, see e.g., Bargrave-Weaver (1959); Stokes (1965).

85 The motive function of Mind is emphasised in Arist. *De an.* 404a25–405a15 (divided up in 59 A55 DK, A99 DK and A100 DK) and *Phys.* 256b24 = 59 A56 DK and *Phys.* 250b24 = A59 Lanza. There are dissenting voices: Hussey (1972) 140 expresses some doubts; according to Schofield, Anaxagoras himself might have been in doubt: 'Conceivably Anaxagoras could see no way of deciding whether the primordial mixture was static or moving, and so made no pronouncement on the subject' (Schofield (1980) 154 n. 45).

although it is not clear in exactly what way, air and aither 'dominated' or held all the other things in 'subjection'.[86]

Once again, the initial situation in the Derveni author's cosmogony, as reconstructed in the previous chapter, shows important similarities with the Anaxagorean account. We start the cosmogony with a jumbled mixture of matter. Furthermore, even though it is not obvious what the divine air was doing in that phase of the cosmos, it is quite likely that the two dominant ingredients were air and fire already at that stage.[87] On the other hand, if the original mixture was indeed static in Anaxagoras, we have a momentous difference: it is quite clear that the mixture in the Derveni cosmogony was characterised by the excessive motion of matter.

The parallelism continues in the respective reports about the emergence of separate entities. In both accounts, it is described as the result of two complementary processes: the separating off from the rest of the matter of the future entity, and the conjunction and combination through which the ingredients of the new entity get compounded. Given that the physical process has an important role in both accounts and that both Anaxagoras and the Derveni author use an elaborate and complex vocabulary, it is worth while comparing the respective terminologies.

In Anaxagoras, in the main, ἀποκρίνεσθαι (B4; B6; B7; B9; B13; B16) refers to the two-stage process by which a new entity is 'separated out': (i) διακρίνεσθαι (B5; B13; B17) refers to the breaking up of the original material as lumps of matter get separated from the rest; and (ii) συμμίσγεσθαι (B12; B17; σύμμιξις B4), προσκρίνεσθαι (B14), συγκρίνεσθαι (B4), and συμπαγῆναι (B4; B16) refer to the different aspects of the conjunction and compounding of a new entity. Besides, συγχώρειν (B15) and ἐκχώρειν (B15) mean the re-grouping of different masses in space; whereas χωρισθῆναι (B6; B8) signifies the counterfactual absolute separation of physical characteristics within any portion of matter.[88] According to the analysis of Lanza, followed and developed by Sider, Anaxagoras employs here a technical vocabulary, which shows important correspondences with the medical terminology of the age.[89]

86 59 B1 DK. For a discussion of the meaning of κατεῖχεν, see, e.g., Sider (1981), ad loc.; Schofield (1980) and KRS 372.

87 It is interesting to note that in discussing the primal role of air and aither in the Anaxagorean cosmogony, Sider (1981) 46–7 refers to theo-cosmogonies, like that of Epimenides, where the primordial divinities are Aer and Night. If my reconstruction of the Derveni theogony is right, we get an even closer parallel with the primordial pair Night and Aither.

88 It would be hardly possible to capture all the shades of meaning obtained by the different combinations of verbs and prefixes. Obviously, there are as many renderings and interpretations as there are commentators.

89 Lanza (1966) 230–1; Sider (1981) 78 and 103.

The table that follows compares the outcome of this survey with the Derveni author's terminology as reconstructed in the previous chapter.

		Anaxagoras	Derveni author
(i)	separation from the rest	ἀποκρίνομαι	–
		διακρίνομαι	διακρίνω
		–	διαλύω
	gaining distance in space	ἐκχωρέω	–
		–	χωρίζομαι (the sun)
		–	διίςτημι
(ii)	compounding	ςυμμίςγομαι	–
		–	μίγνυμι
		ςυμπήγνυμι	ςυμπήγνυμι
		–	πήγνυμι
		προςκρίνομαι	–
		ςυγκρίνομαι	–
		–	εἴκω
		–	ἁρμόζω
	coming together in space	ςυγχωρέω	–
		–	ςυνέρχομαι
		–	ςύνειμι
(iii)	formation of a new entity as a result of (i) and (ii)	?ἀποκρίνομαι	–
		–	ςυνίςτημι

It seems to me that apart from the common usage of the most obvious verbs and prefixes in such a context, and in view of the underlying similarity of the processes envisaged, the two lists are strikingly different. This consideration actually raises the problem that, even though the importance of Anaxagoras' influence on the whole is beyond any doubt, we cannot know how immediate this influence was.

The structure of the cosmos

As direct information on the Derveni author's picture of the geometrical structure of the cosmos is insubstantial, and the reconstruction I have proposed in the previous chapter is highly conjectural, the discussion in this section is bound to remain largely speculative.

Although, as far as I can see, the extant text of the papyrus does not contain any indication concerning the shape of the earth, I would venture the hypothesis that the author did not depart from the standard Ionian view, namely that the earth is cylindrical, maintained without variation by Anaxagoras, Leucippus and Democritus, and, at least according to the majority of commentators, also by Diogenes of Apollonia.[90] Furthermore, it is reasonable to suppose that the Derveni author's earth is supported by air, once more in accordance with the mainstream Ionian tradition, and with Anaxagoras in particular.[91] All the more so because it can be a further indication of the power of the divine air.

The heavenly bodies

As we come to the astronomical theories of the two authors, the differences become more pronounced, and indeed irreconcilable. First, the formation of the heavenly bodies is explained in fundamentally different terms in the respective theories. True, there are some slight uncertainties regarding the account given by Anaxagoras. One part of the problem is caused by Plutarch's unparalleled testimony in A12 according to which the stars were formed when a solid celestial vault broke up into smaller pieces. This is in contrast with the generally accepted version reported by Hippolytus in A42 (probably going back to Theophrastus) according to which the rocks forming the heavenly bodies were originally at the centre with the other earthy stuff and then received their place closer to the periphery.[92] The other difficulty is that, in this latter version, it is not entirely clear why these rocks left their natural central position. One possibility is that they were simply torn off and taken up by the extreme force of the rotation,[93] the other is that there was some aither trapped in them and they were blasted upwards by these portions of aither.[94] Yet no matter which scenario we eventually opt for, it is clear that the formation of the celestial bodies is explained in mechanistic, physical terms in Anaxagoras' cosmogony. This is not the case with the Derveni author. In his theory, Mind itself fashioned the sun (and probably also the stars) in an immediate way, and this deed is Mind's most essential direct contribution to cosmogony.

[90] The evidence for Diogenes' view on the shape of the earth is circumstantial at best. Laks (1983) 175 challenges the standard interpretation and maintains that a spherical earth fits better with the logic of the cosmogonical process. This is not accepted e.g., by KRS. The major piece of evidence against Laks' hypothesis is that Aëtius in 59 A67 DK ascribes also to Diogenes the Anaxagorean theory that the earth got somehow tilted at one end with the emergence of terrestrial life.

[91] Hippolytus in *Ref.* 1.8.3–10 = 59 A42 DK says that διὰ τὸ τὸν ἀέρα ἰσχυρότατον ὄντα φέρειν ἐποχουμένην τὴν γῆν. In the identification of Okeanos as air, Okeanos' strength is the kernel of the allegorical explanation (col. 23.1–9). It would be tempting to think that the air *qua* Okeanos encircles and keeps up the earth, as has been suggested by Burkert (1968) 97. Yet, we apparently have the entire argument about Okeanos and this role of the air is certainly not spelt out here.

[92] Cf. Stokes (1965) 225–9. [93] Cf. *HGP* II, 302. [94] Cf. Sider (1981) 118.

The other contrasting point is that the Derveni author did not follow the more modern view advocated by Anaxagoras (and followed with some modifications by Diogenes of Apollonia and the Atomists), that the heavenly bodies are formed of earthy stuff. Quite the contrary, it is a vital part of the Derveni author's cosmogonical theory that the sun and the stars are made of the fire which was originally mixed with the rest of matter.

I have already discussed the more specific point whether the Derveni author's moon has her own light or only reflects the light of the sun.[95] The question is relevant in the present context as well, as according to some of our sources Anaxagoras pioneered a specific theory of lunar light. It has to be noted that it is not immediately obvious exactly what Anaxagoras' innovation was.[96] At any rate, some of our sources claim that Anaxagoras' moon not only reflects the sun's light but also absorbs and emits some of it even when she is not directly lit by the sun.[97] In a recent discussion of Anaxagoras' contribution to the theory of lunar light Dmitri Panchenko, developing a hypothesis advanced by Denis O'Brien, suggested the following: the solid body of the moon on the one hand reflects the light of the sun, but on the other hand it absorbs some of the heat emitted from the sun and on account of this absorbed heat, it has a dim light on its own, gleaming like a hot piece of charcoal.[98] Daniel Graham, on the other hand, has argued that Anaxagoras took from Parmenides the theory that the moon receives its light from the sun, and Anaxagoras' innovation was to use the theory to explain eclipses.[99]

The reconstruction of Anaxagoras' lunar theory is certainly not easy: to begin with, it is not obvious why the moon is not ignited by the same effects as the other celestial bodies. But irrespective of such remaining problems, I cannot see in what way Panchenko's tempting interpretation is supposed to be explanatory of the Derveni author's contention that 'Those things out of which the moon is (composed) are the whitest of all, divided according to the same principle (*or* measure), but they are not hot' (col. 25.1–3), as has been more recently maintained by Richard Janko.[100]

[95] Cf. 246–9 above.

[96] Cf. Plato *Crat.* 409a–b; Hippol. In 59 A42.10 DK; Plut. *De E apud Delph.* 391a–b. The majority of scholars emend the name of Anaximenes to Anaxagoras in the text of Theon of Smyrna in 13 A16 DK; see, e.g., Whörle (1993) ad loc.

[97] Cf. Plato *Crat.* 409a–b; see also Sider (1981) 122.

[98] Panchenko (2002) 328–9. [99] Graham (2002) 364–5.

[100] Unfortunately, Janko does not give his reasons for the contention that 'The same doctrine [sc. the one ascribed to Anaxagoras by Panchenko] fully explains this passage in the papyrus' (Janko (1997) 66). The problem, as it seems to me, is aggravated by the fact that Janko originally translated τὰ δ' ἐξ ὧν ἡ cελήνη [λ]ẹυκότατα μẹ̀ν τῶν ἄλλωγ as 'those of which the Moon consists are the brightest

As I have said in discussing this passage of the papyrus, I would tentatively accept that the Derveni author's moon has reflected light – or, at least, that the author emphasises that lunar matter has some kind of visual specificities as compared to solar and stellar matter. As the evidence is scanty in the one case, and vexed in the other, I suspend judgement on the nature of the relationship between the Derveni author's theory and that of Anaxagoras.

I would nevertheless call attention to one point, which certainly marks a difference. No matter how we interpret the Anaxagorean material, it is quite clear that the visual characteristics of the moon are explained as supervenient on the post-cosmogonical astronomical situation. The ignition – and hence the heat and brightness – of the sun is largely due to the sun's present position in the whirling aither and not to the inherent characteristics of the stone it is. It follows, then, that the observable visual features of the moon, dependent on the thermal and visual features of the sun, are causally linked to these astronomical factors. As opposed to this conception, the formulation in the papyrus (τὰ δ' ἐξ ὧν ἡ cελήνη [λ]ε̣υκότατα μ̣ὲν τῶν ἄλλωγ) makes it clear that the visual characteristics emphasised by the author are inherent in the lunar matter, and not dependent on the specific post-cosmogonic astronomical position either of the sun or of the moon herself.

There is, however, a passage in Euripides which might constitute a remarkable, albeit admittedly indirect and slight, link between Anaxagoras' astronomical theories and those of the Derveni author. In *Orestes* 982–6 Electra sings:

τὰν οὐρανοῦ
μέσον χθονός τε τεταμέναν
αἰωρήμαcιν
πέτραν ἁλύcεcι χρυcέαιcι φερομέναν
δίναιcι βῶλον ἐξ Ὀλύμπου

Although Electra only speaks about a rock, the context shows that she most probably means the sun.[101] This is also evinced by the imagery of the golden chain, a reference to *Il.* 8.19–22 and 15.19–20, which apparently became part of the solar symbolism in allegorical practice at an early stage. Incidentally, Pierre Boyancé has maintained that the allegorical explanation of Zeus's

of all'. How can Anaxagoras' doctrine explain – on any interpretation of it – that the moon is composed of a stuff which is inherently brighter than that out of which the sun and the stars are composed? Janko (2001) already had 'whitest of all', which is in my view the correct translation, but does not come back to the theory of lunar light. Janko (2002) has now emended the superlative into a comparative, and translates 'are whiter than the other elements'.

[101] So Willink (1986) ad loc. Cf. also Richardson (1975) 70. Contra West (1987) ad loc.

golden chain in the *Theaetetus* goes back to the cosmological speculations of Oinopides of Chios.[102] More important, this clod, βῶλος, on golden chains is echoed by the expression χρυσέα βῶλος which Diogenes Laertius quotes from the now lost *Phaethon* of Euripides, and exactly in connection with Anaxagoras' astronomical theories:

> He [sc. Anaxagoras] is said to have predicted the fall of the stone at Aegospotamoi, which he said would fall from the sun. Hence Euripides, who was his pupil calls the sun a 'golden clod' in the *Phaethon*.[103]

Now, in the *Orestes* passage we have the δίνη together with the mention that this clod is 'from Olympus', which recalls Anaxagoras' idea that the clods and rocks which were to become the heavenly bodies were torn off from the protruding parts of the earth and held up by the vortex.[104] Thus Euripides apparently epitomises here in a poetic form the basic features of the Anaxagorean astronomical theory. On the other hand, it is remarkable that Euripides uses the word αἰωρήμασιν which recalls the Derveni author's predilection for the verb αἰωρέω to describe the way the heavenly bodies are located and held firm in and by the air.[105]

Before concluding this section, I have to discuss Walter Burkert's recent thesis which has an immediate bearing on our present concern.[106] In his contribution to the volume edited by Laks and Most, Burkert focuses on column 25 of the papyrus, and tries to establish parallels between Anaxagoras' cosmology and that of the author of the papyrus on the basis of this passage. The hypothesis is built up of two separate but interrelated claims. (i) Speaking about 'other things now in the air, floating far away from each other' (col. 25.3–4), the Derveni author means not the stars in general, but the Milky Way in particular. (ii) When the author says that these heavenly bodies are kept apart by force to prevent their coming together in one mass, what the author envisages is the formation of a second sun, and hence the redoubling of the cosmos. Burkert then transfers this theory back to the long-standing debate on Anaxagoras' multiple worlds, claiming, although

[102] Boyancé (1952) 347–8.

[103] D.L. 2.10. Cf. Diggle (1970) and Collard, Cropp and Lee (1995) ad loc. A scholiast on the passage just quoted from the *Orestes* is of the same opinion as Diogenes: Ἀναξαγόρου δὲ μαθητὴς γενόμενος ὁ Εὐριπίδης μύδρον λέγει τὸν ἥλιον· οὕτως γὰρ δοξάζει. μύδρον δὲ καλοῦσι τὸν πεπυρακτωμένον σίδηρον. πέτραν δὲ καὶ βῶλον κατὰ τοῦ αὐτοῦ εἴρηκεν.

[104] 59 A42 DK. Cf. above. West (1987) ad loc. interprets the Euripides passage along these lines.

[105] If we accept the standard interpretation of ἀπολαμβανομένου ἐμ μέσωι in col. 15.3–4 (which I do not), we have a further notable parallel, since Electra says that the sun is in the middle (μέσον) between earth and sky.

[106] Burkert (1997).

with some measure of tentativeness, that Anaxagoras might have had a similar strategy: adumbrating the possibility of the emergence of rival worlds just to rule this possibility out.[107]

The evidence for (i), it seems to me, is extremely weak. In the *Meteorology* Aristotle says that:

> Anaxagoras, Democritus, and their schools say that the Milky Way is the light of certain stars. For, they say, when the sun passes below the earth some of the stars are hidden from it. Now the light of those on which the sun shines is invisible, being obscured by the rays of the sun. But the Milky Way is the peculiar light of those stars which are shaded by the earth from the sun's rays. (*Meteor.* 345a25–31, trans. Webster)

What weakens the validity of the alleged parallel with the Derveni author is that many of Anaxagoras' heavenly bodies are invisible even at night. This is why we specifically need the earth's shadow – a negative counterpart, as it were, of the ray of light which makes Leucippus' small particles visible – in the scope of which these lesser heavenly bodies of feeble light can manifest themselves. The statement of the Derveni author, on the other hand, is unambiguous: 'during the day they are invisible being dominated by the sun, whereas during the night it is visible that they exist' (col. 25.4–6). There is absolutely no indication that we need the beam of the earth's shadow on the night sky to make them visible.[108]

As far as (ii) is concerned, my main objection is that Burkert assumes that the counterfactual scenario the Derveni author has in mind is the formation of a *second* sun. But this is unlikely. Why would a second, separate sun form? As the author makes clear, the stars τὴν αὐτὴν δύναμιν ἔχει, ἐξ ὧν ὁ ἥλιος cυνεcτάθη (col. 25.8–9). So why would the force of 'like to like' be operating between the stars, but not between the stars and the sun? Therefore, it is more probable that – irrespective of the question of whether Anaxagoras raised the possibility of multiple worlds, and if so, how he dealt with this possibility – the cosmic catastrophe envisaged by the Derveni author is that the existing sun would become exceedingly large with the addition of the stars, and this inflation of the sun would make the present cosmic order collapse. This is enforced by the final contention of the author: 'If the god

[107] As Burkert notes (p. 171), this interpretation of the Anaxagorean material was already suggested by Fränkel (1960). On the problem of multiple worlds in Anaxagoras, see now Louguet (2002).

[108] Burkert also adduces a similarity in wording between the Derveni author's ἐπικρατεῖται δὲ διὰ cμικ[ρ]ότητα (col. 25.6) and Anaxagoras' καὶ πάντων ὁμοῦ ἐόντων οὐδὲν ἔνδηλον ἦν ὑπὸ cμικρότητοc (B1). Yet, as Burkert himself acknowledges, the contexts are entirely different. In the papyrus, these smaller things, far apart from each other, are invisible because their more feeble light is overpowered by that of the sun. In Anaxagoras, on the other hand, things are invisible – or more exactly not manifest – because everything is mingled together, and there is nothing separated out.

had not wished the things which are now to exist, he would not have made the sun. But he made it of *such a sort and of such a size* (τοιοῦτογ καὶ τ[ο]coῦτον) as is explained in the beginning of the account' (col. 25.9–12). The danger is not the formation of a rival cosmos, but the destruction of the present one.

THEORY OF MATTER

Burkert was the first to suggest that the sentence in col. 19.1–2 εκ[. . .]τὰ ἐόντα ἓν [ἕκ]αcτογ κέκ[λητ]αι ἀπὸ τοῦ | ἐπικρατοῦντοc should be understood in the light of the last sentence of Anaxagoras' B12 (ἕτερον δὲ οὐδέν ἐcτιν ὅμοῖον οὐδενί, ἀλλ' ὅτῳ πλεῖcτα ἔνι, ταῦτα ἐνδηλότατα ἓν ἕκαcτόν ἐcτι καὶ ἦν), taken together with the Theophrastean rendering of the point (πάντων μὲν ἐν πᾶcιν ἐνόντων, ἑκάcτου δὲ κατὰ τὸ ἐπικρατοῦν ἐν αὐτῷ χαρακτηριζομένου Simpl. *Phys.* 27.2 = 59 A41.19–20 DK).[109] Burkert elaborated this interpretation in his subsequent paper,[110] and the suggestion has been referred to and treated by various commentators – both on Anaxagoras and on the Derveni author – as a received fact ever since.[111]

As already argued above,[112] I think there are strong reasons to challenge this interpretation. First, it is misleading to treat the sentence quoted from the papyrus without its immediate context. The whole passage runs as follows:

> . . . existing things have been called each single name by reason of what dominates (them); all things were called Zeus according to the same principle. For the air dominates all as far as it wishes. (col. 19.1–4)

The most important counter-argument, which I have already sketched out, is that there is no suggestion in the text of the papyrus to the effect that the air dominates *from within* the individual things.[113] As a matter of fact, other references to domination suggest that things are dominated by the air as they are in the air. On the other hand, it is obvious that Anaxagoras' theory requires that the dominant element is a quantitatively predominant *ingredient* within the individual entity. And this must be the interpretative background of the tendentious translations of the sentence of the Derveni text, starting with Burkert's rendering 'nach dem *in ihm* vorherrschenden

[109] Burkert (1968) 98. [110] Burkert (1970) 444–7.

[111] Just to mention a few, Laks (1997) 128; Janko (1997) 64–5; Furley (1989) 62 n. 61; with more caution Baxter (1992) 138. Sider refers to it in the discussion of B12 (Sider (1981) 108), and elaborates the point in Sider (1997) 136–7.

[112] 270–1. [113] The plural of Tsantsanoglou's reading πάντων (col. 19.3) is important.

benannt sei'[114] down to the translations by Laks and Most 'has received its name from what dominates *in it*' and by Janko 'has been called after the dominant (element) *in it*' (italics are mine).

Besides, I do not think that it is merely due to the physical condition of the papyrus that domination is only mentioned with regard to air and fire. It is much more likely that it is in accordance with the author's physics that we have two active elements, air and fire, which can dominate the others. It seems to me improbable that the author would speak about, for instance, bone dominating the others. Air and fire are special cases in the Derveni author's theory, thus the assertion with regard to air should not encourage us to generalise the principle.

Furthermore, as Laks has rightly pointed out,[115] Anaxagoras' principle of predominance is part and parcel of Anaxagoras' complex ontology and theory of matter. Indeed, it can only work in such a system where we have as many different elements as the many different types of matter we see in the world.[116] Take Empedocles, for example. Had Empedocles held a similar principle of predominance, we would have a very limited number of types of bodies around us, because either there is a balance of elements, or it is either earth or water or air or fire which quantitatively predominates in an entity. What counts in Empedocles is not the preponderance of one of the elements over the others, but the proportion of the ingredients. It is noteworthy that Laks infers that the Derveni author was probably closer to Anaxagoras' richer ontology than to Empedocles' four-element theory only because the principle of predominance, allegedly maintained also by the author of the papyrus, would not make sense otherwise.

Moreover, the Anaxagorean theory of predominance requires that the lumps of matter – any lumps of matter – have a complex internal structure containing all kinds of ingredients. Excepting the original situation, when things are mingled under the force of fire, I can see no sign of such a view in the Derveni text. Once again, commentators have maintained that the Derveni author endorsed this Anaxagorean theory only because they have taken for granted that the author had accepted the Anaxagorean principle of predominance. Yet I do not think that, for example, when the author says that τὰ δ' ἐξ ὧν ἡ cελήνη [λ]ẹυκότατα μ̣ὲν | τῶν ἄλλωγ κατὰ τὸν αὐτὸν λόγον μεμερισμένα | θερμὰ δ' οὔκ ἐστι (col. 25.1–3), we should understand this as meaning that although these lumps of matter contain

[114] Burkert (1968) 98. [115] Laks (1997) 128.

[116] Admittedly, the picture would be somewhat different with the interpretation sketched in 289 n. 55 above.

hotness – just as everything else contains everything else – they are not hot because coldness prevails in them over their internal but concealed hotness.

Finally, we can raise a further highly important point – yet again made by Laks – against the hypothesis that we can reconstruct the Derveni author's theory of matter on the basis of Anaxagoras. As Laks has pointed out, the author's cosmological and physical account is economical in the sense that it focuses only on those elements of cosmology which evince the working of the intelligent supreme being.[117] To all intents and purposes, to accommodate the highly complex Anaxagorean theory of matter, possibly the most complex in the whole Presocratic period, would be entirely superfluous and indeed an unnecessary complication in such a framework.

To sum up, I am more than ready – as should have become clear – to acknowledge Anaxagoras' crucial importance in the doctrine of the Derveni author. I would nevertheless resist the attribution of a highly complex theory of matter to the author on the basis of (as I see it) a misinterpreted and tendentiously translated sentence.

[117] Laks calls the Derveni author's theory a 'truncated cosmology' in this context (Laks (1997) 132).

CHAPTER 8

Diogenes of Apollonia and Archelaus of Athens

Ever since the first studies of Merkelbach and Burkert, Diogenes of Apollonia has always figured among the philosophers whose importance is considered to be a determining one on the Derveni author.[1] Burkert's early study has already established the most obvious points of contact, and further studies on the papyrus have mainly taken over his conclusions.

However, two recent studies, the one by André Laks, the other by Richard Janko, have enlarged the scope of the discussion on the relationship between Diogenes of Apollonia and the author of the papyrus. The two papers, written independently of each other, consider the same basic question: what is the relationship between physics and religion in the papyrus? The two scholars come up with different answers, but Diogenes features in both. The Derveni author, according to Laks' interpretation, is an enlightened priest, who had developed a philosophically valid and interesting thesis by trying to navigate between Anaxagoras' Mind striving towards transcendence and Diogenes' completely immanent Intelligence.[2] Janko, arguing from the opposing corner, tries to show that the Derveni author could well be one of the *physiologoi*, for 'natural philosopher' and *magus* are not mutually exclusive categories. Moreover, Janko claims that the similarities between the author's doctrines and those of Diogenes of Apollonia are so striking[3] that the Derveni author might after all be Diogenes himself.[4]

In what follows I shall propose a systematic comparison between the Derveni author and Diogenes of Apollonia. The focal points and the

[1] Merkelbach (1967) 22; Burkert (1968) 97 and 99–100; cf. also, e.g., Boyancé (1974) 98–9; West (1983) 80; Most (1997) 118.

[2] Laks (1997) 129–30.

[3] Janko (1997) 83: 'In short, only someone who believed in the doctrines of Diogenes of Apollonia could have written the Derveni papyrus' and p. 92: 'From the above arguments I conclude that the Derveni treatise accords perfectly with the doctrines of Diogenes of Apollonia.'

[4] Janko (1997), especially 80–7. Cf. also Janko (1998). Janko's ultimate claim, however, argued more in detail in Janko (2001), is that the Derveni author is probably Diagoras of Melos. I discuss this hypothesis in the Appendix.

structure of the survey will be the same as in the case of Anaxagoras. I shall start by examining the respective intelligent cosmic principles, with special emphasis on their causal role and relation to matter, to continue with the authors' accounts of the constitution of the physical world. My conclusion will be that the convergences are important but far less overwhelming than others have suggested. Although the intelligence and divinity of air is a fundamental point shared by Diogenes and the author of the papyrus, there are other points – especially in the cosmic role of the intelligent principle and in the details of the respective cosmological theories – in respect of which Diogenes and the Derveni author maintain a different view.

It is customarily held that, together with Anaxagoras, Diogenes had a considerable effect on contemporary intellectual circles. The majority of scholars accept that Diogenes had a certain renown at least in Athens and at least for a certain time during the second half of the fifth century. Apart from a reference to identifiably Diogenean ideas in a fragment of Philemon,[5] the major evidence for this claim is that Socrates in Aristophanes' *Clouds* appears to hold Diogenes' doctrines.[6] It is not clear, however, whether the historical Socrates actually held all those doctrines[7] or whether Aristophanes just merged the two figures. If the later option is the right one, it is somewhat problematic what Aristophanes' objective could be in portraying Socrates professing doctrines which the audience could recognise as the tenets of another philosopher.

Besides, we find a Τῶν Διογένους cυναγωγή in one book in the list of Theophrastus' works.[8] In some specific doctrines Diogenes' influence is detectable in Aristotle.[9] His influence on professional doctors, and on the Stoics, is also observable.[10]

The central problem in evaluating Diogenes' impact on later philosophers concerns the role we ascribe to him in the emergence of teleological

5 Fr. 95 *PCG*.

6 For a recent discussion of the point, see Vander Waerdt (1994) 61–6. Kahn (1994) 106–7 had previously maintained, however, that the doctrines in question are not specific to Diogenes, and hence that the evidence is overestimated. At any rate, I find Vander Waerdt's contention that 'Diogenes exercised an influence in popular thought unparalleled among pre-Socratic philosophers' (p. 61) exaggerated.

7 So Janko (1997); Vander Waerdt (1994) 64 n. 58 is more cautious.

8 D.L. 5.43.

9 E.g., his theory of the role of *pneuma* in the semen, which informed Stoic thinking, perhaps through Aristotle's mediation or through some medical writings. Cf. Hahm (1977) 68–71. In more general terms, I would agree with those who say that Diogenes might have had a more profound influence on Stoicism than usually agreed. On this question see the dated but still instructive analyses of Theiler (1965) 57–61; and Diller (1941) 375–81.

10 E.g., the concept and role of ἰκμάc, moist excretion from the earth, recurrent in Hippocratic writings. Cf. also the material collected in Deichgräber (1935) 27–30 and Jouanna (1965).

thinking. It is a matter of dispute whether or not Diogenes held a more clear-cut teleological cosmology than his predecessors, and if the answer is positive, whether or not he had an immediate influence in this respect on later thinkers, most notably on Socrates, Plato, Aristotle and the Stoics. The comparison with the Derveni author might add further material to the general discussion of this point.

MIND AND INTELLIGENCE

The first point, minor but not to be overlooked, is a difference in terminology. The Derveni author follows Anaxagoras in calling his intelligent cosmic principle νοῦς, and thus, at least in this respect, joins the mainstream of post-Anaxagorean philosophers. Diogenes, by contrast, chooses and consistently uses the cognate word νόησις. Quite apart from the question whether or not this is indicative of any significant difference in doctrine,[11] the point certainly establishes an opposition and renders more problematic the hypothesis claiming an immediate dependence of the Derveni author on Diogenes. At any rate, I shall mark the terminological difference by using the word Intelligence for Diogenes' νόησις, while retaining the word Mind for the Derveni author's νοῦς.

This difference in usage notwithstanding, the three fundamental characteristics of Diogenes' cosmic principle and that of the Derveni author correspond remarkably well: both are (i) intelligent, (ii) divine, and (iii) identified as air. In what follows, I attempt to analyse the similarities and differences of the two cosmic principles with regard to these qualifications.

The primary qualifications of Diogenes' Intelligence

(i) Having previously argued for material monism in B2,[12] Diogenes deduces in B3 that there must be an intelligent cosmic principle by the application of what would later be called the 'argument from design'. We see that the cosmos shows order, and such an order could not have come into being without the intervention of intelligence. Therefore, there must be an intelligent principle. Apart from the not particularly informative claim that it 'knows much', which figures in the list of the principle's most important qualifications in B8, the remaining fragments do not give any further indication about the principle's intellectual capacities. We can assume thus that

[11] Cf. Laks (1983) xxxix.

[12] For the order of the fragments and the general structure of Diogenes' argument, see Laks (1983) 11–13.

its main mental power consists in providing the cosmos with its rational order. So far Diogenes' Intelligence accords perfectly with the Mind of the author of the papyrus.

(ii) The theological implications of Diogenes' intellectual principle are also clear. Diogenes not only describes the cosmic intelligence by adjectives traditionally associated with divine beings and praises it in hymnic style,[13] but, as opposed to Anaxagoras, he seems also ready to call it a god: 'for this itself seems to me to be a god'.[14] Thus in this respect too, the Derveni author appears to be very close to Diogenes, indeed, closer than to Anaxagoras.

In this context, commentators call attention to a further notable point of contact between Diogenes and the author of the papyrus: both identified their divine intelligent cosmic principles with Zeus.[15] For Diogenes, our evidence comes from Philodemus' *De pietate*:

> Diogenes commends Homer that he spoke about the divine not in a mythical way but according to the truth: he says that he (sc. Homer) believes that the air is Zeus, for he said that Zeus knows everything.[16]

So Diogenes claimed that in so far as the Zeus of Homer knows everything, he can be identified with the intelligent air.[17] One should not forget, however, that in the Derveni text, even though Zeus is the main character of the poem, he is only one of the gods whom the author identifies with the cosmic principle. The parallel would certainly be more powerful if Diogenes had also identified the other gods of Homer with air. At any rate, the evidence from Philodemus makes it likely that Diogenes not only was ready to theologise his air, but also tried to allow some communication with the traditional representations of gods and accordingly agreed to some degree of allegory.[18]

13 Cf. B7 DK and the first part of B5 DK. Cf. Deichgräber (1933) 353–5.

14 B 5.3–4 DK: αὐτὸ γάρ μοι τοῦτο θεὸϲ δοκεῖ εἶναι. It is true that this reading, proposed by Usener, requires two emendations to the received text. But, as Laks ad loc. notes, it also accords well with Theophrastus' testimony in A19 DK: ὅτι δὲ ὁ ἐντὸϲ ἀὴρ αἰϲθάνεται μικρὸν ὢν μόριον τοῦ θεοῦ. Note also that this idea corresponds with the doctrine ascribed to Hermotimus or Anaxagoras by Aristotle in the *Protrepticus* (fr. 61 Rose): 'For mind is the god in us – whether it was Hermotimus or Anaxagoras who said so – and mortal life contains a portion of some god.'

15 Burkert (1968) 97–8; Boyancé (1974) 98; Janko (1997) 66 and 80.

16 Philod. *de piet.* 6b p. 70 Gomperz: Διο[γέ]νηϲ ἐπαι[νεῖ] τὸν Ὅμηρον ὡς [οὐ] μυθικ[ῶς] ἀλλ' ἀληθῶς [ὑ]πὲρ τ[οῦ] θείου διειλε[γ]μένον. τὸν ἀέρα γὰρ αὐτὸν Δία νομίζειν φηϲίν, ἐπειδὴ πᾶν εἰδέναι τὸν Δία λέγει καὶ [. . . .

17 It is nevertheless interesting to see that the ground for Diogenes' allegorical interpretation is the omniscience of the Homeric Zeus, which is not all that evident in the epic. Laks ad loc. refers to *Il.* 24.88: Ζεὺϲ ἄφθιτα μήδεα εἰδώϲ.

18 Laks (1983) 102 expresses his doubts whether Diogenes was indeed in the business of allegorical interpretation.

(iii) The third point concerns the physical determination of the intelligent principle. Having shown in B3 that there must be a cosmic intelligence, Diogenes urges in B4 that the cosmic intelligence is air. For this claim he applies, on the one hand, what we could call the argument from breathing, and, on the other, implicitly calls on the analogy between microcosm and macrocosm. Hence the intelligent and divine cosmic organising force must be the air. This conclusion is drawn in the first sentences of B5. With the identification of the cosmic intelligence as air we have, yet again, a glaring accord between our two authors.

The remainder of B5 may nevertheless cause some complication. For it states that air has many forms, and specifies that what is life and intelligence in animals and human beings is air when it is within a certain range of temperature: warmer than the atmospheric air but much colder than the air around the sun.[19] It is not immediately obvious which of the following two renderings is the right one: (*a*) the air as such is intelligence, but it is the intelligence of living beings when it is in the specified range of temperature, or (*b*) the temperature of air *qua* intelligence in general must be within this range, therefore also when it functions as cosmic intelligence. Even though the less restrictive interpretation might seem more plausible, many commentators, both ancient and modern, opt for (*b*).[20] If (*a*) is true, the correspondence between Diogenes and the Derveni author remains undisturbed. If (*b*) is the right reading, the parallel needs some qualification. For it follows from the reconstruction I have proposed that, for the author of the papyrus, the air encircling and controlling the sun and the stars must also be intelligent, divine air. Indeed, it is the air's primary cosmic function to enclose and bridle the fire of the heavenly bodies.

Besides, Diogenes' argument concerning the physical representation of the divinity draws attention to a notable difference in perspective. The argument of B4 shows that Diogenes started from the microcosmos and extrapolated to the macrocosmos. On the basis of the physiological function of breathing, he concluded that air is the conveyor of life and intelligence

[19] 64 B5 DK: καὶ πάντων τῶν ζῴων δὲ ἡ ψυχὴ τὸ αὐτό ἐστιν, ἀὴρ θερμότερος μὲν τοῦ ἔξω ἐν ᾧ ἐσμεν, τοῦ μέντοι παρὰ τῷ ἡλίῳ πολλὸν ψυχρότερος. The next sentence makes it clear that within this range the air's temperature shows differences, as the different individuals do not have the same intelligence.

[20] So, e.g., Barnes (1982) 582; KRS 44. It has often been assumed that this interpretation forms the basis for Nicolaus' claim that Diogenes' *arche* is an intermediary between air and fire – on this point, see n. 27 below. I find Guthrie *HGP* II 369 difficult to follow: 'Finally, it is natural to infer from fr. 5 that the purest form of air, which is sheer intelligence and God, is warmer than that which gives life and less perfect powers of cognition to the animals and ourselves.' As far as I can see, there is nothing in the fragment to substantiate the claim that the divine air is warmer than our breath, nor does it speak about degrees of purity.

in living beings. From this he deduced that air must have a similar role also in the cosmos. What is the most valuable and important in us must also be the supreme cause and the divine itself in the cosmos as well. There are some indications that the same line of thought had already been applied by Anaximenes. Although we do not possess any evidence for the Derveni author, it is not unlikely that he also made a connection between the role of air and breathing for human beings and the intelligence and cosmic role of Mind/air.[21]

Although the inference is not so obvious, Heraclitus might have been extrapolating from the macrocosmos to the microcosmos. Fire has no apparent role in us, but it does in the cosmos. Yet what is the most dignified element in the cosmos must also be the intelligent element in it; and the intelligent element in the cosmos must also be the intelligent element in us. According to the first view, what can be recognised as divine in us must also be divine in the cosmos;[22] while according to the second view, what can be recognised as divine in the cosmos must also be the divine in us.[23]

Intelligence and matter

So far our comparison has exhibited an impressive accordance between the two authors. Yet divergences will soon appear. The first difference concerns the relationship between the ordering principle and the rest of the world. On the standard interpretation of Diogenes, it follows from the argument of B2 that not only the intelligent cosmic force, but also all things are modifications of the same stuff, or simply are the same,[24] i.e., air. It entails that that which is commonly recognised as air governs and arranges (cf. B5.1–2) other, non-manifest forms or modifications of air. Or, by the application of the restrictive definition of intelligence, one form of that which is commonly recognised as air governs and arranges other forms of commonly recognised air and also the non-manifest forms of air.

As opposed to this concept, I have argued in the reconstruction of the Derveni author's ontology that, even though monism cannot completely be discarded, and there are even some indications that for instance earth was formed out of air, there are also strong arguments to show that the Derveni author did not subscribe to material monism. For it appears that

[21] Cf. Arist. *De an.* 410b27–30: τοῦτο δὲ πέπονθε καὶ ὁ ἐν τοῖς Ὀρφικοῖς καλουμένοις ἔπεσι λόγος· φησὶ γὰρ τὴν ψυχὴν ἐκ τοῦ ὅλου εἰσιέναι ἀναπνεόντων, φερομένην ὑπὸ τῶν ἀνέμων . . . See also 346 below.

[22] Cf. e.g., 64 A19 and 42 DK. [23] Cf. e.g., Heracl. 22 A16 DK.

[24] B2.1–2: πάντα τὰ ὄντα ἀπὸ τοῦ αὐτοῦ ἑτεροιοῦσθαι καὶ τὸ αὐτὸ εἶναι.

the air-fire opposition is at the heart of the author's physics, and that at least air and fire are not reducible to each other.[25]

We get a different view of the relationship between the cosmic organising force and the other constituents of the cosmos from a dissenting interpretation of Diogenes, propounded by Jonathan Barnes.[26] Barnes convincingly shows, I think, that there is nothing in the original fragments from which to conclude that the underlying stuff in Diogenes is air. Moreover, we can assume on the basis of B7, Barnes claims, that Diogenes' underlying stuff is simply body (cῶμα), not qualified by any perceptible properties: an approximation of Aristotle's prime matter.

The discussion of the Diogenean material in Simplicius actually makes it likely that the treatise did not contain any explicit arguments or statements to the effect that the underlying stuff, which can take the form of earth, water, air and fire and thus provide the unity of the cosmos as required by the argument of B2, is air itself.[27] Simplicius himself draws the conclusion that, according to Diogenes, the cause of life and intelligence in animate beings 'is from *this* arche *which is air*'[28] on the basis of B4. But, as shown above, this fragment only states that the air is the source of intelligence for living beings – it does not say that this is also the *arche* in the sense of material substrate.

Two options remain. Either Diogenes only showed that the intelligent principle causing order in the cosmos is air, tacitly implying thus that it

[25] Janko (1997) 65 with n. 40, following a suggestion by Robert Sharples, states that 'the expression οὐκ ἐξ ἑτέρ[ων] | ἕτερ' ἄλλ' ἑτε[ρ' ἐκ τῶν αὐτῶν] (col. 15.9–10) recalls the material monism of Diogenes of Apollonia.' Yet the expression does not entail that the things are of the same stuff; it can simply mean that the same constituents enter into different arrangements from one cosmic phase to the other.

[26] Barnes (1982) 574–6.

[27] B3 establishes that the order of the cosmos calls for a cosmic intelligence. Then some steps of the argument between B3 and B4 are apparently missing (cf. the ἔτι δὲ πρὸc τούτοιc καὶ τάδε introducing the argument of B4), and B4 is already arguing for the identification of the organising, intelligent principle as air. Hence B5 can already start from this conclusion and continue by praising this divine air and, further on, show that it can appear in many forms. Yet when Diogenes draws attention to the different manifestations of air in B5, he only mentions those which are commonly recognised as air – 'that which is called air by people'. He certainly does not say that, for example, not only the air around the sun but the sun itself is of air. This is the reason why commentators have to assume that the passages not quoted by Simplicius from between B3 and B4 must have argued to this effect (so e.g., Laks (1983) 39; KRS 443). But this would be highly surprising in view of the fact that Simplicius starts to discuss Diogenes in the context of the interpretation of Aristotle's *Physics* 187a12–26, and in order to decide which is right: the traditional Peripatetic interpretation according to which Diogenes' *arche* is the air, or the alternative reading of Porphyry and Nicolaus of Damascus according to which Diogenes' *arche* is an intermediary substance between air and fire. Thus Simplicius would be quoting all these interesting passages from Diogenes, but would leave out those arguments which could give the solution to his original problem: the identification of the unifying material substrate as air.

[28] Simpl. *in phys.* 152.16–7: ὅτι καὶ ἄνθρωποι καὶ τὰ ἄλλα ζῷα ἐκ τῆc ἀρχῆc ταύτηc, ἥτιc ἐcτὶν ὁ ἀήρ, καὶ ζῇ καὶ ψυχὴν ἔχει καὶ νόηcιν λέγων οὕτωc.

must also be the basic material substance required by the argument of B2. Or Diogenes argued on the one hand for a basic form of matter out of which and into which earth, water, air and fire can turn, without further specifying this stuff, and showed, on the other hand, that there must be something intelligent, and this intelligent thing must be (a specific form of) air.[29] On this latter interpretation, the Peripatetic authors recognised, just as we have apparently seen Simplicius do, that air has a privileged position in Diogenes' philosophy and, in their urge to identify Diogenes' *arche*, took it as natural that this privileged element must also be the basic stuff.[30] The same logic could also account for the alternative interpretation advanced by Nicolaus: he reserved the privileged status – intelligence – not for air *simpliciter*, but for hot air. Therefore, we must assume that his *arche* was some intermediary between air and fire. Hence Nicolaus' interpretation would also try to determine the identity of the basic stuff solely on the basis of the physical description of intelligence.[31]

This reading of the Diogenean material would certainly leave more distance between the organising principle and the other physical constituents of the cosmos. The air would still be the most important of the elements, but it would no longer be true that air is the material substrate or the archetypal form of matter. However, the air *qua* intelligence would obviously not be exempted from the argument of B2, and it would still be true that all the elements – including air and fire – can transform into each other. Consequently, the difference between Diogenes and the Derveni author, pointed out above, would still not disappear.

Intelligence as cosmic cause

The cosmological role of Intelligence is dealt with in two or perhaps three fragments. In B3 Diogenes draws attention to the fact that we can observe a certain orderliness in things, and this applies not only to the more obvious cases of meteorological phenomena showing periodicity. What is more,

29 Barnes (1982) 574–6 takes the latter view. I would, nevertheless, leave Diogenes' position vaguer than Barnes does, since I do not think that either B2 or B7 entitles us to ascribe to Diogenes a stronger thesis, i.e., that the basic stuff is entirely unqualified by any perceptible properties and thus approximates to Aristotle's prime matter. Even Barnes inclines to agree that air is the starting point in cosmogony (p. 582). I wonder, however, if there is anything in the text to suggest that Diogenes distinguished between the logically prior material substrate, which never actually appears in nature as such, and the temporally prior air. Moreover, the last sentence of B2 seems to imply that the cosmogonical starting point and the archetypal form of matter providing the possibility of interaction in the cosmos are the same.

30 This can also be taken as a response to the objection raised against Barnes' hypothesis in KRS 439 n. 1.

31 According to Laks (1983) 9 n. 2, this had already been suggested by Panzerbieter; see also *ibid.* p. 98.

careful observation can discover that the cosmos displays the best possible arrangement. Yet this arrangement would not be achievable without intelligence:

For, he says, without intelligence it could not be divided in this way, so that it has measures of all things, of winter and summer and night and day and rains and winds and fair weather. The other things, too, if one is willing to consider them with intelligence, one would find to be disposed in the most beautiful possible way.[32]

B5 can already claim that it is the air which has or is the intelligence required by the orderliness of the cosmos. Furthermore, the first part of the fragment gives a description of the 'deeds' of air in hymnic terms:

And it seems to me that that which has the intelligence is that which is called air by men, and that all men[33] are governed by this and <this>[34] has power over everything. For this very thing seems to me to be a god and to have reached everywhere and to dispose everything and be present in everything. And there is no single thing that does not have a share of this; but there is not a single thing that has a share in it similarly to something else, but there are many fashions of air itself and also of the intelligence.[35] (trans. based on KRS)

It is notable that after this description the fragment does not continue with the cosmic role of air/intelligence, but follows with the psychology of man and animals by showing that all of them have a different intelligence.

B7 might add something to the cosmic function of air, by showing that the intelligent air is responsible for the generation and destruction of the individual entities of the world:[36]

[32] 64 B3 DK: οὐ γὰρ ἂν οἷον τε ἦν οὕτω δεδάcθαι ἄνευ νοήcιος, ὥcτε πάντων μέτρα ἔχειν, χειμῶνός τε καὶ θέρους καὶ νυκτὸς καὶ ἡμέρας καὶ ὑετῶν καὶ ἀνέμων καὶ εὐδιῶν· καὶ τὰ ἄλλα, εἴ τις βούλεται ἐννοεῖcθαι, εὑρίcκοι ἂν οὕτω διακείμενα ὡς ἀνυcτὸν κάλλιcτα.

[33] Laks ad loc., in explaining the reference to air's governing men (πάντας as opposed to πάντα) calls attention to PDerv. col. 19.3, where the *ZPE* text had πάντᾳς γὰρ ὁ ἀὴρ ἐπικρατεῖ. This parallel obviously disappears in view of Tsantsanoglou's new reading πάντῳṿ.

[34] The subject for κρατεῖν is problematic. I would prefer to supply a τοῦτο with Calogero than accept with Laks that it is the people who govern everything.

[35] 64 B5.1–7 DK: καί μοι δοκεῖ τὸ τὴν νόηcιν ἔχον εἶναι ὁ ἀήρ καλούμενος ὑπὸ τῶν ἀνθρώπων, καὶ ὑπὸ τούτου πάντας καὶ κυβερνᾶcθαι καὶ <τοῦτο> πάντων κρατεῖν· αὐτὸ γάρ μοι τοῦτο θεὸς δοκεῖ εἶναι καὶ ἐπὶ πᾶν ἀφῖχθαι καὶ πάντα διατιθέναι καὶ ἐν παντὶ ἐνεῖναι. καὶ ἔcτιν οὐδὲ ἓν ὅ τι μὴ μετέχει τούτου· μετέχει δὲ οὐδὲ ἓν ὁμοίως τὸ ἕτερον τῷ ἑτέρῳ, ἀλλὰ πολλοὶ τρόποι καὶ αὐτοῦ τοῦ ἀέρος καὶ τῆς νοήcιός εἰcιν.

[36] This aspect is lost if we accept the correction of τῷ into τῶν proposed by the Aldine, retained by Diels in DK (but not in his edition of Simplicius), and followed, e.g., by KRS. On this reading, the fragment only sets against each other the 'divine' form of air and the 'mortal' forms of it (so KRS 443), or possibly the air as such and the individual entities formed of it. For a defence of τῷ, see Laks (1983) ad loc. On Barnes' interpretation, the first clause is not even about the air, but describes the unspecified material substrate.

And this very thing is both eternal and deathless body, but due to it some things come into being and others pass away.[37]

These are the texts which prompted the debate on the question whether or not Diogenes propounded a teleological cosmology, and if so in what form. For, notably, our doxographical sources do not give any further positive information on the cosmic role of intelligence. Some points are certainly clear: first, that the cosmos shows the best possible arrangement, and, second, that the intelligence of air is the cause of this arrangement. If this is enough for a teleological cosmology, then Diogenes' is no doubt teleological. If, on the other hand, for a teleological view we need an active agent forming and arranging the cosmos in view of a well-defined end,[38] it is not obvious whether we can ascribe such a view to Diogenes. The question is all the more interesting for us as we can certainly ascribe the latter view to the author of the papyrus. As we have seen, the Mind/air of the author actively intervenes in the cosmic processes by 'removing' the fiery particles, and the supreme cosmic principle created the sun and thus started the cosmogony in view of the ensuing cosmological developments. Can we find a similar view in Diogenes' cosmology?

The description of the divine Intelligence is certainly heavily loaded with anthropomorphic language, especially in the hymnic part of B5. The Intelligence commands or steers (κυβερνᾶcθαι) all men, has power over all things (κρατεῖν, cf. also ἰcχυρόν from B8), and, in accordance with the argument of B3, disposes or arranges (διατιθέναι) everything. This stratum of Diogenes' language would certainly suggest a personal cosmic agent. From the statement of B8 that it is 'much-knowing' (πολλὰ εἰδόc ἐcτι), one might deduce that – as in the case of Anaxagoras' Mind – this knowledge is primarily about the calculable outcome of cosmic processes. This line of thought might lead to the conclusion that Diogenes' Intelligence arranges the cosmos according to a plan.

However, the distribution of Intelligence becomes important at this juncture. B5 explains that it 'has reached everywhere' but adds immediately that it is 'present in everything' and that 'there is no single thing that does not have a share [of it]'.[39] This development strongly suggests that the intelligent

37 B7 DK: καὶ αὐτὸ μὲν τοῦτο καὶ ἀίδιον καὶ ἀθάνατον cῶμα, τῷ δὲ τὰ μὲν γίνεται, τὰ δὲ ἀπολείπει.

38 Cf. Laks' critique of the interpretation of Theiler (1965): 'L'erreur est de conclure de ce constat que tout être est le produit d'une technique travaillant la matière d'après la représentation d'une fin' (Laks (1983) XXVIII). See also Hüffmeier (1963). It is quite another problem whether or not we can ascribe to Diogenes the teleological explanation of the constitution of the human body as a sign of the gods' care for man as elaborated in Xenophon's *Memorabilia* 1.4 and 4.3. I shall not tackle this latter question here, for it is not immediately relevant to the remaining text of the Derveni papyrus.

39 Barnes (1982) 580 translates already ἐπὶ πᾶν ἀφῖχθαι as 'to have penetrated everything'.

air can arrange things not because it is a demiurgic cosmic agent, or even an immanent physical constituent of the cosmos as such, but because it is a component in all the individual entities of the cosmos. The individual entities of the cosmos can show a rational structure because they are organised according to a rational component in them. Orderliness is thus not imposed on the phenomena from the outside by an efficient cause trying to realise a plan but things develop in this orderly way because they comprise some intelligence: the organising principle is immanent in them.[40]

In view of this consideration, it becomes understandable that, as far as I can see, there is no actual cosmogonical act to be ascribed to Intelligence. There is certainly no indication of such an act either in the fragments or in the testimonies. Consider the case of Anaxagoras. In his theory, the main cosmogonic act of Mind consisted in starting motion in the primordial mixture. Although it is reasonable to suppose that Intelligence has a motive function also in Diogenes, its motive function cannot be regarded as the cosmogonical act of a cosmic agent for, also as a consequence of the immanence of air, matter has always been in motion.[41]

This interpretation is strengthened by strong circumstantial evidence. As is commonly maintained, the revisions that Archelaus of Athens introduced into the system of Anaxagoras converge at important points with the amendments proposed by Diogenes.[42] One of these is Archelaus' attempt to get Mind closer to the other components of the cosmos. He maintained that 'right from the beginning some mixture inheres in mind'.[43] As according to our sources he accepted the main principles of Anaxagoras' theory of matter,[44] a correlate of this thesis is that mind also inheres in all forms of matter as well. This tenet is explicitly stated in the testimony of Augustine.[45] Thus Archelaus maintained that all the entities of the cosmos contain some portion of Mind – and not only animate beings as Anaxagoras had claimed. We can also assume that Mind's inherence in entities could account for their ordered structure.[46] Yet Archelaus apparently did not assign any immediately cosmogonical role to Mind. As a short notice of

[40] Cf. Jaeger (1947) 166: 'we obviously must think of the Mind [of Diogenes] as working from within outward'.

[41] Cf. e.g., 60 A6 DK.

[42] It is almost impossible to establish the relative chronology of the two thinkers, but most probably their lifetimes overlapped at least to some extent.

[43] 60 A4 DK = Hippol. *Ref.* 1.9.1: οὗτος δὲ τῷ νῷ ἐνυπάρχειν τι εὐθέως μῖγμα.

[44] 60 A4 and A5.1 DK.

[45] 60 A10 DK = August. *De civ. d.* 8.2: 'etiam ipse de particulis inter se similibus quibus singula quaeque fierent ita putavit constare omnia, ut inesse etiam mentem diceret, qua corpora aeterna, id est illas particulas, coniungendo et dissipando ageret omnia'.

[46] Cf. 60 A11 DK = Clem. *Protrept.* 5.66: τούτω μέν γε ἄμφω (sc. Anaxagoras and Archelaus) τὸν νοῦν ἐπεστησάτην τῇ ἀπειρίαι.

Aëtius informs us: 'Archelaus [said that] god is air and Mind, but Mind is not creator of the cosmos.'[47]

The other parallel comes from Heraclitus. As Hans Diller forcefully argued, Heraclitus' influence is pre-eminent in Diogenes. This is so especially in the characterisation of Intelligence/air, where Heraclitus' Logos/fire is more dominant than Anaxagoras' Mind.[48] Now, on the standard interpretation of Heraclitus, the fire, the physical aspect of Logos, is one of the elements partaking in physical transformations, but is also somehow the archetypal form of matter. It is endowed with intellectual features, and is the conveyor of soul and intelligence in individuals. Moreover, it also maintains cosmic order by regulating the measures of physical, cosmological processes.[49] All these features accord well with Diogenes' conception of air.[50] Yet, as B30 of Heraclitus underlines, there is no cosmogonic agent involved – indeed there is no cosmogony either. Fire is an immanent structuring principle, which continually governs both the individual entities and ultimately the cosmos itself; or more exactly, the individual entities and the cosmos itself can show order, for fire inheres in them.[51] Finally, to the material collected by Diller, we might also add that the idea that the cosmos shows the most beautiful arrangement (ἀνυϲτὸν κάλλιϲτα), stressed by Diogenes in B3, and usually treated as a novelty on his part, is perhaps an echo of Heraclitus B124: ὥϲπερ ϲάρμα εἰκῇ κεχυμένων ὁ κάλλιϲτοϲ, φηϲὶν Ἡράκλειτοϲ, [ὁ] κόϲμοϲ.[52]

I would maintain, then, that both the internal and the circumstantial evidence make it likely that the cosmic Intelligence in Diogenes functions as an organising or structuring force permeating all the entities of the cosmos. It is thus not an independent demiurgic cosmic agent which imposes order on the entities from outside. Accordingly, there is no definite cosmogonical act and cosmogonical moment either. By contrast, the Derveni author's

47 60 A12 DK = Aëtius 1.7.14: A. ἀέρα καὶ νοῦν τὸν θεόν, οὐ μέντοι κοϲμοποιὸν τὸν νοῦν.

48 Diller (1941) 370–81; cf. also KRS 441; see also Janko (1997) 83–4.

49 It is noteworthy that, suspending the more rigid classification of Presocratic 'causes', occasionally even Aristotle can treat the Heraclitean fire together with Empedocles' Love and Strife and the Anaxagorean Mind, standard examples of Presocratic approximations of the efficient cause, cf. *Phys.* 196a17–19.

50 The argument in KRS 441 trying to show the difference between the organising and regularising function of the Heraclitean Logos on the one hand, and the Diogenean Intelligence on the other, is not very convincing to me.

51 It is true that Heraclitus' fire is sometimes treated as demiurgic in later doxographic literature (cf. Aëtius 1.3.11 in A5 and 1.7.22 in A8 DK), but in these texts the terminology betrays Stoic influence.

52 For a recent discussion of the fragment both in the context of Theophrastus' *Metaphysics* and in Heraclitus' philosophy, see Polito (1997). Notably, the double role of fire (material substrate and formal/efficient cause) constitutes the core of the interpretation of the Heraclitean idea of 'the most beautiful cosmos'. Just as in Diogenes, the complete immanence and omnipresence of the primary element is the guarantee of the beauty of the cosmos.

Mind, as we have seen above, retains more traits of a personal god who actively changes the course of events at his own will. Correspondingly, there is a demiurgic agent who has a plan and carries out a specific cosmogonical act in order to realise its plan.

Moreover, when Diogenes claims that the air has power over and can control everything, he highlights the fact that everything has air in it, or has a share of air. As opposed to this, there is no indication in the remaining text of the papyrus that the air would be immanent in all the individual entities. On the contrary, when the Derveni author speaks about the power of air, he claims not that the air is in the other things, but that the other things are in the air. The power of air is coercive not through its immanence but due to the fact that it contains, encircles and delimits the other things.

COSMOLOGY AND PHYSICS

The evidence on Diogenes' cosmology is scanty. It has often been assumed that the relative silence of our sources can be explained by Diogenes' lack of originality. The doxographers found hardly anything original, for Diogenes had taken over the main lines of Anaxagoras' physical theories, amending them in some respects, while simplifying them in others.[53]

There is some debate on the question whether or not Diogenes posited an unlimited number of worlds, and if so, whether these worlds are supposed to follow successively upon each other over time, or to exist simultaneously, or both.[54] At any rate, it seems likely that Diogenes had a cyclical cosmogony.[55] We have seen that, as far as we can judge on the basis of a highly problematic sentence in col. 17.6–9, the author of the papyrus also subscribed to such a view. On the other hand, it is also probable that Diogenes, perhaps following Leucippus, posited an infinite number of worlds scattered in unlimited space.[56] If this is the right interpretation, then the point constitutes a difference, since, even though the text of the papyrus is not explicit on the question, I have argued that it is unlikely that the author's universe comprised more simultaneously existing worlds than one.

[53] So, e.g., KRS 445–6; *HGP* II 370–1; Diller (1941) 360 (stressing also the importance of Leucippus). On the question whether or not Diogenes exposed his cosmological doctrines in the same treatise from which Simplicius quotes the B fragments, see Laks (1983) App. 2.

[54] The evidence of 60 A1, A6 and A10 DK goes for an unlimited number of worlds, whereas Simplicius in 13 A11 DK ranges Diogenes with those who had one cosmos periodically destroyed and reorganised.

[55] Simpl. *in phys.* 1121.12–15, and also 64 B2 DK.

[56] So, e.g., *HGP* II 371; Laks (1983) 184 also trying to account for the conflicting testimony of Simplicius.

According to some interpreters Diogenes held the doctrine of a cosmogonic vortex.[57] In the absence of direct evidence, this interpretation is supported by contextual arguments. Anaxagoras and the Atomists certainly hypothesised a cosmogonic vortex. Moreover, Socrates in the *Clouds* also teaches that there is a cosmic vortex, and Socrates in Aristophanes' portrayal holds the doctrines of Diogenes; therefore, so the argument goes, Diogenes must have espoused the doctrine of the cosmic vortex, which must also have a role in the formation of the cosmos. The only relevant piece of information in the doxography on Diogenes is surely not about a cosmogonic motion. According to the last sentence of A1

> The earth is rounded, fixed in the middle, getting its structure (σύστασιν) according to the circular motion stemming from the hot, and its solidification (πῆξιν) by the cold.[58]

This means that, in the only place where circular motion is mentioned in the context of Diogenes' cosmogony, it is apparently not a primeval vortex.[59] The vortex in Anaxagoras differentiates portions of matter showing different qualities, whereas the circular motion in the sentence just quoted is explicitly said to be 'from the hot'. It is thus supervening on hot which is already separated from cold.

The interest of this passage lies elsewhere for us. It shows that the force of the hot and the cold is evoked in the explanation of the coagulation of the earth. It is a notable point of convergence – ignored by previous commentators – for we have seen that the Derveni author explained the formation of separate entities, and presumably that of the earth as well, along similar lines: the motive force of fire helps the getting together of smaller parcels of matter (sometimes using the derivatives of συνίστημι in this context), whereas some cooling is necessary for their compounding (where the derivatives of πήγνυμι are repeatedly used). It is also noteworthy that, in Diogenes, the end of the world is apparently marked by the gradual drying up of the earth and the homogenisation of thermic conditions.[60] Growing entropy might mean the end of this cosmic phase for the Derveni author as well.

As far as the formation of individual entities is concerned, our sources are extremely meagre. The doxographies reported by Diogenes Laertius

[57] So, e.g., Diller (1941) 360–1; KRS 445.

[58] τὴν γῆν στρογγύλην, ἠρεισμένην ἐν τῷ μέσῳ, τὴν σύστασιν εἰληφυῖαν κατὰ τὴν ἐκ τοῦ θερμοῦ περιφορὰν καὶ πῆξιν ὑπὸ τοῦ ψυχροῦ.

[59] συμπεριφέρεσθαι in Stob. *Ecl.* 1.24.1d = Aëtius 2.13.5 and 9 refers to the revolution of the fixed stars in the present cosmic situation.

[60] A18 DK.

and Ps.-Plutarch suggest, however, that Diogenes returned to the theory of Anaximenes which operated with rarefaction and condensation, although he also referred to other qualitative changes of air as well. As opposed to this, there is no mention of such qualitative changes in the papyrus: the Derveni author certainly seems closer to the separation model used by Anaxagoras.

Another point of contact might be that air holds up the earth. Yet, on the one hand, it can only be suspected in the case of the Derveni author, and, on the other, Diogenes is far from the only one to hold this view.[61]

There is, however, a fundamental difference. Diogenes apparently followed Anaxagoras in defining the substance of the heavenly bodies: he also accepted the evidence provided by the meteorite of Aegospotami and maintained that the heavenly bodies are stones. However, he amended Anaxagoras' original theory by specifying that they have pores like the pumice stone.[62] Moreover, the sun has a distinctive feature. It is not inflamed in the same way as the stars, but its heat and light are due to the fact that it concentrates the rays coming from the aither. That the sun is not inherently fiery is confirmed by another piece of information according to which it is periodically quenched by the effect of the cold.[63] It is also possible to assume with KRS that the sun's orbit is close to the periphery of the cosmos.[64] The moon is imagined to be of the same substance as the sun and the stars.[65]

It follows that their respective astronomical theories constitute a major contrast between the Derveni author and Diogenes. They do not agree either on the matter of the heavenly bodies, or on their formation. Moreover, as opposed to Diogenes, the Derveni author thinks that the sun and the stars differ from each other only by their sizes, but maintains on the other hand that the moon is not of the same substance. And these features are far from negligible, since the theory of celestial bodies lies at the very core of the Derveni author's doctrine: the formation of the sun and the stars out of the surplus of fire is the main manifestation of the god's purposeful cosmogonical activity.

In conclusion, then, I would say that the doctrines professed by the Derveni author converge at certain fundamental points with the tenets of Diogenes of Apollonia. The most notable of these is the intelligence and

61 For Diogenes, see A16a DK, a scholion to Basil.

62 A12 DK, with the different sources as shown under T 26 a, b and c (Laks). 63 A13 DK.

64 KRS 446 on the basis of the last sentence of A6: τὰ <δὲ> κουφότατα τὴν ἄνω τάξιν λαβόντα τὸν ἥλιον ἀποτελέσαι.

65 A14. Laks ad loc. suggests that the difference between the luminosity of the moon and that of the other heavenly bodies is due to their respective locations.

divinity of air, to which we might add the cosmogonical role of hot and cold. There are nevertheless points no less important in respect of which the two authors do not agree. Such is the question of material monism, the functioning of the intelligent principle as cosmic cause, and the cosmogonical role and nature of the heavenly bodies.

ARCHELAUS OF ATHENS

The name of Archelaus is not normally mentioned in connection with the Derveni text. One principal reason for this might be that Archelaus' name is rarely mentioned in modern discussions at all.[66] It seems to me, however, that as far as one can judge on the basis of the insubstantial evidence on his teaching, the mild modifications and shifts of emphasis he imposed on the Anaxagorean tenets touch the doctrine of the Derveni author at important points. Actually, if we assume that on the majority of those points which are left in silence by the doxographers he simply agreed with Anaxagoras, it can even be claimed that the Derveni author is closer to Archelaus than he is either to Anaxagoras or to Diogenes.

To begin with, it seems clear that Archelaus followed Anaxagoras in positing an intelligent principle, which he, too, called Mind. Yet Archelaus, as briefly discussed above, evidently tried to get Mind closer to matter, just like Diogenes. He extended Anaxagoras' doctrine about the universal mixture and claimed that Mind also contains mixture.[67] The other facet of this amendment is that Mind also inheres in other things.[68] This means that Mind has lost its purity, and its *chorismos* from matter which was so much emphasised in Anaxagoras.

There is also strong evidence to show that the air has gained privilege over the other physical elements; some of our sources even identify the air as Archelaus' *arche*.[69] Moreover, there are some indications that air and Mind get remarkably close to each other, if, indeed, they are not identified with each other. At least this is what an entry in Aëtius, already quoted above, suggests:

[66] See, however, now the discussion of all the ancient evidence on Archelaus in Tilman (2000), and Panchenko (1999) on the shape of the earth in Archelaus.

[67] 60 A 4.1DK, see 316 above. It is true that the wording is curious and not without ambiguity: οὗτος δὲ τῷ νῷ ἐνυπάρχειν τι εὐθέως μῖγμα. See also Tilman (2000) 80–2.

[68] 60 A10 DK.

[69] 60 A7 DK = Sext. Emp. *adv. math.* 9.360 and Aëtius 1.3.6. Tilman (2000) 76: 'Il semble cependant avoir conçu ce mélange primordial comme identique à l'air.' Tilman (2000) 82, however, goes, I think, too far when she claims that 'l'*air* d'Archélaos serait le nom générique donné à la totalité des substances matérielles – dont la substance-Intellect chez Archélaos – que reconnaissait Anaxagore.'

Archelaus (said that) god is the air and Mind, but Mind is not creator of the cosmos.[70]

This notice is all the more important for us since it shows that Archelaus also divinised the Mind and the air.[71]

Other testimonies evince that Archelaus followed, at least in the main lines, Anaxagoras' theory of matter.[72] Thus, in sum, he did not reach the more transcendental concept of Mind pioneered by Anaxagoras. He raised the air to a privileged position, and apparently even identified it with Mind, yet did not go so far as Diogenes: he resisted the more radical solution of an air-based monism and retained the pluralist ontology of Anaxagoras. Even though we can only conjecture the philosophical reasons for such a mediating position, all this accords remarkably well with the teaching of the Derveni author.

So far the difference lies in the point which Aëtius finds important enough to attach to the entry just quoted: '. . . but the Mind is not creator of the cosmos.' It is difficult to see what the import of this assertion is. It might be connected to the immanence of Archelaus' Mind, or, on the assumption that the details of his cosmogony were explained in mechanistic terms even more than in Anaxagoras, it might even have something to do with Socrates' criticism of the Anaxagorean Mind in the *Phaedo*.[73] It has to be noted, however, that Clement has a contrasting claim: both Anaxagoras and Archelaus 'impose Mind on the infinity [i.e., matter]'.[74] Nevertheless, if we accept Aëtius' emphatic affirmation at its face value, it certainly makes a point of contrast between Archelaus and the author of the papyrus.

Further significant parallels emerge as we reach the sphere of cosmology. One of the most notable innovations of Archelaus seems to be that he explained much in terms of the effect of hot and cold. According to Diogenes Laertius 'he says that the causes of generation are two: the hot

[70] 60 A 12 DK = Aetius, i, 7, 14: A. ἀέρα καὶ νοῦν τὸν θεόν, οὐ μέντοι κοσμοποιὸν τὸν νοῦν.

[71] In view of these considerations, I wonder why some of the texts which are usually treated as demonstrating the influence of Diogenes of Apollonia could not show the effect of Archelaus. For example, Hecuba's prayer from Euripides' *Trojan Women* is quoted in DK as a C text (*Nachwirkung*) for Diogenes. However, I cannot see anything in it which could not be ascribed to Archelaus. And we must not forget that Diogenes' influence in Athens (or even that he spent a long period of time there, as Jaeger has claimed) is only extracted from these texts, whereas Archelaus was Athenian, perhaps the only Athenian philosopher before Socrates. Moreover, we have every reason to suppose that Socrates' teacher was not unfamiliar to Euripides, as, indeed, according to the Suda (*s.v.* Archelaus), some even said that Euripides was a disciple of his.

[72] 60 A4.1; 5 and A10 DK. [73] So, e.g., KRS 388.

[74] 60 A11 DK = Clem. *Protrept.* 5.66: τὸν νοῦν ἐπεστησάτην τῇ ἀπειρίᾳ. Tilman (2000) 84 maintains, on the basis of D.L. 2.17 and Hippol. *Ref.* 1.9.2–6, that Archelaus followed Anaxagoras in thinking that Mind started the cosmogonic vortex which began the separation of hot and cold. I do not think that the evidence quite warrants this interpretation.

and the cold'.[75] The more detailed statement of Hippolytus is of special interest for us:

> The source of movement is the separation of the hot and the cold, and while the hot moves, the cold stays still.[76]

The meaning of the first clause is not immediately clear, but, in the light of the parallel text of Diogenes Laertius, one possible reading of it is that the generation (of the cosmos and of everything in it) can start when the hot and the cold get separated from each other.[77] As we remember, cosmogony starts in the Derveni text when Mind separates the excess of fire from the rest: with the separation of the hot and the cold the generation of entities can start.

The one verbatim fragment of Archelaus might even go further. According to Plutarch, Archelaus said that 'the cold is a bond',[78] where the context makes it likely that Archelaus was speaking about the cold being 'the bond' of the earth, i.e., being responsible for the fixed position of the earth. If one takes this piece of information together with the testimony on Archelaus' cosmogony in A1, one gets the impression that Archelaus connected the air's ability to hold the earth still with the air's coldness.

A no less interesting point of contact is that in the short summaries of Archelaus' cosmology, the one in Diogenes Laertius, the other in Hippolytus, there are two elements which are said to dominate (κρατεῖν) the other elements: the air and the fire.[79] There is some disagreement between the sources as to whether there is a further hierarchical relationship between these two. In the account of Diogenes Laertius, we read that while the air dominates the earth, the air is dominated by the circular motion of the fire.[80] This is obviously contradicted by the Derveni author's belief that,

75 60 A1 DK = D.L. 2.16: ἔλεγε δὲ δύο αἰτίαϲ εἶναι γενέϲεωϲ.

76 60 A4 DK = Hippol. *Ref.* 1.9.2: εἶναι <δ'> ἀρχὴν τῆϲ κινήϲεωϲ <τὸ> ἀποκρίνεϲθαι ἀπ' ἀλλήλων τὸ θερμὸν καὶ τὸ ψυχρόν, καὶ τὸ μὲν θερμὸν κινεῖϲθαι, τὸ δὲ ψυχρὸν ἠρεμεῖν.

77 This reading might get additional support from another short statement by Aëtius (2.4.5 = A14): Ἀ. ὑπὸ θερμοῦ καὶ ἐμψυχίαϲ, where Meineke has conjectured ἐμψυχρίαϲ instead of the transmitted ἐμψυχίαϲ. ἐμψυχρίαϲ would give good sense but is not attested elsewhere, whereas ἐμψυχίαϲ is somewhat problematic. One solution would be to argue from the identity of Mind with soul (cf. A18); yet this would contradict the idea of a non-cosmogonic Mind. Alternatively, we could also fall back on the identity of soul and air (cf. A17). Tilman (2000) 85 suggests that the primordial opposition between hot and cold expresses all the fundamental physical oppositions.

78 60 B1a = Plut. *De primo frig.* 21 954f: ἡ ψυχρότηϲ δεϲμόϲ ἐϲτιν.

79 This can easily be regarded as a development from Anaxagoras' idea that in the original mixture air and aither 'held things in subjugation' (59 B1 DK: πάντα γὰρ ἀήρ τε καὶ αἰθὴρ κατεῖχεν).

80 60 A1.17 DK: τηκόμενόν φηϲι τὸ ὕδωρ ὑπὸ τοῦ θερμοῦ, καθὸ μὲν εἰϲ τὸ <κάτω διὰ τὸ> πυρῶδεϲ ϲυνίϲταται, ποιεῖν γῆν· καθὸ δὲ περιρρεῖ, ἀέρα γεννᾶν. ὅθεν ἡ μὲν ὑπὸ τοῦ ἀέροϲ, ὁ δὲ ὑπὸ τῆϲ τοῦ πυρὸϲ περιφορᾶϲ κρατεῖται. Tilman (2000) 88–9 maintains, mainly on the basis of this fragment, that Archelaus identified Mind with the cosmogonic fire.

although fire is also able to dominate the other elements, air ultimately takes the upper hand, even over fire. Nevertheless, there is a generally accepted conjecture by Roeper to supplement the cosmological summary in Hippolytus, according to which 'the air dominates the whole'.[81] At any rate, just as in the Derveni text, air and fire are the active elements, and, as we have seen, there is also ample evidence that air is more privileged in Archelaus' system than fire.

The creation of the four world masses in Archelaus, as KRS remark, 'is somewhat obscure and very peculiar'.[82] It nonetheless displays a feature which might be connected to an element of the tentative reconstruction suggested above for the Derveni author's theory. There, we have seen that on the one hand, air and fire are not likely to be able to transform into each other, but, on the other hand, there are some indications that earth and perhaps also water developed from air. In Archelaus' cosmogony, first the hot and the cold get separated, then under the influence of the hot the cold goes through different phases of transformation and turns into water, earth and air. Thus, of the two primary components one produces solely fire, whereas the other is transformed into the remaining three elements.[83]

The last point in our survey, once again, concerns the substance of the heavenly bodies. Here, yet again, Archelaus offers some novelty. According to Hippolytus, the most detailed of our sources, the heavenly bodies, of which the sun is the biggest, are produced 'from a first combustion of the air'.[84] Besides, an entry in Aëtius claims that the stars are of red-hot metal.[85] Even though it would be difficult to reconstruct the details of this astronomical theory, it seems that Archelaus followed the basic idea of Anaxagoras, and would not assent to the alternative view according to which the heavenly bodies are of fire or the 'hot', i.e., the other of the two original world masses.

[81] <τὸν δ' ἀέρα κρατεῖν τοῦ παντός>. At all events, the gender of the relative pronoun and participle in the next sentence (ἀφ' οὗ πρῶτον ἀποκαιομένου) necessitates some supplement.

[82] KRS 388.

[83] Obviously, this also shows a strong Anaxagorean influence: the 'cold' is the functional equivalent of Anaxagoras' primordial 'air', while the 'hot' correlates to Anaxagoras' primordial 'aither'.

[84] 60 A4 DK: ἀφ' οὗ πρῶτον ἀποκαιομένου τὴν τῶν ἀςτέρων εἶναι φύςιν κτλ. Admittedly, the subject of the participle, the air, only comes from Roeper's emendation.

[85] 60 A13 = Aëtius 2.13.6: Ἀ. μύδρους ἔφηςεν εἶναι τοὺς ἀςτέρας, διαπύρους δέ.

CHAPTER 9

Physics and eschatology: Heraclitus and the gold plates

Heraclitus' place in the Derveni author's intellectual landscape is demonstrated by direct textual evidence: apart from Orpheus, Heraclitus is the only author mentioned by name in the surviving text of the papyrus. Besides, col. 4 contains the text of fr. B3 + B94, which – if we accept the standard dating of the Derveni text – is the earliest known quotation from Heraclitus. This quotation has, of course, raised much attention and has been extensively discussed in the literature; the main concern has been to offer a satisfying interpretation of the sentence – the two parts of which were previously known as independent fragments – both in itself and in the framework of Heraclitus' philosophy.[1] More recently, David Sider in his contribution to the volume edited by Laks and Most has not only proposed an interesting interpretation of B3 + B94, but has also given an excellent general overview of the points of contact between Heraclitus and the text of the papyrus.

The above-mentioned papers, however, almost exclusively concentrate on what the papyrus adds to our knowledge of Heraclitus. This is the main focus of Sider's paper too, although he does indeed enlarge our understanding of the Derveni author by pointing out certain parallels with Heraclitus.[2]

[1] Cf. Burkert (1983); Mouraviev (1985); Sider (1987) and (1997); Tsantsanoglou and Parássoglou (1988); Lebedev (1989); Tsantsanoglou (1992) and (1997); Schönbeck (1993); (1998a) and (1998b); Betegh (1994). See also the conjectures proposed in Mansfeld (1987) 266. The three papers written before 1988 were based on the *ZPE* transcript, and a substantial part of the conjectures and interpretative hypotheses proposed in them were definitely ruled out by the new readings provided by Tsantsanoglou and Parássoglou. Lebedev (1989) proposed further conjectures which have, in turn, been refuted by Tsantsanoglou (1992). Schönbeck, for his part, repeatedly challenges the reading εἰ γά̣[ρ in line 8 as proposed by the Greek editors, maintaining that ει̣κ[or ειπ̣[cannot be ruled out. Schönbeck (1998a) gives an extensive survey of Heraclitus' ideas on the sun, with an elaborate methodology and a large scientific apparatus. He also raises some important questions regarding the validity of the interpretative framework ordinarily applied in the discussion of astronomical fragments.

[2] Sider (1997) 130: 'Although my chief aim is to determine the contribution made to our knowledge of Heraclitus by the Commentator, I will also touch upon the reasons for his having quoted Heraclitus in the prologue to his text and commentary on the Orphic poem.'

On the other hand, the studies on the Derveni author's intellectual background have had surprisingly little to say about Heraclitus' place in the Derveni author's theories.

Accepting a substantial part of Sider's conclusions (with modifications and qualifications suggested by Schönbeck) the following survey is meant to be a complement to his approach: my primary aim is to examine Heraclitus' contribution to our knowledge of the Derveni author. My main conclusion will be that Heraclitus is indeed fundamental for our general understanding of the Derveni text. I shall try to show that, even though the Heraclitean influence is not as tangible as that of Anaxagoras on the level of specific cosmological tenets, a conscious and probably critical reflection on Heraclitus is nonetheless a highly important component in the Derveni author's doctrine on the cosmological and eschatological role of fire. This is also the context in which I shall evoke the evidence of the gold plates.

THE QUOTATION

The most obvious starting-point is the quotation of B3 + B94 in col. 4:

ἥλι̣[οc . . .]τ̣ου κατὰ φύcιν ἀν̣θρω[πηΐου] ε̣ὖροc ποδόc [ἐcτι]
τοὺ̣[c οὔρου]ϛ̣ οὐχ ὑπε̣ρβάλλων· ει.[. . . . .]ρουcε[
[ἐ]κ̣[βήcετα]ι̣, Ἐρινύε[c] νιν ἐξευρήcου̣[cι, Δίκηc ἐπίκουροι.]

Some way or other it must constitute a point of contact between Heraclitus and the author of the papyrus. Since my present concern is Heraclitus' reception by the Derveni author, I can, for the time being, ignore some otherwise important questions. For it is not immediately relevant in this context whether or not the text in the papyrus is a verbatim quotation, and, correspondingly, whether or not we should henceforth print B3 (known from Aëtius) and B94 (known from Plutarch) as a continuous sentence in future editions of Heraclitus.[3] What matters is that the author of the papyrus took it to be by Heraclitus. It is also not my primary purpose here to examine what the meaning of this reunited sentence could be in

[3] My inclination is to think that the papyrus should be given preference over the evidence of Aëtius and Plutarch. It certainly remains a possibility, as e.g. Most (1997) 126 n. 31 notes, that the Derveni author conflated two separate passages of Heraclitus. It is, however, much more difficult to explain why the author found it important also to quote B3 than to explain why Plutarch did not mention that part of the sentence, or why Aëtius quoted only B3 (cf. Betegh (1994) 73). It should nevertheless be noted that – *pace* Sider (1997) 130 with n. 4 – the *paragraphos* at the beginning of line 7 does not necessarily indicate a direct quotation, for I see no reason to believe that the *paragraphoi* before coll. 10.11; 11.8; 13.7; 15.11; 20.11; 23.8 and 26.3 indicate quotations either. Cf. 95–6 above.

the context of Heraclitus' own philosophy.[4] What I am now interested in is the way the Derveni author interpreted it.

The Derveni author found Heraclitus' statement important enough to quote it. The question naturally arises: why? Commentators agree that the immediate reason is not entirely obvious.[5] More precisely, there appear to be two equally good reasons for the quotation; yet these reasons are from two different strata of the author's discourse – or so it seems – and, therefore, we have to choose one of the two alternatives.

One option is that the author adduces the sentence because Heraclitus is speaking here about the size or 'limits' of the sun. As we have seen, the creation of the sun constitutes the central episode in the author's cosmogony.[6] Indeed, the specific nature and appropriate size of the sun is the *sine qua non* of the emergence and maintenance of the presently existing entities and cosmic order. Commentators assume that there is even textual evidence to show that Heraclitus is invoked in this context. When the author in col. 25 recapitulates his central tenet about the cosmogonical role of the sun, he makes a back reference:

> If the god had not wished that the things which are now should exist, he would not have made the sun. But he made it of such a sort and of such a size as is explained in the beginning of the account. (col. 25.9–12)

This reference to 'the beginning of the account (*logos*)' is usually understood as pointing to the Heraclitus quotation in col. 4, which could, in principle, be somewhere around the beginning of the author's text.[7] This interpretation is open to challenge, however. For it seems much more probable that διηγεῖται does not refer to the author's own text, but to Orpheus' account. For a back reference to one's own text one would surely use not a present but a past tense; the present tense, on the other hand, is perfectly normal to refer to some part of the text commented on. Besides, the only other

[4] A brief remark about Sider's recent interpretation: Sider connects B3 + B94 with B43 (ὕβριν χρὴ cβεννύναι μᾶλλον ἢ πυρκαιήν), Plato *Rep.* 498a, where Heraclitus' sun is said to be quenched, B6 (ὁ ἥλιοc νέοc ἐφ' ἡμέρῃ), and the 'moon illusion' (i.e. that the sun appears to be bigger at the horizon than at the zodiac). The outcome would be that the Erinyes punish the sun's attempt to transgress its measures at each sunset by quenching it. To this I would object that the moon illusion applies not only to the setting but also to the rising sun: the Erinyes should quench the sun already at dawn. Cf. also Schönbeck (1998a) on the problems concerning the moon illusion. I am not, however, entirely convinced by Schönbeck's own solution, as proposed on pp. 151 and 168–9, that the fragment is ironical, expressing a counterfactual absurd case which could be paraphrased as 'ΗΛΙΟΣ will not overstep his borders. Or it might be that the Erinyes will track HIM down . . . and therefore deflect from HIS trajectory from fright'.

[5] Cf. e.g. Laks (1997) 133 n. 35. Laks consistently speaks about col. 5, but this must be a slip.

[6] Cf. esp. 230–1 above.

[7] So e.g. Tsantsanoglou and Parássoglou (1988) 126 and Laks (1997) 132–3.

occurrence of the same verb in the remaining text of the papyrus (col. 15.8) refers to Orpheus' account and not to the author's own explanation. Therefore, I find it much more likely that the reference in col. 25 is to ***L2***=col. 8.4–5 as interpreted by the author in col. 9. Here the author interprets the lemma about Zeus's ascent to power as speaking about the creation of the sun; in the extant part of the column he also mentions in this context that the divinity placed the fiery particles in the most appropriate position.[8] Furthermore, ***L2*** was most probably at the very beginning of the narrative part of the poem, possibly immediately after the proem. Thus, when the author says 'But he made it of such a sort and of such a size as is explained in the beginning of the account', he is thinking about ***L2***, the beginning of Orpheus' account as understood by him.[9] Nor should it cause surprise that the author calls Orpheus' poem a *logos*; especially if one considers that, for him, Orpheus' text was most probably a *hieros logos*.

However, even if it is agreed that the reference in col. 25 is not to the Heraclitus quotation, this point does not rule out the possibility that the author adduces B3 + B94 in relation to his doctrine about the cosmological importance of the sun. As a matter of fact, the hypothesis that the author treated questions pertaining to cosmogony already in col. 4 might receive further support from the otherwise desperately fragmented first part of the column. They might suggest that the passage not only treated the size of the sun with the help of the Heraclitus quotation but also the creation of the sun, or, more generally, the creation of order in the cosmos, which, in the author's view, depends on the creation of the sun.[10] Even if the supplements and interpretation of the previous lines are highly conjectural, the plausibly restored rhetorical question in line 4 (ἆρ' οὐ τά̣[ξιν ἔχει ἐκ τῶ]ν̣δε κόcμοc;) certainly goes in this direction.[11] Taken together with the phrase . . . παρ]ὰ τῆc τύχηc γ[ὰρ] | οὐκ εἴ̣[α λα]μμάνει̣[ν.] immediately preceding it, it forms something like an argument from design: mere chance could not

[8] See pp. 230–1 and 233–4 above.

[9] It should also be noted that if the reference had been to col. 4, then the text of what we have as col. 4 could be referred to as 'the beginning of the account'. On the assumption that there was no other text preceding this one on the roll, this remark could then be taken as an indication that there were just a few columns preceding our col. 1. On the other hand, judging from the blank space left after it, col. 26 appears to be the last column on the roll. From these premises it would follow that the entire roll contained hardly more than 30 columns. Although we do not have sufficient comparative evidence from the same age, one should note that this is considerably less than the average number of columns in third-century rolls.

[10] This is the general line of interpretation offered by Tsantsanoglou in his detailed commentary on the column (see Tsantsanoglou (1997) 106–10).

[11] The letters]μηνιτακ[from line 13 may also speak about the order of the months as corresponding to astronomical phenomena. I find Tsantsanoglou's hypothesis that the reference is to 'the month appointed for the offerings to the Erinyes' (Tsantsanoglou (1997) 97) less likely.

produce such an order in the world. And, as the other part of the text shows clearly, the source and guarantee of the cosmic order is the creation of and constant constraints on the sun by the divinity. Thus, Heraclitus would be adduced for his view that the sun must not transgress its limits, for otherwise the cosmic order would be endangered.[12]

Alternatively, it can be argued that not the sun and its size, but Heraclitus' mention of the Erinyes prompted the quotation in col. 4. For in the ritual, eschatological context of the first seven columns the Erinyes and the Eumenides turn up repeatedly. The supplement Ἐ ρι]ν̣ύ̣ων in line 7 of the badly damaged col. 1 is not unlikely, while the fragments Ἐ̣ριν[υ | . . . Ἐ̣ρινύω̣[ν in lines 3 and 4 of col. 2 make the appearance of these mythological figures certain. Moreover, there is some indication that the author also equated both the Eumenides and the Erinyes with souls.[13] Thus, we have a discussion of the Erinyes in the columns preceding col. 4 and a similar discussion of the Eumenides in the columns following it. Therefore, the argument goes, we have every reason to suppose that Heraclitus' dictum is quoted because it speaks about the Erinyes, the auxiliaries or guards of Dike.[14]

In my own view, both arguments are too strong to be dismissed. The cosmological role of the sun is evidently fundamental to the author's physical theory, while the eschatological discourse, in which the Erinyes apparently have a pre-eminent part, is undoubtedly prevalent in the first seven columns. I suggest, then, that we should consider the possibility that the Heraclitus quotation somehow constitutes the link between eschatology and cosmology, the two main parts of the Derveni text.[15]

But this point brings us immediately to an even more pressing issue. Quite apart from the interpretation of the Heraclitus quotation, such a link is still missing: we do not know what keeps the author's eschatology and physics together. Or, in other words: what is the author's overall argument (assuming that he had one)? This problem has been articulated in three recent papers by Dirk Obbink, André Laks and Glenn W. Most respectively.

[12] So, e.g., Sider (1997) 133; cf. also Tsantsanoglou and Parássoglou (1988) 126.

[13] On these questions, see 85–9 above.

[14] So also Tsantsanoglou (1997) 109. For interesting remarks on the role of the Erinyes within the Heraclitus fragment, see Schönbeck (1998a) 103–6 and Johnston (1999) 265–7. Independently of each other, and on slightly different grounds, both authors maintain that, contrary to what has often been claimed, the story of Achilles' horse whom the Erinyes made stop speaking in *Iliad* 19, cannot provide a parallel for the function of the Erinyes in the Heraclitus fragment. Moreover, both of them point out, independently of each other, that 'maintaining the natural order' was not part of the Erinyes' function in other texts. These are important points that should not be neglected in future discussions of the Heraclitean text.

[15] So already Laks (1997), with somewhat different emphases.

All three authors point out convincingly that we cannot claim to have arrived at a satisfying interpretation of the text without proposing an answer to this question.[16] Moreover, all three authors agree that the answer should be sought in the direction set in Richard Seaford's pioneering study: in the connection – evinced also in other texts – between a theory of elements and a doctrine of salvation.[17]

It is also true that the link between eschatology and cosmology might be on a more general level. In a sense, all eschatological doctrines necessarily imply a cosmology, for if the soul survives the death of the individual, one has to know where the soul goes after it leaves the body.[18] Therefore the understanding of the organisation of the cosmos can be fundamental for the understanding and acceptance of a doctrine which speaks about the afterlife. Even though I think this is a valid general interpretation, I would go further and examine whether we can find a closer relationship between a particular cosmology and the particular eschatology it is connected with. In other words, I should like to examine whether or not specific eschatological tenets can govern or require specific cosmological doctrines, and, conversely, whether or not a specific cosmological theory can be explanatory of the eschatology related to it.[19]

The trouble is that the surviving text of the papyrus offers very little help regarding the author's religious, eschatological tenet(s). As we have seen in chapter 2, apart from the fact that the author's discourse focuses on souls, daimones, Erinyes and Eumenides, sacrifices and propitiatory rites, hardly anything can be said about the exact course of the argument. Moreover, apart from the Heraclitus quotation, there is no clear trace of any connection with a physical theory, and there is no mention of the physical elements or the celestial bodies. Thus, the remaining desperate fragments of the first columns are not sufficient in themselves to establish a connection with the author's physics. Given that the internal evidence at this point is so scanty, we are obliged to have recourse to some external evidence.

Such external help can primarily be expected from two sources: first, from the sepulchral context of the find, and, second, from the independent evidence on Orphic and other eschatological lore. Such abstruse matters need to be broached at this point, *nolens volens*. The most obvious denominator for all these texts and contexts is fire. Laks tentatively raises the possibility that the author's physical doctrines about fire are somehow related to the ritual context, i.e. that the papyrus was designed to be burnt with the

[16] Obbink (1997); Laks (1997) 122–4; Most (1997) 119; cf. also Funghi (1997) 31.

[17] Seaford (1986). [18] This was pointed out to me by Luc Brisson.

[19] This set of questions is obviously relevant also to authors such as Empedocles and Plato.

corpse on a funeral pyre.[20] Glenn Most goes further and suggests a more specific hypothesis by bringing in the evidence of the texts inscribed on gold plates.[21] He tries to connect the author's physics and cosmology primarily with group B of the gold plates, taken together with the Hipponion and the Thessalian gold leaves.[22] In these texts, the thirst of the dead person's soul appears to be the central motif; a thirst which the soul is permitted to quench only by the refreshing water of the fountain of the lake of Memory, whereas another fountain, tempting as it may appear, is to be strictly avoided. Most suggests that the image of this netherworldly thirst can stem from the observation that the corpse, no matter whether cremated or inhumed, dries out and turns into bones and ashes: the observable dehydration of the corpse might be expressed by the tormenting thirst of the newly arrived soul in Hades. In the final step Most connects the netherworldly thirst of the soul with the Derveni author's theory about the destructive effects of fire: the Derveni author would console his folk by pointing out that fire, the cause of drying out, can ultimately be overcome and controlled by the divinity.

Although this line of interpretation seems promising at first, it does not turn out convincing enough at the final stage. An immediate objection to Most's conjecture is that there is a basic discord between the two groups of evidence in their reference to elements: the accent is on water in the underworld situation described on the gold plates, and, on the whole, there is no question of air and fire, whereas the Derveni author focuses on air and fire and is not at all interested in water – for example, he does not seem to show any interest in Orpheus' verse about ποταμοὶ καὶ κρῆναι ἐπήρατοι in col. 16, and even interprets Okeanos as air in col. 23. More important, Most's interpretation cannot account for the fact that for the Derveni author fire is not only, and not even primarily, destructive; indeed, once it is properly mastered by the divine intelligence, it becomes the source of all generation.

In what follows I shall try to offer – just as tentatively as Most – an alternative hypothesis. Given that Heraclitus B3 + B94, as shown above, has immediate links with both parts of the text of the papyrus, I will try to tackle this problem via the Heraclitus quotation.[23]

[20] Laks (1997) 140; Most (1997). [21] Most (1997) 131–4.

[22] On the gold plates discovered before 1971, together with their classification in groups A and B, see Zuntz (1971), esp. 277–393. On the more recent finds and their classification, see Graf (1993b) and now Bernabé and Jiménez San Cristobal (2001). On the Hipponion leaf, see Pugliese Carratelli (1974) and Cole (1980) and Bernabé and Jiménez San Cristóbal (2001) 25–86 with further bibliography. On the gold leaf from Thessaly, see Tsantsanoglou and Parássoglou (1987).

[23] The possibility is mentioned but not carried further by Most (1997) 127.

There is a further consideration to show that this may be the right track. If we accept that the author's cosmology is in some way thematically related to his eschatological ideas, then it is reasonable to suppose that the core element of the cosmological theory has something to do with his religious, eschatological tenets. If, moreover, it is true that the physical theory should be seen as an underpinning of the author's position on certain religious and eschatological matters, then I find it inescapable that the author's most important doctrine in physics is precisely the point which is meant to create the link between the two domains: the author comes up with and develops this physical, cosmological point because he thinks that it can further explain, rationalise, and render convincing his specific religious, eschatological doctrine.

As argued above in chapter 6, the author's most remarkable and possibly original physical doctrine concerns the importance of the creation of the sun and the stars by the intelligent cosmic principle, and, more generally, the dynamic interplay between the intelligent air and the brute force of fire in the formation of the cosmos. If my reasoning is valid, this is the physical doctrine which is meant to explain or rationalise a religious or eschatological dogma.

Furthermore, although I do not find Most's specific hypothesis conclusive, I also think that the gold plates might offer a clue towards a more economical interpretative hypothesis. Thus, I shall first turn to these texts with the hope that after this detour we may find a connection between the author's eschatology and physics, and may even be in a better position to find a proper place for the Heraclitus quotation in the context of the first columns of the papyrus.

THUNDERBOLTS AND PYRES

My suggestion is that we should expect a clue not from group B of the Thurii plates as Most did, but from group A. The reason is obvious: fire is directly involved in the text of this latter group.

Yet before turning to these documents, an argument in defence of this general procedure is in order. For one could easily object that by adducing the evidence of the gold plates we are just mixing everything with everything: the tumuli of Thurii are very far from that small burial ground at Derveni, and the fact that both contain nebulous eschatological stuff is no sufficient ground for combining the two groups of evidence. I would maintain, however, that there is some – and as I see it quite strong – textual evidence to justify such a comparison. The link, I suggest, is to be found in the text of Tablet C.[24]

Tablet C was found in the Timpone Grande, inside a coffin of a cremated person, close to the head of the skeleton. The tablet in question is a tiny gold plate (81 × 23 mm) which was folded and into which Tablet A4 was wrapped.[25] The text of C is notoriously difficult to read and interpret. Ten lines of writing are squeezed onto a tiny surface, while the wrinkles caused by the folding of the plate further obscure the original carving.[26] But what makes the decipherment practically hopeless is that the text seems to contain not only some scribal errors (inversions of letters, reduplications, omissions) and a mixture of different dialectal forms, but also manifestly unintelligible letter combinations. Various hypotheses have been proposed to account for this feature. According to Diels these strings of letters stem from complete carelessness, while others have suggested that they are abbreviations, or cryptic codes, or Greek transcriptions of words of another language, or magical abracadabra.[27] Alberto Bernabé has recently proposed a modified version of the hypotheses first suggested by Comparetti: unintelligible strings of letters were intentionally inserted between meaningful words and shorter phrases, in order to make the text more obscure to those who did not know how to use it and what to look for in it. The text, according to this interpretation, is comparable to the puzzle in newspapers known as 'word search': those who know how to approach such a seemingly random collection of letters, and know what to look for in it, can elicit the meaningful linguistic units. They are not bothered by the rest because they know that unintelligible groups of letters are just part of the game. Others, by contrast, see only a perplexing jumble of letters. Of course, the function of the gold plate is not to offer entertainment, but to convey some serious

[24] I presented the main lines of this interpretation in Betegh (1999) 330–3 and (2002b) using the text of Tablet C as printed by Zuntz. Since then Bernabé and Jiménez San Cristóbal (2001) have published their new readings and interpretation of Tablet C = fr. 492 B. I am to a great extent in agreement with their proposed interpretation, while some of their suggestions certainly corroborate the correspondences I am trying to establish between Tablet C and the Derveni papyrus. I do not intend to enter here into the debate whether or not the gold plates are 'Orphic'. The first editors and commentators were convinced that they were. Then came a long period of denial, culminating in the work of Zuntz. However, on the force of the new evidence from Hipponion, Pelinna and elsewhere, scholars nowadays tend to give justice to the first view. See most recently Bernabé and Jiménez San Cristóbal (2001) 18–21.

[25] For the description of the excavation and the archaeological details, see Zuntz (1971) 287–93, esp. 290.

[26] On the palaeographical difficulties, see Zuntz (1971) 345.

[27] 'A text more corrupt than this will not easily be found . . .' Zuntz (1971) 345. For an overview of the different hypotheses, see now Bernabé and Jiménez San Cristóbal (2001) 184–7.

religious message, while secrecy, exclusive access to information, can play an important role in such a context.[28]

Indeed, even though longer intelligible syntactical units cannot be found on Tablet C, certain words and shorter phrases are legible or can be reconstructed with a fair degree of confidence. The recognisable words and phrases that ensue show a rather clear thematic unity. Diels defined it as an Orphic hymn to Demeter,[29] whereas Zuntz, more convincingly, tried to explain it as Kore's prayer to her mother Demeter.

To make reference easier, I reproduce here the text as printed by Bernabé and Jiménez San Cristóbal. Following them, I print the meaningful units with lower case characters, leaving the unintelligible or not yet deciphered strings of letters in upper case.[30]

1. Πρωτογόνῳ<ι> ΤΗΜΑΙΤΙΕΤΗ Γᾶι̣ ματρί ΕΡΑ Κυβελεία<ι> Κόρρα<ι> Ο̣ΣΕΝΤΑΙΗ Δήμητροϲ ΗΤ
2. ΤΑΤΑΙΤΤΑΤΑΠΤΑ Ζεῦ ΙΑΤΗΤΥ ἀέρ ΣΑΠ̣Τ̣Α Ἥλιε, πῦρ δὴ π̣άντα ΣΤΗΙΝΤΑΣΤΗΝΙΣΑΤΟΠΕ νικᾶι Μ̣
3. ΣΗΔΕ Τύχα ΙΤΕ Φάνηϲ, πάμν̣ηϲτ̣οι Μοῖραι ΣΣΤΗΤΟΙΓ̣ΑΝΝΥΑΠΙΑΝΤΗ ϲὺ κλυ̣τὲ δαῖ̣μον ΔΕΥΧΙ
4. Σ πάτερ ΑΤΙΚ παντο̣δαμάϲτα ΠΑΝΤΗΡΝΥΝΤΑΙ ΣΕΛΑΒΔΟΝΤΑΔΕΝ ἀντα̣μοιβή ΣΤΛΗΤΕΑΣΤΛ
5. ΤΗΜΗ ἀέρ Ι πῦρ ΜΕΜ μᾶτερ ΛΥΕΣΤΙΣΟ̣ΙΛ̣-ΕΝΤΑΤΟ Νῆϲτ̣ι Ν νύξ ΙΝΗΜΕΦ ἡμέρα ΜΕΡΑΝΕΓΛΧ̣ΥΕΣ
6. ἐπ̣τ̣ῆμαρ ΤΙ νήϲτιαϲ ΤΑΝ Ζεῦ ἐνορύττιε(?) καὶ πανόπτα. αἰέν ΑΙΜΙΥ* μᾶτερ, ἐμᾶϲ ἐπ-
7. ἀ̣κο̣υϲον ΕΟ εὐχᾶϲ ΤΑΚΤΑΠΥΑΡ̣ΣΥΟ̣ΛΚΑΠΕΔΙΩΧΑΜΑΤ̣ΕΜΑΝ καλ{η}ὰ Δ ἱερά ΔΑΜΝΕΥΔΑΜΝΟΙ
8. Ω̣ΤΑΚΤΗΡ ἱερά ΜΑΡ Δημῆτερ, πῦρ, Ζεῦ, Κό̣ρη Χ̣θονία ΤΡΑΒΔ̣ΑΗ̣ΤΡΟΣΗΝΙΣΤΗΟΙΣΤΝ
9. ἥρῶϲ ΝΗΓΑΔΥΝΗ φάοϲ ἐϲ φρένα ΜΑΤΑ̣ΙΜΗΤΝΝΤΗΣΝΥΣΧΑ̣ μή̣ϲτωρ εἷλε Κούρην
10. αἶα ΦΗΡΤΟΝΟΣΣΜΜΟ̣-ΕΣΤΟΝ ἀέρ ΤΑΙΠΛΝΙΑ̣ΛΥἐϲ φρένα ΜΑΡ*ΤΩΣ̣

Obviously, I do not pretend to understand much of this. And, of course, it would be irresponsible to argue that this 'reading' is any more definitive

[28] Bernabé and Jiménez San Cristóbal (2001) 187–8.

[29] Diels (1902). It should be noted that Diels' textual conjectures were severely criticised by all subsequent commentators, see e.g. Kern ad *OF* 47.

[30] For the transcript and critical apparatus, see Bernabé and Jiménez San Cristóbal (2001) 273–7 and Zuntz (1971) 346–8. Cf. also the alternative reconstruction, short discussion and facsimile (drawn by M. Puccetti) in *OF* 47.

than the ones proposed before; others may prefer to find other meaningful units in the strings of letters set in upper case by Bernabé, or construe the remainder differently. But no matter what solution one prefers, the tablet, I submit, shows undeniable points of contact with the Derveni text. Ironically, my case is best stated by Zuntz himself. So before presenting some suggestions of my own, and taking over some further suggestions advanced by Bernabé and Jiménez San Cristóbal, let me quote Zuntz' analysis of line 2, based on a reading slightly different from the one printed above, *in extenso*:

> In between these two passages [line 1 on the one hand, and the last word of line 2 and its continuation in line 3 on the other], the combination Ἥλιε πῦρ, however precarious syntactically, may suggest an echo of philosophical allegorism, and a corresponding equation of Zeus and air might be suspected immediately before. Confirmation of a mythology of elements seems to emerge in l. 5 ἀέρι πῦρ (μεμ<ειγμενον>?),[31] but this impression is shattered on reaching l. 8, where Demeter appears equated with fire; she whom the wildest syncretism could never identify with the Sun-god. Even so, the impression that fire and air played some significant part in this text, and with them, probably, earth (l. 1; but I find no indication of the fourth element), is not easily rejected (further traces may or may not be in l. 7 ΠΥΡΑϹ and l. 10 ΑΕΡ), but the wording is too corrupt to admit of any specification.[32]

To appreciate fully the force of these lines one has to bear in mind that Zuntz does not treat the Derveni papyrus,[33] and, more importantly, one of his major objectives was to deny any Orphic traits in the gold plates. Yet he unwittingly points out momentous similarities with the papyrus. First there is the connection between the sun (or Sun) and fire – true, not surprising in itself. Much more significant is the supposed equation between Zeus and air, and, in general, the 'mythology of elements'. As a matter of fact, not even Demeter's assumed equation with fire seems completely unthinkable once one has seen, for example, Okeanos being equated with air in the papyrus. But it is also possible that line 8 presents a list and not equations. Bernabé, for example, has proposed reading a list of the four elements in this line: Demeter would be the earth, Zeus the air, while the chthonic Kore would be equated with water through the figure of Nestis, who comes up in line 5 in a similar list of elements/gods. But other possibilities are not excluded either. Another telling observation mentioned by Zuntz is that from all the elements air (probably equated with Zeus) and fire (mainly

[31] Actually, the text has μη (μή?) before ἀέρι.

[32] Zuntz (1971) 351; cf. also Seaford (1986) 22.

[33] The only reference to the papyrus is on a palaeographic point for the dating of the gold plates (p. 295 with n. 3 – not in the bibliography in L&M).

in the form of the sun) seem to play a major role, while earth appears in a secondary place, and water does not turn up as such at all. It may be sheer coincidence, but this is exactly the distribution we have found in the papyrus.[34] (The picture may be somewhat different if one accepts Bernabé's hypothesis that Nestis and the chthonic Kore represent water.)

To this, we can add further parallels, some more, some less specific. In line 1, we have Πρωτογόνῳ<ι> ΤΗΜΑΙΤΙΕΤΗ Γᾷ ματρί, retained with some variation by Kern and most other editors. Again, the syntactic relationship and meaning is unclear. However, according to the interpretation I eventually opted for, the genealogical scheme of the Derveni poem has Ge and Ouranos in the first generation born from the primordial pair Aither and Night, and Ge's companion Ouranos bears the epithet 'the first-born king'. It seems possible to me that the term 'First-born' refers to Ouranos in the gold tablet as well.[35] Further, there is the equation of Δημήτηρ and Γῆ μήτηρ suggested for this line e.g. by Diels – and not denied by Zuntz – which explicitly turns up in the papyrus.[36]

More interesting is the importance allotted to Moira, and especially the rare epithet πάμμηστοι or παμμήστο<ρ>ι in line 3, as proposed by Murray and Diels respectively (which I find preferable to πάμνηστοι proposed by Colli and retained by Bernabé). It can be compared with the verb ἐμήσατο used in the Derveni poem to describe Zeus' creative power,[37] and, further, with the Derveni author's reasoning about Moira's being the wisdom of Zeus, ordaining 'how the things that are and the things that come to be and the things that are going to be must come to be and be and cease.'[38]

On the whole, the papyrus might even offer some help in overcoming Zuntz' ultimate worry: 'All this is hypothetical, but I suppose, not inconceivable [i.e. that the text originally presented Kore's prayer to her mother on Kore's abduction to the underworld], and it is clear that what can be made out with some confidence fits into hexametric rhythm. On the other hand I confess myself unable to fit the "allegorization of elements" into this frame – if indeed this motif has to be acknowledged (. . .)'[39] Now, the papyrus is a perfect parallel for a text from a tomb containing parts of a mythological hexameter poem together with an 'allegorization of elements' apparently with some kind of eschatological significance.

[34] Cf. 260–5 above.

[35] Bernabé has now proposed the same interpretation independently of me (see Bernabé and Jiménez San Cristóbal (2001) 189–90).

[36] Cf. col. 22.7 and 9–11. [37] Cf. col. 23.4 and ***L16*** = col. 25.14 with 129–30 above.

[38] Col. 19.4–7. Cf. also col. 18.2–10. [39] Zuntz (1971) 352.

To conclude, the fact that most of the intelligible words of this gold tablet, found in the tomb of a cremated person, are indeed catch-words in the Derveni text (air, fire, sun, Zeus, Moira, Ge, mother, Demeter), makes it at least defensible to evoke the evidence of the gold plates in the interpretation of the papyrus. I would even venture to say that Tablet C might shed some dim light on a religious lore operating with a theory of elements that lays special stress on air and fire, and from which the cosmological theory developed in the Derveni papyrus may have emerged.

Be that as it may, the folded Tablet A4, as mentioned above, was wrapped in C. This arrangement secures that C and A4 were somehow related in the ritual; and, as has been shown by Zuntz, A4 belongs to the same family as A1–3. The most complete and least corrupt of group A is A1, found in the neighbouring Timpone Piccolo.[40] The first two lines of A1 contain an invocation of the queen of the underworld together with the claim that the speaker is 'pure out of the pure' (Ἔρχομαι ἐκ κοθαρῷ<ν> κοθαρά, χθονί<ων> βαcίλεια, | Εὐκλῆc Εὔβο<υ>λεύc τε καὶ ἀθάνατοι θεοὶ ἄλλοι·). In line three, the speaker boasts of being from the same race as the gods (καὶ γὰρ ἐγὼν ὑμῶν γένοc ὄλβιον εὔχομαι εἶμεν). In line four, connected in A1 to the previous by ἀλλά, the speaker says that his death was by lightning and due to Moira. (ἀλ<λ>ά με μο<ῖ>ρ' ἐδάμαc<c>ε καὶ ἀc{c}τεροβλῆτα κεραυνοῷ̣ι̣). Line 5 hails the leaving behind of the 'woeful and grievous cycle' (of rebirths?), and says that the speaker won the wreath at a – surely metaphorical – footrace (κύκλο<υ> δ' ἐξέπταν βαρυπενθέοc ἀργαλέοιο | ἱμερτο<ῦ> δ' ἐπέβαν cτεφάνο<υ> ποcὶ καρπαλ̣ίμοιcι). In the next line the speaker rushes to the breast of the chthonic queen, and then is greeted and proclaimed deified (δεc{c}ποίναc δ' ὑπὸ κόλπον ἔδυν χθονίαc βαcιλείαc. | "ὄλβιε καὶ μακαριcτέ, θεὸc δ' ἔcηι ἀντὶ βροτοῖο"). Finally, line 8 contains the much debated reference to the kid falling in or rushing to the milk (ἔριφοc ἐc γάλ' ἔπετον).[41]

Most important for us is the central line 4. There has been some discussion in the literature as to whether or not all the three buried in Timpone Piccolo with a similar text by them actually died from lightning. Jane Harrison maintained that they did not, and that the reference is purely mythical,[42] whereas Zuntz finds it more likely that the people buried in this tumulus

[40] Graf (1991) 96–7 argues convincingly that the Pelinna and Hipponion plates bridge the gap between Zuntz' A and B groups showing that, in fact, all these tablets are closely connected and must have belonged to the same religious movement. It has to be noted that the gold plates from other locations still do not parallel or explain the text on Tablet C.

[41] For a critical discussion of the text, with the differences as compared to the other tablets of the same group, see Zuntz (1971) 299–327 and now Bernabé and Jiménez San Cristóbal (2001) 135–78.

[42] Harrison (1922) 587.

were actually struck by thunderbolts, and thereby 'sanctified'.[43] Of course, this question is quite impossible to answer definitely; and I agree with those like Peter Kingsley who say that the main interest of this body of evidence does not lie in this factual matter.[44] It concerns the much more general theme of heroisation or even deification (cf. line 7) by lightning and fire.

As has been pointed out most recently and ardently by Kingsley, death by fire was a principal way of attaining heroic, or even divine, status for a Greek.[45] Heracles' death surely has a prominent place in this tradition. A salient feature of this story – or at least the most widely known version of it – is that the pyre and Zeus's thunderbolt are related to each other in it. As soon as Heracles' pyre was eventually lit, a thunderstorm came, and the hero was lifted into the heaven by lightning.[46] The fire of the thunderbolt and that of the pyre thus become symbolically connected.

The pyre and the thunderbolt of Zeus are connected also in the story of Asclepius. Apollo falls in love with the Thessalian Coronis. But the girl, already pregnant with the child of Apollo, sleeps with a mortal man. Artemis punishes the unfaithful Coronis with death. When the girl is already on the funeral pyre, Apollo decides to save the unborn baby, Asclepius. The centaur Chiron brings up the child and teaches him the art of medicine. But Asclepius transgresses the limits of his art, and not only heals the ill, but also raises the dead. For this excess, Zeus strikes him with the thunderbolt. Thus Asclepius, who was born on a funeral pyre, receives divine punishment and at the same time heroisation by Zeus' thunderbolt. Fire is the element of birth, death and heroic rebirth, the element that sets the limits of mortal life. In the figure and story of Asclepius, I maintain, we find together numerous elements that are relevant to the Derveni papyrus and its author. Asclepius is a Thessalian hero just like Orpheus. Apollo has a distinguished role both in the cult of Asclepius and in the Thessalian cult of Orpheus. Just as the figure of Orpheus, the figure of Asclepius combines chthonic elements with the cult of the Sun. Moreover, the art of Asclepius corresponds in many ways to that of an Orphic initiator priest: it brings together healing – primarily magical healing in the Thessalian version of Asclepius – with oracles, mainly dream oracles.[47]

[43] Zuntz (1971) 316, contra Bernabé and Jiménez San Cristóbal (2001) 150–5.

[44] Kingsley (1995) 257–8.

[45] Kingsley (1995), esp. 252–8, with ample bibliography; from the earlier literature, see esp. Rohde (1925) and Cook (1914–40) vol. 2, esp. 11–29.

[46] See, e.g., Diodorus Siculus 4.38.4; Lucian, *Hermotimus* 7, Apoll. 2.7.7 with further references in Fraser's note ad loc. in the Loeb edition of the *Bibliotheca*.

[47] One may also add that, as Henrichs (1984) 258 has shown, the combination of sacrificing cakes and wineless libation that we find in col. 6. is closely reminiscent of the sacrifice to Asclepius in Athens.

Some of our sources make the cathartic role attributed to fire in this situation more explicit: the fire (of the pyre or of the thunderbolt) destroys what is earthly and mortal and thus helps the liberation of the divine part in man.[48] On Tablets A1–3, the reference to death by lightning on the one hand, and the deification of the dead man's soul on the other, are, in all likelihood, connected. It is highly probable that the pyre could also assume the role of lightning in such a context.[49]

Another interesting piece of evidence comes from Plato. At the end of the final book of the *Republic*, Socrates relates the story of Er of Pamphylia. Conspicuously, the account of the brave Er, chosen by the gods to be their messenger, brings together some of the central themes of the papyrus: we get not only the description of the journey and lottery of the souls – that is a colourful presentation of a retributive eschatology that props up the ethical tenets of Plato – and a detailed topology of the underworld to give the settings for the story, but also a cosmological account.[50] Now it is generally agreed that Plato incorporated many Orphico-Pythagorean elements in the story of Er. And, quite notably, Er, whom everybody thought to be dead, revived to transmit the divine message by disclosing the fate of the soul and the major motive forces that govern the cosmos on the pyre (ἐπὶ τῇ πυρᾷ κείμενος ἀνεβίω *Rep.* 614b7).

Immortalisation by fire has a prominent place also in the myth relating the foundation of Demeter's temple in Eleusis. Demeter wanted to bestow immortality on Demophoon, the son of Celeus, king of Eleusis, by putting him into the fire of the hearth every night. All went well and the child was growing as an immortal god until the mother of Demophoon decided to spy on this unusual way of childcare. When she discovered what was going on during the night, she, quite naturally, got frightened, fearing that the fire would destroy the child. But this natural reaction raised the anger of Demeter. She stopped caring for the child

48 Lucian *Hermotimus*, 7; Cf. Kingsley (1995) 255, with bibliography to the earlier literature.

49 The three corpses buried in Timpone Piccolo were accompanied by the gold plates stating that they were struck by lightning. In the excavation documentation there is no mention of ashes, carbon, or coffin in these tombs. Zuntz (1971) 292 infers from this detail that these corpses were not cremated but inhumed. On the other hand, the person buried in the neighbouring Timpone Grande was cremated, but the text accompanying him – otherwise parallel in many ways to the ones mentioning death by lightning – does not say that this person was struck by a thunderbolt. Is a pyre needed only for those who did not have the privilege of being struck by lightning? At any rate, the importance of fire is evident in the text of Tablet C, found in the tomb of the cremated person of the Timpone Grande.

50 Even though this part of the account also is expressed in mythical terms (especially because of its reference to the four goddesses, Ananke, Lachesis, Clotho and Athropos), it gives the first approximation of a much more sophisticated explanatory model for the heavenly motions.

and ordered the Eleusinians to build a temple where her rites could be performed.[51]

Zeus's thunderbolt – not as a means of heroisation, but as a tool of punishment – appears also at a crucial point in the Orphic mythological narratives. According to a generally accepted reconstruction, the story goes as follows. Dionysos was born from the intercourse of Zeus with Kore, Zeus's daughter from the rape of Rhea/Demeter. Zeus gave royal power to the young Dionysos; but Hera, heated by jealousy, plotted with the Titans against the infant king. On Hera's instigation, the Titans killed and ravaged Dionysos and ate from his flesh. But they could not escape Zeus's wrath: he struck them with the thunderbolt, and Dionysos could then be reborn with the help of Athena.[52] It seems certain that at least this much of the story was already known in pre-Platonic times.[53] It is not implausible that the now lost part of the Derveni poem contained this story – it is only regrettable that our text is cut off at the episode of the rape of Rhea/ Demeter.

It is possible, but far less certain, that the ancient version of this story already culminated in the creation of mankind: man was born from the ashes – or soot – of the Titans struck by Zeus's thunderbolt. The ultimate aim of this account is to explain in mythical terms the double nature of man. We are of the ashes of the wicked Titans, hence evil; but the same ashes also contained something of Dionysos just consumed, and hence there is also something of the god in us. As a matter of fact, Murray and Harrison interpreted the claim about divine origin (καὶ γὰρ ἐγὼν ὑμῶν γένοϲ ὄλβιον εὔχομαι εἶμεν) followed by the reference to death by lightning in line 4 of Tablet A1 in the light of this narrative. The problem is that the only explicit evidence for this anthropogony comes from the *Commentary on Plato's Phaedo* attributed to the sixth century Alexandrian Neoplatonist

[51] *Hymn. Dem.* 239–75.

[52] On the symbolic level, this episode seems to be the antithetical pair of the episode in which Zeus swallows Ouranos' phallus (or Protogonos-Phanes-Eros). The first act of engulfment is not only θέμιϲ, sanctioned by Night's oracle, but also the most important act towards the establishment of a stable world-order. It is also a climax in which the world attains the highest degree of unification inside the god. As opposed to this, the devouring of Dionysos by the Titans is king-slaughter, the most outrageous revolt against the world-order; a dramatic anticlimax, in which the god itself gets torn apart and dispersed. (The connection is reinforced by the fact that Dionysos could easily be identified either with the phallus or with Phanes.) Nonetheless, both engulfments result in the rebirth of something that previously existed. Empedocles' cosmology shows striking similarities with this pattern: total unification in the divine as the climax, then multiplication, extreme separation as anticlimax, followed by rebirth through unification. For more on this point, see Betegh (2001). For some new evidence vindicating the cyclical model in Empedocles, see Martin and Primavesi (1999) 87–97.

[53] For the evidence, see e.g. Brisson (1992) 494–7.

Olympiodorus, and, as Luc Brisson has recently argued, earlier references to the Titanic origin of mankind seem not to know about the account concerning the soot of Titans.[54] Fortunately, I do not have to take a position on this question, since a version containing only the punishment of the Titans and the resurrection of Dionysos – accepted as early also by Brisson – is enough for my argument; it is true, on the other hand, that the anthropogony would strengthen rather than weaken it.[55]

Death by lightning appears also in Orpheus' mythical biography. According to a version recorded by Pausanias, Orpheus himself died by Zeus's thunderbolt: the king of the gods punished him for having revealed things people had not heard before about the mysteries.[56] Just as in the case of Semele and Asclepius, death by lightning does not seem to entail total annihilation, but rather a means of heroisation.

This survey is far from exhaustive, but it is enough to show that fire, in the form of the thunderbolt or of the pyre, had a crucial eschatological role in an impressive number of myths, often related to the mysteries. The connection between these stories, which may or may not have been explicit in the minds or writings of those involved, is that at a more general level, not specific either to the eschatology of the gold plates or to the Orphic narrative, the thunderbolt is the means by which Zeus's divine justice is maintained and reaffirmed.

The bone plates of Olbia may provide further evidence concerning the importance allotted to the thunderbolt in Orphic circles.[57] In the upper part of plate 1 we have the words ΒΙΟϹ ΘΑΝΑΤΟϹ ΒΙΟϹ, then ΑΛΗΘΕΙΑ a little below, and ΔΙΟ(. . .) ΟΡΦΙΚΟΙ. This last line finally assures us that we are in Orphic circles – actually, it is assumed to be the first explicit evidence for the existence of Orphic communities.[58] Now in between these lines, in the middle section of the plate, there is an isolated letter Α, and two

[54] Brisson (1992). Contra e.g. Kahn (1997) 58–9. Burkert (1999b) 101 n. 67 points out, against Brisson's argument, that the reference to four (and not six) monarchies in Olympiodorus attests that the writer preserves an old tradition. To my mind, this is not sufficient to discredit Brisson's analysis, for he, too, accepts that the account in Olympiodorus is based on an ancient tradition, but he tries to point out that there is something in it – the specific details of the anthropogony – which is not attested in earlier documents.

[55] As, e.g., KRS 30 suggests, the statement of the dead person 'I am the son of Earth and starry Sky' on the Hipponion gold plate can be taken as a reference to the Titanic origin of man. This evidence might also reinforce the conviction of the early editors of the Thurii gold plates that the dead man's claim to be of the gods' race, followed by the statement about death by lightning, is to be connected with the Orphic anthropogony, or at least with the story of the Titans devouring Dionysos.

[56] Paus. 9.30.5.

[57] For the find and the text, see Rusjaeva (1978) and Tinnefeld (1980); for the interpretation, see West (1982), with the corrections at specific points by Vinogradov (1991) and Zhmud' (1992).

[58] See esp. Zhmud' (1992).

zigzag symbols. Other plates contain either some designs the significance of which is difficult even to guess (a ship, a horse, an oblong figure divided into seven sections, with a circle in each of the sections, etc.), or some words and abbreviations, such as ΕΙΡΗΝΗ ΠΟΛΕΜΟϹ | ΑΛΗΘΕΙΑ ΨΕΥΔΟϹ | ΔΙΟ on Tablet 2, and ΔΙΟ | [ΨΕΥΔΟϹ] ΑΛΗΘΕΙΑ | ϹΩΜΑ ΨΥΧΗ on Tablet 3.[59] The zigzags, often together with the solitary A, appear on most of the plates. West in his suggestive commentary on the bone plates proposes that the zigzags represent either a snake symbolising rebirth, or, more probably, lightning. One must agree with Vinogradov that these suggestions do not exhaust the possibilities, but a better solution, to the best of my knowledge, has not yet been proposed.[60]

Whether or not the zigzags are to be interpreted as thunderbolts, the bone plates of Olbia bring us finally back to Heraclitus. For the antithetical pairs carved onto the plates are often put in relation with Heraclitus. In the discussion of the words ΒΙΟϹ ΘΑΝΑΤΟϹ ΒΙΟϹ Heraclitus B62 (ἀθάνατοι θνητοὶ θνητοὶ ἀθάνατοι, ζῶντεϲ τὸν ἐκείνων θάνατον τὸν δὲ ἐκείνων βίον τεθνεῶτεϲ) is adduced, while the pair ΕΙΡΗΝΗ ΠΟΛΕΜΟϹ recalls B67 (ὁ θεὸϲ ἡμέρη εὐφρόνη, χειμὼν θέροϲ, πόλεμοϲ εἰρήνη, κόροϲ λιμόϲ).[61] But not only is Heraclitus called to help in the interpretation of the bone plates, but the converse happens also. Kirk, Raven and Schofield, for example, propound on the basis of the carving ΒΙΟϹ ΘΑΝΑΤΟϹ ΒΙΟϹ that '[in B62] Heraclitus, therefore, may be giving a special interpretation of an alternation between life and death that was broadly accepted in exotic mystery-cults of Bacchic or Orphic flavour.'[62]

There is the obvious problem of deciding the direction of influence. Was Heraclitus influenced by the antithetical style as used on the plates, and the eschatological beliefs about the fate of the soul after death as propounded in the mysteries, or did some Orphics draw inspiration from Heraclitus' writings? As David Sider argues,[63] the most probable scenario is that influence was mutual: Heraclitus took over something, both in form and content,

[59] For the reading ϹΩΜΑ, see Vinogradov (1991) 79.

[60] West (1982) 19; Vinogradov (1991) 78 n. 6. West also adduces the famous story of the Scythian king Skyles who participated in the Bacchic mysteries at Olbia and who saw his house struck by lightning (Hdt. 4.79). It is also true, however, that Zeus' thunderbolt is not commonly represented as a zigzag in Greek iconography. See Cook (1914–40) 2.764–85. It is sheer speculation, but I could envisage that the solitary A next to the zigzags stands for ΑΗΡ.

[61] Cf., e.g. West (1982) 18–9, 22–3 and 26.

[62] KRS 208 n. 1; see also Seaford (1986) 17 and Burkert (1999a) 71. Also to be considered now in this connection are the first lines of the gold plates from Thessaly: νῦν ἔθανεϲ | καὶ νῦν ἐγ|ένου, τριϲόλβ|ιε, ἄματι τῶιδε. Cf. Tsantsanoglou and Parássoglou (1987) ad loc.

[63] Sider (1997) 147–8.

from some mystic movements,[64] and, as the Derveni papyrus also evinces, people connected with Orphism got interested in Heraclitus.[65]

Yet, although they certainly provide welcome cumulative evidence, we are not dependent on the bone plates in connecting Heraclitus with the eschatological and mythological matters previously discussed. For it does not need a detailed demonstration to make plain that fire has an eschatological role – closely linked to its cosmological, epistemological and psychological ramifications – in Heraclitus' philosophy.

Let me just mention the most obvious points, without pursuing an analysis of the wider implications in the Heraclitean context. That Heraclitus was indeed interested in eschatological issues – parallel to his critical attitude towards traditional views on these matters – is evident from fragments such as B122: 'Such things await men when they die that they neither expect nor suppose' (ἀνθρώπους μένει ἀποθνόντας ἅσσα οὐκ ἔλπονται οὐδὲ δοκέουσιν).[66] As far as the eschatological role of fire is concerned, the most explicit is perhaps B66: 'The fire will judge and convict all things when it comes upon [them]' (πάντα γὰρ . . . τὸ πῦρ ἐπελθὸν κρινεῖ καὶ καταλήψεται). As noted by Marcovich ad loc., the parallel use of καταλαμβάνω in B28 ('Dike will convict the fabricators of lies and those who bear witness to them'; Δίκη καταλήψεται ψευδῶν τέκτονας καὶ μάρτυρας) reinforces the feeling that divine justice is involved here. It is equally noteworthy for our discussion that B66 is immediately followed by B64 in Hippolytus' text: 'The thunderbolt steers all things' (τὰ δὲ πάντα οἰακίζει κεραυνός). In this fragment, the thunderbolt offers an easy mediation between the Heraclitean fire and Zeus, the traditional representation of the supreme divine power. B11 ('Every beast is driven to pasture by blows', πᾶν (γὰρ) ἑρπετὸν πληγῇ νέμεται) probably also belongs in this context. As Diels and Kranz suggest, θεοῦ or Διός is to be understood with πληγῇ, and the whole concept evidently alludes to Zeus' thunderbolt (cf., e.g., Hes. *Thg.* 853–5, adduced by Marcovich ad loc.).

Fire and divine justice, from which there is no escape, return in a somewhat more cryptic manner in B16: 'How could one escape the notice of

64 On this point see also the brief but, as I see it, circumspect and well-balanced analysis in Seaford (1986) 14–20. For a certainly much exaggerated ancient opinion see Clem. *Strom.* 6.27.1: σιωπῶ δὲ Ἡράκλειτον τὸν Ἐφέσιον, ὃς παρ' Ὀρφέως τὰ πλεῖστα εἴληφεν.

65 Commentators agree that, whatever the exact meaning of it might be, the pair ṭὰ κοινὰ . . . τὰ ἴδ̣[ι]α̣ in col. 4.5–6, immediately preceding the quotation, is an allusion to the Heraclitean use of these concepts (Tsantsanoglou and Parássoglou (1988) 130; Sider (1997) 134–5). To put such an allusion to work, the author must assume that his audience is familiar with Heraclitus' philosophy. Cf. also e.g. the Orphic appropriation of B36 in *OF* 226; the two are quoted side by side in Clem. *Strom.* 6.17.1–2.

66 This fragment, by the way, is sometimes mentioned in relation to P. Derv. col. 5.6–7.

what never sets?' (τὸ μὴ δῦνόν ποτε πῶc ἄν τιc λάθοι). This fragment is yet again a corrective reappropriation of a traditional concept, this time that of Ἥλιοc πανόπτηc, 'the Sun who sees all'. Heraclitus' general line of thought, as generally agreed by commentators, is well captured in the passage in Plato's *Cratylus* where the etymology and meaning of δίκαιον is discussed. The sun in itself cannot fulfil the role of 'the eye of Zeus' or 'the eye of Justice', because it sets every night – but the same function can easily be assumed by the cosmic divine fire, the sun which never sets.[67] Thus, the cosmic fire, yet again, appears in the role of the supreme cosmic Watchman, this time assuming the role traditionally ascribed to the sun.

Some of these fragments may even lead us back to Tablet C of the Timpone Grande, reinforcing thus the feeling that the web of common themes and possible cross-references is indeed tightly woven. There are at least three passages in that most obscure of the gold plates to which the Obscure of Ephesus can provide interesting parallels.[68] First, we have seen in line 2 the strongly related terms 'Sun, fire overpowers all' (Ἥλιε, πῦρ δὴ π̣άντα [. . .] νικᾶι). Second, accepting Bernabé's convincing reading, we have the relatively rare term ἀνταμοιβή in line 4, which can mean 'interchange' or 'requital'. The same term is applied to fire in fr. B90 of Heraclitus where it is said that 'All things are requital for fire, and fire for all things, just like goods for gold and gold for goods.' Third, we find the adjective πανόπτα in line 6 either referring here to Zeus, as Colli suggests, or, in a more usual manner, but with a slight emendation, to the Sun (Ἄ<λ>ιε), as Diels has proposed. Notably, all three Heraclitean parallels are related to the theme of fire, and two of them to the sun.

Indeed, the sun has appeared at a prominent place in many of the texts we have been discussing. Let us recall that we have started this whole discussion with the suggestion that the sun, mentioned in the Heraclitus fragment quoted in col. 4, and allotted a crucial role in the Derveni author's cosmological interpretation of the Orphic poem, creates a link between the two parts of the papyrus. Then we have seen that the sun, with possible Heraclitean echoes, appears in the text of the Tablet C. To this we may also add an inscription by a 'northern thiasos' (βορεικοὶ θιαcῖται) from around 300 BC from Olbia: ΒΙΟC ΒΙΟC ΑΠΟΛΛΩΝ ΑΠΟΛΛΩΝ ΗΛΙΟC ΗΛΙΟC ΚΟCΜΟC ΚΟCΜΟC ΦΩC ΦΩC.[69] I would not venture to

[67] Plato *Crat.* 413b6–c2: (reflecting on the former hypothesis that the δίκαιον is the sun) . . . καὶ ἐρωτᾷ εἰ οὐδὲν δίκαιον οἶμαι εἶναι ἐν τοῖc ἀνθρωποιc ἐπειδὰν ὁ ἥλιοc δύῃ. λιπαροῦντοc οὖν ἐμοῦ ὅ τι αὖ ἐκεῖνοc λέγει αὐτό, τὸ πῦρ φηcιν . . . .

[68] All three are listed in the commentary to Tablet C by Bernabé and Jiménez San Cristóbal (2001).

[69] Dubois (1996) no. 96.

propose any elaborate hypothesis about the significance of this inscription, but it is remarkable that we are back to Olbia, in a group of initiates, and we find that the sun and the cosmos are mentioned in a clearly religious, mystic context. The cumulative effect of these texts is that the sun functions as an interface between religious and cosmological matters not only in the Derveni papyrus but in its wider, but relatively well-defined, cultural context extending from Heraclitus to mystic communities from Macedonia to Magna Graecia.

The general image we get from all the above Heraclitean passages is that cosmic and eschatological guidance, justice and surveillance are closely related, or even identical, functions of the supreme divine power, fire. If I interpret him correctly, the Derveni author would agree with much of this – except that last word, fire. First and foremost, he would certainly concur with Heraclitus on the point that cosmic justice and eschatology are just two facets of the same divine order. He would also assent to the view that fire's role is central in both cosmology and eschatology. Yet he would not concede that the supreme divine power is fire. True, it is the force of fire whose effect we can immediately perceive; it is fire which can dominate (ἐπικρατεῖ) and has the ability to determine or intervene in the course of events. But fire is neither divine, nor intelligent in itself: it is not much more than a brute force. It is only the apparatus through which the cosmic intelligence communicates its will to the world. *Pace* Heraclitus, it is not the thunderbolt which steers the world, but Zeus, and the thunderbolt is but a tool in his hand.

As a matter of fact, the Derveni author could point out that in B3 + B94 Heraclitus apparently contradicts his own principal doctrine. For is not the sun the most magnificent cosmic manifestation and concentration of fire? And if the fire of the sun is intelligent and divine in itself – indeed that which steers and oversees the rest – why should it be under the surveillance of Dike and the Erinyes? Is it not true that Dike is, or is a function of, the cosmic intelligence? Conversely, if the sun, this huge mass of fire, fundamental to cosmic order and life, needs some coercive force to check that it keeps its measures, then fire cannot be the supreme cosmic intelligence.[70] By referring to Heraclitus, the author could insist on the (cosmological and

[70] These questions which I am putting in the mouth of the author indicate genuine problems – quite apart from the Derveni text – since the (Fire-Logos)-Dike-Erinyes-sun relationship is certainly problematic. Dike, as universal Justice, must have a close relationship to the cosmic balance (the Logos) and the ordering fire (cf. e.g., Kahn (1979) 273). Schönbeck (1998a) 103–14 lists important questions concerning the sun (HΛIOC)-Erinyes relationship. I agree with Schönbeck that the role the Erinyes play in connection with the sun/Helios in the Heraclitus fragment depends on the reading and interpretation of the conditional particle in line 8 of column 4 of the papyrus.

eschatological) importance of fire; whereas by quoting B3 + B94 he could point out that, as even Heraclitus is obliged to admit, the fire concentrated in the sun is *not* the ultimate power.

Pursuing this line of thought, there may be even more in Heraclitus' B3 + B94 that the author could easily turn to his own use. As far as I can see, in all those systems where there is a cosmic intelligence posited, the individual soul (or the most valuable, rational, divine part of it) shares in the basic nature of the cosmic intelligence. This pattern is certainly present in the respective systems of those philosophers who had the most impact on the Derveni author: Anaxagoras, Diogenes of Apollonia and Heraclitus. Moreover, the most evident form of this scheme is where the cosmic intelligence manifests itself in one of the physical elements; in this case the soul (or the most elevated form or part of it) is characterised by the same element. This is the case in Diogenes and Heraclitus. I see no reason why the Derveni author would not subscribe to this view. If so, the author's souls must have at least a component of air in them. And Aristotle's testimony according to which in (at least some of) the Orphic writings the soul is described as something airy is of considerable value at this point.[71]

On the other hand, as mentioned above, the Eumenides are expressly identified as souls (the souls of the righteous?) in col. 6,[72] and there is some indication that the Erinyes are also identified with (some) souls.[73] If we combine the two points, it follows that, for the Derveni author, the Erinyes are airy souls. On the author's interpretation Heraclitus' sentence would entail, then, that these airy souls can join the divine air and assist in the maintenance of the cosmic order.[74] Here we are: unwittingly even Heraclitus agrees that the sun is controlled by the divine air and its airy helpers.

This admittedly conjectural interpretation would, by the way, accord well with the tone of the remarks introducing the quotation from Heraclitus. As far as one can see from the fragments preceding the quotation, the author is critical but at the same time approving of Heraclitus' saying. The opening κατὰ̣ [ταὐτ]ὰ̣ Ἡράκλ̣[ε]ιτος . . . seems to imply that Heraclitus can be adduced as confirmation of the view just expressed. On the other hand, the

[71] Arist. *De an.* 410b27–30: τοῦτο δὲ πέπονθε καὶ ὁ ἐν τοῖς Ὀρφικοῖς καλουμένοις ἔπεσι λόγος· φησὶ γὰρ τὴν ψυχὴν ἐκ τοῦ ὅλου εἰσιέναι ἀναπνεόντων, φερομένην ὑπὸ τῶν ἀνέμων κτλ. Burkert (1999a) 97–100 discusses the possible Iranian background of the idea documented e.g. in Epicharmus B9 and B22 DK, in Euripides (*Suppl.* 532 and *Erechtheus* 532) and on an inscription at Potidea from 432 BC (*IG* I² 1179) that the soul is some kind of *pneuma* which goes to the sky to join the godhead after death. The text in D.L. 1.7 about *eidola* (which I argued 79 above refers to the soul of the dead) moved in the air by exhalaisons, and perceived by the magi, may count as further evidence.

[72] Col. 6.9–10. [73] Col. 2, with 87–8 above; cf. also Tsantsanoglou (1997) ad loc.

[74] Cf. the phrase θ̣εῶν ὑπηρέται in col. 3.7, probably referring to the souls of the righteous.

plausibly restored κατ̣[αςτρέ]φει τὰ ἴδ̣[ι]α̣ appears to mean that Heraclitus, on the author's interpretation, ended up confuting his own teaching.

On this basis, let me offer a tentative free paraphrase of the author's main tenet. Fire – the lightning in a thunderstorm, the burning flames of the pyre or the fire of Hades – can appear terrifying; but there is nothing to be feared.[75] True, fire has an immense and possibly destructive power, yet it does not act on its own. On the contrary, it is the principal means through which divine justice and intervention can manifest themselves, both at the cosmic level and at the level of the individual souls. Fire, this tremendous force, is well under the control of the supreme divine intelligence, be it called air or Zeus. Just look at the heavenly bodies. That vast amount of fire concentrated in them could, in principle, destroy everything. For example, the stars, driven by the force of 'like to like', could collapse into each other and join the fiery mass of the sun, and thereby annihilate this cosmic order. But once fire is mastered and fashioned in a proper form by the cosmic intelligence, it becomes the source of all becoming. For fire is nothing but the device with the help of which the god, the intelligent air, keeps justice and governs the world. Thus, they who have no reason to be afraid of the divinity should not fear fire either. The fire of the pyre, or that of the thunderbolt, destroys only the corpse, but the (superior) soul is airy, thus fire has no power over it. Moreover, as far as the individual soul is assimilated to the cosmic intelligence and is thus airy, fire becomes its helper: by destroying something inferior, it promotes the birth of something superior.[76] The ultimate message, therefore, is the proper explication of the relationship between air and fire; and the main clue is the exegesis of Orpheus' word: 'Zeus the king, Zeus who rules all with the bright bolt', Ζεὺς βαςιλεύς, Ζεὺς δ' ἀρχὸς ἁπαντων ἀργικέραυνος.

Besides, if the last part of this interpretation is accepted, we might finally be able to propose an answer to a further problem, internal to the text of the papyrus. In col. 20, the Derveni author interrupts the exegesis of the Orphic poem, and launches a forceful attack on rival initiators and initiation ceremonies. His main criticism consists in the fact that in these ceremonies the initiands, although they formally accomplish the rites, cannot attain genuine understanding because they are offered no proper explanation of what they have seen and heard. And without such understanding, the author affirms, the whole ceremony is worthless.

[75] As commonly agreed, the main aim of mysteries was to provide relief from some kind of anxiety connected with eschatology; see, e.g., Burkert (1987a) 21–9.

[76] The first lines of col. 4, preceding the Heraclitus quotation, seem to express something in this vein: something/someone does something positive rather than doing harm ( ]δ̣οῦναι | μᾶλλ[ον ἢ] ς̣ίνεται̣[ ). I wonder if in ὁ κείμ[. . .] μ̣ετ̣α̣θ̣[there could be a reference to the corpse, cf. LSJ *s.v.* κεῖμαι I. 4–5.

It has often been remarked that this column is difficult to find a place for in its immediate context. As André Laks has pointed out, this column belongs rather to the discourse of the first six columns, and seems to be an intrusion in the allegorical commentary.[77] In a way, this problem mirrors the problem concerning the place of the Heraclitus quotation in col. 4: a seemingly misplaced passage, which would pertain rather to the complementary part of the text. Although the general significance of the passage has been well explained by Dirk Obbink, it remains unclear why this passage appears at this particular point in the course of the commentary on the Orphic poem.[78]

A momentous detail: the last lemma adduced in the bottom part of the extant text of the previous column is precisely the one just mentioned: Ζεὺ̣ς̣ βαcιλεὺc, Ζεὺc δ' ἀρχὸc ἁπάντων ἀργ̣ικέραυνοc (col. 19. 10). The scanty fragments after the quotation show only that the author starts to elaborate on the word ἀρχόc; unfortunately, we cannot decipher what he has to say about this word and about ἀργικέραυνοc. A plausible suggestion is that it is precisely this lemma which prompted the diatribe against those other initiations and initiators. If people go home after the rites without understanding the real significance of this verse, indeed the condensed expression of Orpheus' central teaching, they have really spent their money in vain, and their expectations can get frustrated. This is all the more disastrous as these people, thinking that this is all an initiation rite can offer to them, have to continue their lives bereft even of hope.[79]

Admittedly, the above reconstruction, as I tentatively submit it for further discussion, is predominantly speculative. But the nature of the evidence certainly requires some degree of conjecture. As I see it, the main force of this hypothesis is that it unites in a reasonably coherent but not excessively complex framework a relatively large number of the elements directly or indirectly involved in the Derveni text: a theory of elements, eschatological concerns, the sepulchral context, the author's cosmological theory, Heraclitus' text, other texts connected with mystery religions and the Orphic poem itself.

77 Laks (1997) 124. See also the valuable discussion of the column in Obbink (1997) esp. 42–8. Previously, Rusten (1985) 138–40 went so far as to argue that the first 10 lines of col. 20 are not even by the Derveni author. For a criticism of this view, see now Obbink (1997) 42–8.

78 Obbink (1997) 52 n. 21 suggests that the word τε̣[λ]έ̣cαι occurring at the end of col. 19 occasioned the digression on the telestic rites in col. 20; he notes however that 'the connection remains obscure'.

79 Col. 20.3–12.

CHAPTER 10

Understanding Orpheus, understanding the world

Who is, then, the Derveni author? As I am unable to answer this question with a name, in what follows I shall try to identify him by describing the specific nature of his project.

Since the first reports and partial transcripts of the papyrus it has been customary to call the text a commentary and its author a commentator.[1] I myself have argued in chapter 3 that the Derveni author offers a systematic commentary on the Orphic poem. But does this make the text as a whole a commentary? The trouble, of course, lies in the first columns. As has been made clear by the publication of the first seven columns by Tsantsanoglou, the commentary on the Orphic poem only starts in column 7. This fact makes it at least problematic to continue to define the whole text as a commentary;[2] the first columns certainly have a bearing on how best to describe the overall subject matter and genre of the text. Let me list some of the emerging options:

1. The subject matter of the text as a whole is the interpretation of the Orphic poem. The first columns explain some external features or *realia* relevant for the understanding of the poem.
2. The subject matter of the text as a whole is not the Orphic poem, but the description and explanation of some ritual practices. The exegesis of the Orphic poem is subordinate to this, because it is in some way relevant for the understanding of the ritual practices in question.
3. The subject matter of the text as a whole is some eschatological or other lore in relation to which the Derveni author describes and explains both

[1] So already Kapsomenos (1963) 22; this has been argued for most recently by Lamedica (1990) and (1992).

[2] As a matter of fact, some scholars were already noting this problem on the basis of the *ZPE* text. See e.g. West (1983) 78 and Burkert (1986). On the other hand, it remained usual to stick to the previous description of the text. So e.g. Sider (1997) refers to the author as 'the Commentator', even though a substantial part of his paper is about col. 4, which precedes the commentary on the Orphic poem. See also Burkert (1999a) 78.

some ritual practices and the Orphic poem. The ritual practices and the Orphic poem are not otherwise related.

4. The subject matter of the text as a whole is a specific ritual, for example a certain funerary or initiatory ritual. The explanation of ritual actions in the first columns and the commentary of the Orphic poem complement each other, because they represent the two aspects of this ritual, the *dromena* and the *legomena*.

The list can be continued. We regrettably lack crucial information – both internal and external to the papyrus – for establishing which of these options is to be preferred. For instance, we do not know whether or not the Orphic poem commented on in the papyrus had a role in a specific ritual, and if so what that ritual was. Moreover, the articulation of the two parts of the surviving text, where the author probably said something about the shift of subject matter, is lost in the lacuna between columns 6 and 7. One should not forget either that we do not know how many columns, or for that matter rolls, preceded the surviving column 1, and we cannot know whether or not the text continued after our column 26 on another roll. For these reasons, I am quite sceptical about the possibility of deciding once and for all among the above options, or defining the genre and subject matter of the text as a whole with certainty. Personally, I would be most inclined to vote for the last option on the list above, but only, admittedly, because I find it the most economical hypothesis.

Yet no matter which of the above alternatives one finds the most appealing, it seems to me that the most probable general characterisation of the author – especially in view of the first columns – is that he was a priestly figure with Orphic allegiance.[3] Moreover, the two parts of the text have a common denominator in the author's manner of procedure: in both parts he provides explanation and interpretation.[4] He explains the true significance and meaning of certain ritual actions and eschatological terms in the first columns, just as he explains the true significance and meaning of Orpheus' words in the remainder of the text.

EXPLANATION AND PROFESSIONAL RIVALRY

In view of the role of explanation in his text, the single most promising way to characterise the Derveni author is to apply the well-known words of

[3] I think this remains true despite the recent challenge by Janko (1997). See also Laks (1997) and Obbink (1997). For a detailed critical appraisal of Janko's hypothesis, see the Appendix.

[4] This aspect of the text has been stressed by the excellent study by Obbink (1997). As the present chapter will show, I am in agreement with Obbink's general presentation of the Derveni text to a great extent.

Plato: he could well be one of those priestly persons 'who have made it their concern to be able to give an account of their practices' (*Meno* 81a10–b1). Indeed, these words have often been quoted in attempts to describe the author.[5] Plato's phrase remains vague, probably intentionally, and even the larger context of the passage leaves it undecided who exactly these people are, what their 'practices' consist in, and what kind of accounts they provide for those practices. I shall argue, however, that some other passages in Plato can help us in drawing a more detailed sketch of the Derveni author.[6]

It has often been noted that Plato has an ambiguous attitude towards priests connected with mystery cults. The more sympathetic approach, which we find in the *Meno* passage, is to be contrasted with texts such as Adeimantus' famous description of *orpheotelestai* in his bitter diatribe at the beginning of book 2 of the *Republic*:

> And they present a noisy throng of books by Musaeus and Orpheus, offspring as they say of Selene and the Muses, in accordance with which they perform their rituals. And they persuade not only individuals but whole cities that the unjust deeds of the living or the dead can be absolved or purified through ritual sacrifices and pleasant games. These initiations, as they call them, free people from punishment hereafter, while a terrible fate awaits those who have not performed the rituals.[7]

As a matter of fact, this short description forms the basis of our knowledge of the praxis of the Orphic initiators. Now the practice rebuked by Adeimantus reappears in book 10 of the *Laws* as the third and most dangerous form of irreligion: those who believe in the efficiency of these practices must think that the gods are corrupt, for they can be won over by sacrifices and prayers, without a genuine moral advance and a conscious avoidance of vices.[8] It is clear that Plato in his old age saw the itinerant salvation-mongers as a morally and theologically dangerous lot.[9]

The *Laws* can offer, I suggest, another, less obvious, clue for the characterisation of the Derveni author in connection with Adeimantus' harangue. For, apart from holding morally destructive theological doctrines, what do Adeimantus' initiators do? They prescribe certain rituals, and, on the other hand, they describe in frightful terms 'the terrible fate' of those who do

[5] E.g. Boyancé (1974) 109; Kahn (1997) 55; West (1977) 83 etc.

[6] Funghi (1997) 37 remarks, rightly I think, that 'the only fixed reference point for characterising the author gravitates around Plato's writings.'

[7] *Rep.* 364e3–365a3 (trans. Grube revised by Reeve). Cf. *Phdr.* 244d5–245a1.

[8] *Leg.* 885b8–9 and 905d1–907b4.

[9] It is not clear how the priests and priestesses mentioned in the *Meno* are related to the initiators spoken of by Adeimantus; and, further, how far we can identify the 'account' provided by the priests and priestesses simply with an Orphic eschatological myth.

not follow their instructions. Ignoring for a moment the contents of these orders and threats, this tactic is the same as what Plato ascribes to the inferior type of doctors in the analogy he draws between doctors and legislators.[10] The slave doctors practise their art only on the basis of empirical skills and techniques they have picked up from their masters. Consequently, they do not, and indeed cannot, explain to the patients the nature of their illness and the treatment to be followed; they merely order what seems to them the best cure and threaten the patient by describing what will happen if they do not follow the prescription. Thus, they act much like the initiators who terrify their clientele with horror stories about the fate of lost souls and prescribe the accomplishment of the appropriate rituals as treatment.

The master doctor, by contrast, gives a double treatment. First, he tries to discover the causes of the condition by consulting the patient and the patient's relatives. More important, he then instructs the patient about the illness and the cure, and does not order anything before he ascertains, by means of explanation and discussion, that the patient has understood the nature of the illness, and the reason why the prescribed method is the best cure. Although Clinias hastens to agree that the master (or free) doctor's procedure is certainly much better and more efficient, it is not at all evident why the healing cum explanation method is more efficient than simple healing without explanation. Indeed, when Plato returns to the medical analogy in book 9, we get no clear reply to the slave doctor's cynical remark that the other doctor tutors the patient instead of healing him, even though what the patient wants is not to become a doctor, but to get well.

To sum up, the master doctor (i) does not contend with sheer empirical skills, but wants to set his practices in a more general theoretical framework, involving, most prominently, knowledge pertaining to natural philosophy, and (ii) finds it important to impart knowledge to the patient, to help him or her in understanding what he or she is going through, and to gain the patient's trust in this way. It seems to me that the Derveni author's general stand within his *techne* can be best understood by reference to the professional attitude of Plato's master doctors, these practitioners with higher intellectual standards and didactic aims. The Derveni author's professional opponents – at least as he presents them in col. 20 – are comparable in turn to the slave doctors; those practitioners who, even though they do carry out what is normally expected from the *techne* in question,[11] do not stand

[10] *Leg.* 720a2–e8 and 857c4–e6.

[11] As a matter of fact, in previous dialogues Plato himself maintained that the doctor does not need to instruct and to give an explanation to the patient or to get the patient's consent (see e.g. *Grg.* 521e6–522a7 and *Pol.* 293a9–c2).

up to higher demands, for they do not possess a more thorough theoretical knowledge and do not help their clients to understand their own situation. They prescribe the rituals without explaining either the true meaning of the *legomena* or the way the *dromena* are effective. They prescribe and threaten, but do not explain what is going on.

As Burkert elucidates in his important study significantly entitled 'Craft versus sect: the problem of Orphics and Pythagoreans', *orpheotelestai* can best be considered practitioners of a 'religious craft': professionals of a specific *techne*.[12] Now, col. 20 shows that the Derveni author's text is to be seen also as part of a polemic among representatives of the same craft. We hear about two different groups of professionals performing initiatory rites, and the author claims that what these initiators do is worth nothing; and, consequently, those who let themselves be initiated by these people are inevitably doomed to failure and frustration. However, what he wants to say is not that the possible services of a *mantis* and mystery rites are useless – as I have argued in chapter 2, he identifies himself as a priestly person connected with oracles, initiation and related rites – but that those other practitioners are either unqualified or simply charlatans.

This internal criticism parallels a certain type of restricted criticism of magical practices, well documented by anthropological studies from the Azande to sixteenth- and seventeenth-century England. Declaring this or that witch-doctor a quack, or even saying that most of the practitioners are charlatans, does not necessarily undermine faith in the practice in general, but can even reinforce trust in it: it is only the few true experts who can realise the genuine possibilities of the art.[13] This may well be the Derveni author's tactic. Calculating that his possible clients might have already been deceived by some initiators, or at least could have already heard sceptical or critical objections (for example, something like the speech of Adeimantus), he explains why those others are not authentic representatives of the art. This is also a forceful rhetoric to gain the interest even of those who have already gone through an initiation rite and would otherwise be quite happy with it – they can now realise that, actually, that other initiation was not the real thing.

The most extensive Greek evidence we have about the polemics of practitioners of the same *techne* comes from the Hippocratic corpus. In these

[12] Indeed, as, for Plato, being able to give a *logos* of one's praxis is a distinguishing feature of a *techne* (cf. e.g. *Grg*. 456a6–7), the *Meno* passage can be seen as an acceptance that certain priests and priestesses are professional practitioners also in the restrictive Platonic sense.

[13] This is one of the main upshots of Evans-Pritchard's classic study of Zande magic and witchcraft. See esp. Evans-Pritchard (1937). For the English case, see Thomas (1971). For a very helpful critical overview of the anthropological discussion of the topic, see Tambiah (1990).

professional debates the arguments naturally have two aspects: criticism of rivals on the one hand, and a positive claim for superior expertise on the other. Besides, the polemic has a double objective. On the one hand, the practitioner wants to prove his skills and knowledge in front of his colleagues; professional pride is certainly at stake. But the debate is also directed towards the lay public, the potential market of the *techne*: the doctor has to show his superiority over rival professionals in order to gain the clients' trust.[14] There were some who employed theatrical means to capture the interest of the public,[15] but there is ample evidence to show that there were others who tried the method of Plato's master doctor. These doctors wished to obtain the clients' confidence by presenting a scientific account of the constitution of the human body, the cause of the illness within this framework and an ensuing physical, physiological explanation why this or that curative method should be helpful. Remarkably, such explanations are often put into an even wider context: the doctor's view on the constitution of the human body is presented with reference to a general physical theory and cosmology.[16] The underlying idea is that the human body is part of nature and thus the same physical laws are at work in the organism as govern the cosmos. And these descriptions often reflect ideas developed by the Presocratic philosophers.

Mutatis mutandis this is the strategy of the Derveni author. In order to gain the trust of the public his text is directed to, he claims that he is the one who possesses genuine expertise. To prove this claim, on the one hand he launches a critique of rival practitioners, and, on the other hand, he develops his own, purportedly premium teaching. Again, the substantive teaching places the specific concern of the *techne* – knowledge about the divinity and eschatological, cathartic, telestic matters – into a wider, physical, cosmological context. The relevance of this physical, cosmological framework is in all probability the same as in the case of the medical writers: the phenomena with which the *techne* is concerned are to be understood with reference to the widest possible horizon. These phenomena are integral parts of the cosmos, and the same laws and mechanisms govern the subject matter of the *techne* as are at work in the cosmos in general. The same divinity governs both the development of the cosmos and the fate of the individual souls, and in its activities the supreme principle can rely on the same physical processes. Moreover, to expound the eschatological doctrine requires giving some explanation of the nature of the soul. And if

[14] Lloyd (1979) 89–98.

[15] See e.g. the criticism of such practices in *Praec.* ch. 10 and *Decent.* ch. 2.

[16] Cf. e.g. *De carn.*, *De nat. hom.*, *De victu.*

the soul survives the body, we have to be able to tell to which part of the cosmos it will go and what it will do there.

It has to be stressed, however, that, even though the relevant Greek evidence is scanty, the cosmological frame of reference in both contexts – healing and eschatology – is probably not a new development but an archaic feature. For example, a cuneiform text, presumably used as incantation in primitive dental surgery, relates the story of the worm that causes the pain by presenting a short theo-cosmogonical narrative from the very beginnings (how Anu created the Sky, which created the Earth, which created the rivers etc.) down to the birth of the worm.[17] That eschatological doctrines could easily be presented with reference to a cosmological framework is evinced also by Plato's great eschatological myths. Both in magical healing and in cathartic rites the aim was to reinstall normal order disturbed by illness and/or pollution; and a cosmological-cosmogonical account was supposed to be the most effective means for this purpose.[18]

Nevertheless, the Hippocratic writings and the Derveni papyrus share at least two features that seem to mark a difference as compared to the more archaic tradition. First, we have the obviously dialectical, polemical use of the cosmological and physical doctrines in critical discussions about the theoretical foundations of the craft. Second, and more important, these texts show a conscious use of concepts and explanatory methods developed in the 'inquiry into nature' tradition, among which, first and foremost, there is the unifying concept of nature, and the conviction that the world is an ordered whole within which one can account for the diversity of phenomena and processes with a reductionist, mechanistic, causal explanation, built out of a few basic principles.

On this basis, I suggest, the Derveni text can be seen as an attempt to implement for the *orpheotelestes*' craft a certain type of professional attitude, methodology and argumentative strategy which we can see most notably in the sphere of the medical art. It is not necessary to maintain the stronger hypothesis that the author has actually acquired this approach from the doctors.[19] It has to be noted also that, even though this aspect is not manifest in the Derveni text, there is strong evidence to show that religious practitioners, and Orphic initiators among them, were also occupied with

[17] *ANET* 100–1; cf. Bottéro and Kramer (1989) 484–5.

[18] Cf. e.g. Burkert (1982) 8, (1987b) and (1992) 124–7.

[19] Nonetheless, it is not without interest that some of the specific terms he uses are technical terms of medical writings, e.g. θάλψιϲ in col. 9.7, but cf. also the technical vocabulary describing the mixing and conjunction of entities. Moreover, as noted above (cf. 273 with n. 136), some of his physical explanatory principles – most notably the strong connection between fire and motion – are better documented in medical texts than in the doctrines of the natural philosophers.

magical healing.[20] What is more, the author of the Hippocratic treatise *On the Sacred Disease* claims that 'the magi, purifiers, begging priests and quacks' not only practised ritual cures consisting of purifications and incantations, but also 'added a suitable account', that is, they devised some kind of explanatory theory to prop up their healing methods.[21] It is reasonable to assume not only that Hippocratic doctors knew about their priestly rivals, but also that the Orphic initiators and other magical healers knew about the practices of Hippocratic doctors. This is especially plausible in the case of the Derveni author who certainly had an interest in physical, cosmological theories.

Note also that both incantations and dreams considered as signs can appear within a *physical* framework in Hippocratic writings of a clearly scientific character.[22] Aristotle calls attention to the fact that

> even accomplished physicians tell us that one should pay diligent attention to dreams, and to hold this view is reasonable also for those who are not practitioners, but make theories and do philosophy.[23]

The reason for this, Aristotle explains, is not that dreams are sent by some god, but because dreams are also part of the natural functioning of the organism and can indicate something about physical processes going on in the body. Dreams are significant in so far as they are natural phenomena and obey the general laws of nature. We have no information on the question of how the Derveni author, if pressed, would have explained the origin and function of dreams and in what way he could have tried to integrate them into his general theory of the cosmos. But his theology leaves little room for direct personal divine intervention even in this respect.

At any rate, it is widely agreed that critical professional debates and, closely connected to them, a growing demand for theoretical explanatory frameworks for healing methods were determining factors in the emergence and development of scientific medicine. The Derveni text now shows that something similar could appear in other *technai*. More specifically the Derveni text can show that in the *techne* dealing with initiation and oracular activity, at least in certain cases, this could lead towards natural theology,

[20] Plato *Phdr.* 244d5–245a1; Eur. *Alc.* 966–71 where the parallel reference to Orphic and Hippocratic medicine is especially notable (there is no cure against Ananke: οὐδέ τι φάρμακον | Θρῄσσαις ἐν σανίσιν, τὰς | Ὀρφεία κατέγραψεν | γῆρυς, οὐδ' ὅσα Φοῖβος Ἀσκληπιάδαις ἔδωκε | φάρμακα πολυπόνοις ἀντιτεμὼν βροτοῖσιν). Cf. also Eur. *Cyc.* 646–8.

[21] *Morb. sacr.* 2: μάγοι τε καὶ καθάρται καὶ ἀγύρται καὶ ἀλαζόνες . . . καὶ λόγους ἐπιλέξαντες ἐπιτηδείους.

[22] *De victu* 4 is the most elaborate on the question, but see also e.g., *De hebdom.* ch. 45.

[23] *Div. somn.* 463a (trans. J. I. Beare in the Revised Oxford Translation, modified).

using the conceptual toolbox of natural philosophy.[24] It remains an open question whether the Derveni author was an isolated case or was representative of a trend.

The Derveni author's insistence on a general theory of nature is a notable feature also with regard to magical practice. Magical practice, on a common understanding of the term, involves supernatural intervention. Indeed, as G. E. R. Lloyd has stressed, one distinguishing feature of a scientific, as opposed to a magical, approach is a rigorous demarcation of nature as separate from the sphere of the supernatural.[25] How can we classify the Derveni author within this scheme? The Derveni author certainly makes an effort to provide a general theory of nature. Now in this general theory every event is, or could be, described as the joint result of divine volition and physical causation; divine volition realises itself through physical causation.[26] There is a general framework, involving both the divine and the physical, and, if needed, one can find a causal explanation for every important phenomenon within this framework. This approach is very close, among others, to that of the Hippocratic author of the *On the Sacred Disease*, and – as G. E. R. Lloyd has shown – such a conception of nature rules out the intervention of the supernatural, and can substantiate the Hippocratic author's criticism of magic.[27]

Certainly, in col. 6.2 the Derveni author speaks about the magi's ability to influence the daimones by means of incantations and sacrifices. This can surely qualify as coercive ritual with the aim of manipulating eschatological forces – and this is, at least on some accounts, what magic is. It is also true, however, that the author explicitly speaks about the propitiation of souls, which in the Tylorian-Frazerian account of magic, for instance, is a distinctive feature of religion as opposed to magical manipulation. At any rate, it is not clear how these practices are thought to be related to the theological and physical theory developed in the second half of the Derveni text. The gaps in the papyrus do not entitle us to say anything definite on whether or not the physical and cosmological theory of elements could be deployed in the ritual action described in the first columns. In the previous chapter, however, I tried to formulate a tentative hypothesis that could be developed also in this direction.

[24] The author of *On the Sacred Disease* argues that the religious healers are actually impious because even though they claim that the illness is of divine origin, they treat it by diet and other physical means. The Derveni author could respond that the two types of explanation and causation are not necessarily incompatible if one can provide a framework in which the god also uses physical processes in order to realise his intentions.

[25] See esp. Lloyd (1979) 26–9. [26] On this, see 273–4 and 292 above.

[27] See Lloyd (1979) 26–9.

As well as the possible role of physical causation in ritual action, it would be interesting to see whether the author would establish any connection between his views on language as developed in the second part of the text, on the one hand, and the 'incantations' of the magi mentioned in column 6, on the other hand. This could have important bearings on the topic of the magical use of language.

Be that as it may, in view of the author's general attitude, I do not find it probable that he considered the ritual action of the magi a temporary suspension of the laws of nature, but rather as making use of non-evident possibilities by the application of a thorough knowledge of the underlying causes. The obvious parallel here is Empedocles, and especially his promise in B111 to impart knowledge of controlling nature in an extraordinary, but not *super*natural, way.[28] The case of the Derveni author is a further indication that there is no necessary mutual entailment between a naturalistic mode of explanation and an empirically verifiable, 'scientific' practice.[29]

Besides, it is generally held that the way in which the practitioner deals with empirical evidence that could show the inefficacy of a practice, and which could hence falsify the explanatory theory, is an important indication capable of demarcating science from magic. Surely, this criterion cannot be applied in our case: there could be very little empirical evidence to show how successfully someone is able to secure the post mortem salvation of the soul of an initiate. On the other hand, this is exactly the principal reason, at least in a Popperian framework, why the very aims of such a practice range it outside the possible scope of scientific activity – verification and falsification are hardly possible here.

But the criterion of efficacy of such a practice should be sought elsewhere, one could object, namely whether the religious practice in question is able to cater for psychological needs, to offer relief from anxiety. At least some form of empirical verification is possible as far as its pragmatic and performative function is concerned: whether or not the practice in question is able to fulfil such a psychological and social role. From this point of view, the very popularity of mystery cults shows that these practices were efficacious in this sense. Hence, there is verification, but its criteria are different.

I would respond to such an objection by saying that the objective validity of the explanatory theory and the efficacy of the practice are inherently connected in the medical practice with scientific aspirations, whereas there

[28] Cf. also Lloyd (1979) 34–5.

[29] This formulation can also be taken as a reaction to the general thrust of Kingsley (1995).

is no such intimate relationship between the psychological or social efficacy of the ritual action and the objective validity of the propositions making up the explanatory account for it. However, the explanatory account provided for ritual action can have more persuasive power, which can certainly enhance also the psychological efficacity of the practice joined to it, when it satisfies certain criteria of validity applicable also in the domain of medical theories. And this is especially so in a culture where people have been made sensitive to such features as internal logical consistency, the economy of causal principles and so forth, through the critical discussions and professional rivalry that went on notably in medicine and philosophy. If so, certain representatives of the priestly *techne* could also feel that their explanatory accounts have to satisfy such criteria in order to retain their persuasive power.

THE *EXEGETAI*

The Derveni author's attitude towards the 'meaning' of ritual actions and sacred texts, his stress on the importance of understanding, may be further elucidated through a brief comparison with the professional activities of the *exegetai*. *Exegetai* were religious experts whom people could consult in cases of doubt. They could provide professional interpretation of the relevant sacred laws, and give advice on the correct performance of diverse cult activities, such as sacrifices and purification rituals.[30] *Exegetai*, as for example Cleidemus of Athens, sometimes also wrote professional treatises under the title *Exegetikon*. As far as one can see from the rather scanty evidence, they could evoke and interpret mythical narratives possibly as historical aetiologies, or even use etymologies in their descriptions of cult activities. Yet, these descriptions, and this is where the contrast with the Derveni text can be seen most clearly, were always pragmatic, and certainly did not place the cult activities or the sacred texts in a physical or philosophical framework.

One may also consider in this context Philochorus, who held the official position of a *mantis* and who published a learned work on his art (Περὶ μαντικῆς, frr. 76–9 *FGrHist*), in which he maintained, for example, that Orpheus was also a *mantis*. Philochorus wrote on sacrifices (Περὶ θυσιῶν, 80–2 *FGrHist*), on purification rituals and on dreams (Περὶ καθαρμῶν and Περὶ ὀνείρων fr. 89 no. 17 and 18 *FGrHist*) as well, and thus his sphere

[30] See e.g. Demosthenes 47.68, Jakoby (1949) ch. 1 and Garland (1990) 81.

of interest overlapped to a large extent with the professional activities of an Orphic initiator. If Dirk Obbink is right, he even knew the Derveni author's text.[31] Yet, as Jakoby has made clear, Philochorus' theology was a systematic and rational account based on the traditional, mythic representations of the gods in which there is no trace of philosophical concerns, of physical interpretation of divine powers, of euhemerism, or of the influence of Orphic or Pythagorean religious views. And even though he used etymology for the explanation of a wide range of historical and religious phenomena, his etymologies all remain 'plain and sober', and have nothing to do with physical allegory.[32] If Jakoby's representation of Philochorus is correct, the Derveni author would have surely claimed that one cannot get a genuine understanding of cultic matters either from an *exegetes* like Cleidemus or from a *mantis* like Philochorus, because these people do not have a real grasp of the importance of physical and cosmological factors in religious and eschatological matters.

UNDERSTANDING

There is, however, something more to the Derveni author's attitude. As noted above, we did not get a clear-cut answer from Plato as to why it matters for the success of the cure whether or not the patient obtains an understanding of the illness and of the treatment. The author of the papyrus seems to imply, by contrast, that personal understanding on the part of the initiand is indeed crucial for the success of the ritual. From his assertions about the role and possibilities of understanding transpires an epistemology which shows significant points of contact with Heraclitus. I shall argue that understanding and knowledge are conceived both by Heraclitus and the author of the papyrus as the outcome of a conscious and personal exegetical effort.

As noted already by David Sider,[33] both Heraclitus and the Derveni author claim that they possess a specific knowledge, whereas the great majority of people are in a state of lamentable ignorance. This epistemological criticism is no doubt a recurrent feature in many texts of the Presocratic period. However, Heraclitus' more specific assertion is that people are ignorant despite the fact that they actually have access to all the data which, rightly interpreted, contain the necessary information, and could, in principle, be turned into knowledge. People experience things, they encounter them on a daily basis, nonetheless they remain uncomprehending about

[31] Obbink (1994) and (1997) 49 n. 16 with my remarks 98 n. 20 above.

[32] Jakoby (1954) 225–7 and 233–4.

[33] Sider (1997) 146.

their real significance. Fragments with this implication are plentiful (see in particular B1; B17; B34; B72).

We have a comparable claim in col. 5 of the papyrus:

> . . . terrors . . . consult an oracle . . . they consult an oracle . . . for them we go into the oracular shrine to inquire for oracular answers, whether it is right . . . the terrors of Hades, why do they disbelieve? Not understanding dreams, nor any of the facts, on the basis of what kind of warning would they believe? Overcome by fault and by pleasure as well, they neither learn, nor believe. Disbelief and lack of understanding [? are the same thing]. For if they neither understand, nor do they learn, [it is not possible that they believe] even when they see . . . disbelief . . . appears . . . (col. 5.1–14)

People can, and indeed do, experience the facts (πράγματα), and their own dreams, but this in itself is not sufficient for them to attain genuine knowledge. The author draws attention to the same problem in col. 20, this time with respect to the *legomena* and other elements of initiatory rites:

> . . . those men who, while performing the rites in the cities, have seen the holy things, I wonder less that they do not have knowledge. For it is not possible to hear and at the same time to understand (*or*: learn) what is being said (τὰ λεγόμενα). But all those who (hope to acquire knowledge?) from someone who makes craft of the holy rites deserve to be wondered at and pitied. Wondered at because, thinking that they will know before they perform the rites, they go away after having performed them before they have attained knowledge, without even asking further questions, as though they knew anything of what they have seen or heard or learned; and pitied because it is not enough for them to have spent their money in advance, but they also go off deprived even of their judgement. Hoping before performing the holy rites that they will attain knowledge, they go away after having performed them deprived of hope too. (col. 20.1–12)

It is stated here explicitly that these people did in fact see and hear what is there to be seen and heard during the ritual. They even learned what they had to learn.[34] Yet they have not acquired the crucial thing – understanding – and, as a consequence, they have been completely deceived. What is more, their due is not only deception, but also self-deception: they go home mistakenly imagining that they have actually obtained knowledge from the mystic experience. In a word, they have lost not only their money, but also their good judgement. Although it is applied here to a special situation, the concept of being ignorant of one's ignorance, as well as the image of seeing and hearing without understanding, has an obvious Heraclitean ring.[35]

[34] On *symbola* or *synthemata* to be learned by the initiand, see e.g. Burkert (1987) 46 and 153 n. 14.

[35] See also the parallel, noted by Sider (1997) 133 between εἶδον ἢ ἤκουσαν ἢ ἔμαθον in line 8 and ὄψις ἀκοὴ μάθησις in Heracl. B55 DK. Moreover, some commentators maintain, especially in connection

The same is true with regard to the author's principal concern in the main part of the remaining text: the interpretation of the Orphic poem. The Orphic poem is available to both groups distinguished by the author, both 'the ignorant' and 'those who know'.[36] Yet while the former remain uncomprehending, stuck at the surface level of the text, the latter truly understand Orpheus' meaning. The same text is available to both groups, yet the genuine significance of it nevertheless escapes those who do not understand the language.

The inner tension of the Derveni text stems from the fact that this particular epistemological attitude places the author in a certain sense outside the limits of his profession – at least as far as we can judge it on the basis of the meagre comparative evidence. It has long been recognised by anthropologists that magical actions and mystical rites are primarily to be assessed by their performative and affective side; and, further, that the main criterion of the felicity of these practices is whether or not they are carried out properly, in accordance with the relevant ritual regulations, and, possibly, whether the participants have the right intentions. Clearly, teaching, in the sense of a detailed and falsifiable causal explanation on the part of the practitioners, and a personal and *critical* intellectual effort on the part of the participant, do not have a central role in this context. As Tambiah argues, the ritual use of language is to a great extent performative, and hence it is inappropriate to judge its efficacy in terms of verification statements.[37] What really counts for the participant is not an intellectual awareness through an act of objectification and neutralisation of what is going on and why exactly it is supposed to be working in terms of natural causes, but that he or she is going through an extraordinary experience, and that some of his or her pressing problems will, it is hoped, be solved by the efficacy of the ritual. The two contrasting attitudes, which may overlap to a certain degree in actual experience, can also be termed in the wake of Lévy-Bruhl and Tambiah as 'participation' versus 'causality'.[38] In the Greek context, the relevance of such a difference of attitudes with respect to mystic rites was fully recognised by Aristotle: 'Those who are being initiated into the mysteries are not to be expected to

with B15 DK, that Heraclitus does not disapprove of ritual, mystic practices as such, but criticises the participants for not understanding the true significance of the rites (cf. e.g. Funghi (1997) 32 and the circumspect analysis of Heraclitus' attitude towards popular religion in Adoménas (1999)). Contra e.g. Marcovich ad B15 DK.

36 Tsantsanoglou (1997) 118–19 points out, on the basis of the first words of col. 7, that the Orphic poem apparently did not belong to the ἀπόρρητα or ἄρρητα: everybody had access to it. This, however, appears to be in conflict with the proem of the poem in which the poet orders the profane 'to put doors to their ears' (col. 7.9–10). A possible solution is that, on the author's interpretation, formal, ritual secrecy is here replaced by another sort of secrecy: Orpheus' riddling language excludes 'those who are impure in hearing'.

37 See, e.g., Tambiah (1973) 218–27.

38 See esp. Tambiah (1990) ch. 5.

learn anything, but to undergo something, to be treated in a certain way, that is to say to become adapted.'[39] Quite clearly, Aristotle uses here the word 'to learn' in the sense of 'to understand causally', since he must have known that the initiand had to learn certain magic formulae. Thus, what he says is that the initiands do not have to understand anything, because the stress is on the affective, emotional, participatory side of the ritual,[40] and on the experience of a certain closeness to the divine.[41] Involvement in a rite, in a word, is predominantly an affective and not an intellectual issue.

Not only the general didactic tone of the whole text, but also the explicit second order arguments about the importance of understanding in columns 5 and 20 make it clear that the Derveni author finds this general approach unacceptable. For him, the criterion of validity of the ritual action does not simply reside in the faultless execution of the *dromena* and the correct recitation of the *legomena*. All this may well be a necessary, but certainly not a sufficient condition for the felicity of the action. For it is just as important that the client should leave the scene with an understanding of the meaning of the acts and texts involved in the rites, and what is more, should attain a more thorough awareness of the world and his situation in it not only, and even not primarily, at the level of emotions, but in terms of understanding of physical causes.

Remarkably, according to one of his critical remarks in col. 20.7–8, the author would encourage people to keep asking questions until they really understood what is to be learned. A personal intellectual effort is needed on the part of the initiand. This insistance on understanding is very close to the main tenor of the fragments of Heraclitus. Moreover, it may very well be the case that Heraclitus' criticism of traditional religiosity and the Derveni author's criticism of the populace and rival practitioners come very close: it has been convincingly suggested that Heraclitus' scorn is directed not so much against the cults and rites as such, but rather against people who do not have a clue what they are doing when they partake in the cults.[42]

Assuming that the primary concern of the ritual was eschatological, the ritual action had to effectuate some kind of modification in the individual soul in preparation for its future existence. Given the physicalist assumptions of the Derveni author, one can venture the hypothesis that he

[39] Fr. 15 Rose: καθάπερ Ἀριστοτέλης ἀξιοῖ τοὺς τελεουμένους οὐ μαθεῖν τι δεῖν ἀλλὰ παθεῖν καὶ διατεθῆναι, δηλονότι γενομένους ἐπιτηδείους.

[40] Cf. also Arist. *Polit.* 1342a6–11 describing the psychological effects of music used in rites: 'Some persons fall into a religious frenzy, and we see them restored as a result of the sacred melodies – when they have used the melodies that excite the soul to mystic frenzy – as though they had found healing and purgation' (trans. B. Jowett in the Revised Oxford Translation).

[41] See also Burkert (1987a) ch. 4 entitled 'The Extraordinary Experience'.

[42] See Babut (1975) and Adoménas (1999).

also shared Heraclitus' view that the individual soul may be most effectively transformed, made better and more similar to its divine cosmic counterpart, by attaining a deeper understanding of the world and itself. If so, the Derveni author could claim that the real felicity of a ritual action performed with the aim of preparing the individual soul for a better fate requires not only the correct performance of the rite and the *bona fides* of the participants, but also a properly intellectual component, because this last can bring about a necessary, and also physically expressible, transformation in the soul itself.[43]

LEARNING FROM AN ORACLE

But how can Orpheus' text teach us about the world, and how can one reach a genuine understanding of it? The parallelism between Heraclitus and the Derveni author, I suggest, is not restricted to the criticism of the ignorant, but continues on the positive side, as well. For the paradigm case of understanding, for both Heraclitus and the Derveni author, is the interpretation of oracles. In what follows, I shall try to show that the *hieros logos* of Orpheus is to be interpreted and understood in the manner of an oracular utterance.

To begin with, there is the sacred *logos* of Orpheus which, according to the Derveni author, is enigmatic.[44] What is more, Orpheus is intentionally enigmatic. On the other hand, this *logos* has also an important teaching function. Orpheus wanted to express something momentous in it, something which, if rightly understood, turns out to be of the utmost importance:[45] the constitution of the cosmos, the true nature of the divinity, and the way it created the present world order. Besides, his teaching apparently has an important bearing on the fate of our souls. The situation is paradoxical: how can one teach by intentionally using a cryptic language, reveal and conceal in the same gesture?

As there is no key provided with the poem to break Orpheus' code, prima facie this situation may appear to imply that we can only understand the real meaning of Orpheus' text if we already know what he wants to express. In order to understand the poem, we must already have learnt the

[43] In such a view, the distinction between the 'intellectually' and the 'morally' better soul is missing, without, of course, the Socratic explanation of why this distinction is inappropriate.

[44] I argued above that by the word λόγοϲ in col. 25.11 the author refers to Orpheus' text.

[45] Cf. col. 7.4–7. This claim, by the way, is very much like the τρόποϲ τῆϲ διδαϲκαλίαϲ, the explanation of the author's teaching method by the commentator, a standard feature of latter commentaries, analysed recently by Sluiter (1999) esp. 179–85.

constitution of the cosmos, the true nature of the divinity, etc.; we should, then, interpret Orpheus' text on the basis of our pre-existent knowledge. But this again will not do, for in that case Orpheus' *hieros logos* would lose its point; our aim is not the interpretation of a poem for its own sake but obtaining knowledge about reality.

Thus the constitution of reality and the true nature of the divinity are not self-evident, but nor is Orpheus' text – neither side in itself can be taken as a basis for the understanding of the other. Nonetheless, we can still expect understanding and gain in knowledge *from the proper matching of the two sides.* If we find the right way to make the connection between the text and the cosmos, then the two will mirror and interpret each other. The text will help us in understanding the constitution of the world, while our knowledge about the world will further our understanding of Orpheus' text.[46]

This interpretative scheme echoes the hermeneutic situation implied in oracles.[47] Greeks (as a community or as individuals) consult an oracle when they find themselves in a situation which they feel they cannot solve by their own means. Moreover, more often than not, this will happen in cases when the individual or community considers the situation to be of great importance for their entire existence. They are at an important juncture of their lives, perplexed, asking for divine help, and expecting a key to unravel the real structure of the situation, how it is rooted in the past and how it can affect the future. What they receive, however, is not a clear-cut solution, but a message cloaked in enigma. Indeed, the oracular answer can be just as perplexing as the situation for the interpretation of which they consulted the divinity. The oracle can be enigmatic, indeed up to the point of being ἀδύνατον,[48] but it cannot be lying. The oracular answer, by definition, is true and, rightly interpreted, does provide the key to the solution.

There are thus two sides, the situation to be solved on the one hand, and the oracular answer on the other; and both are puzzling. The solution lies in the right matching of the two. The task is to find how the text describes the world; and as soon as we have understood the way the text mirrors the world, we have also obtained the answer to our question concerning the situation to be solved. The task does not consist in proving *that* the pronouncement is true, but in understanding *how* it is true. As, for example, in the case of the second oracle offered to the Athenians on the occasion of the Persian attack: they had to discover first that the expression 'wooden fortification'

[46] So also Obbink (1997) 42.
[47] For a good discussion of the hermeneutics of oracles, see e.g. Manetti (1987) 27–56.
[48] Cf. Crahay (1974) 213.

ξύλινον τεῖχος did not refer to the ancient wall of the Acropolis, as some of the older generation suggested, but to ships; and, second, it had to be seen that 'the offspring of women' to be killed by 'divine Salamis' does not refer to Greeks but to Persians. The message of the oracle comes from the right mapping of text and reality. Once the mapping is accepted, it presents immediate practical consequences and suggests the outlines of a feasible action plan.[49]

This scheme becomes evident in the story of the quest for Orestes' bones, as related by Herodotus. The Lacedaemonians knew that their task was to find the place to which the description 'where two winds blow under strong compulsion, blow lies upon blow, woe upon woe' might apply. Eventually, all the elements of the oracular answer found their denotata in the blacksmith's shop: 'the two bellows were the two winds, hammer and anvil were blow upon blow, and the forging of iron was woe upon woe'. Moreover, the way Lichas 'by a combination of luck (συντυχίῃ) and skill (σοφίῃ)' discovered the solution shows how text and the elements of reality interpret each other. The oracle in itself might have remained an enigma; but once the necessary additional information was obtained by the sight of the workshop and the blacksmith, every element found its place and meaning.[50]

I suggest, then, that the Derveni author's exegetical attitude towards the Orphic poem is analogous to the hermeneutics of oracles as sketched above. To begin with, there is a text from a divine source. (Even though Orpheus was certainly not considered to be a god, he, like all poets, had a privileged epistemological status.)[51] Moreover, the Derveni author assumes that this text contains not only the truth but also something of supreme importance ([μεγ]άλα): the key to the understanding of the world and our situation in it. Orpheus, however, covered his meaning in riddles.[52] Therefore, the text requires careful interpretation. As a consequence, a mere acquaintance with the text will not suffice; the simple knowledge of it is just as useless as learning the text of an oracular response by heart without trying to interpret it. This is what those criticised in column 20 forget. The poem of Orpheus, just as an oracular response, will provide us with genuine knowledge only if we try to go beneath its surface meaning by making an interpretative effort in all earnest.[53] But once we have broken the code, and found the

[49] Hdt. 7.141–3. [50] Hdt. 1.67–8 (Loeb trans.). [51] Cf., e.g., Bremmer (1987) 4.

[52] Note also that the words αἰνίττομαι and σημαίνω applied by the Derveni author to describe Orpheus' use of language belong also to the vocabulary of oracles. See, e.g., Crahay (1974) 204 and 212–14; Manetti (1987) 30–3.

[53] Indeed, the Derveni author's attitude to the Orphic poem is comparable to Socrates' reaction to the Delphic oracle. The Pythia's reply to Chaerephon's question was plain enough and had an obvious

true denotata of all the words of the text, we have finally obtained access to the true and valuable message. Failing to understand the genuine meaning of the poem is just like failing to understand an oracle which conveys an important message about the fate of our soul, or that of our community. The majority of people do not succeed in solving this exegetical task, and they inevitably run into great troubles; yet the wise (or if you prefer: the properly initiated) get the true message, hence obtain the key to the understanding of their situation, and can thenceforth act accordingly.

It is also worth noting that the hermeneutic procedure involved in interpreting oracles does not depend merely on intuitions: it also needs reasoning. In both examples mentioned above, the members of the community had a debate about the right interpretation and in these discussions rational arguments were used.[54] For instance, the brave Themistocles had to point out against rival interpretations and in support of his own (i.e., that the oracle is propitious for the Athenians) that had the expression 'the offspring of women' to be killed at Salamis referred to the Athenians, the Pythia would not have said "Ὦ θείη Cαλαμίc" but rather "Ὦ cχετλίη Cαλαμίc."[55] Or Lichas realised that 'woe upon woe' can refer to the forging of iron 'since he figured that iron was discovered as an evil for the human race'.[56] Although the interpretative techniques applied there are more varied, this is not all that far from what we see in the papyrus.

Moreover, the oracle about Orestes' bones might help us in understanding another feature of the interpretative procedure of the Derveni author. First, there is a preliminary understanding of the text on the one hand, and the observation of reality on the other. Then, at least in the felicitous case, comes a heuristic moment when the gist of the solution is obtained: the recognition that the text as a whole must refer to such and such a phenomenon, part of reality, etc. The systematic matching of the individual elements of the text with particular entities in the physical world comes only after this heuristic moment. First comes the recognition that the oracle as a whole must refer to the blacksmith's shop, and then Lichas can ponder the denotata of the individual elements of the text. Also, it is only in this phase that, if needed, arguments can have their part, to work out why is it so that this or that phrase of the text must refer to this or that element of

literal sense: 'No one is wiser than Socrates.' Yet Socrates knew that the utterance is both true and enigmatic and thus the real message must lie below the surface: "Τί ποτε λέγει ὁ θεόc, καὶ τί ποτε αἰνίττεται; ἐγὼ γὰρ δὴ οὔτε μέγα οὔτε cμικρὸν cύνοιδα ἐμαυτῷ cοφὸc ὤν· τί οὖν ποτε λέγει φάcκων ἐμὲ cοφώτατον εἶναι; οὐ γὰρ δήπου ψεύδεταί γε· οὐ γὰρ θέμιc αὐτῷ." (*Apol.* 21b3–7.).

54 Cf. Hdt. 7.142: ὡc δὲ ἀπελθόντεc οἱ θεοπρόποι ἀπήγγελλον ἐc τὸν δῆμον, γνῶμαι καὶ ἄλλαι πολλαὶ ἐγίνοντο διζημένων τὸ μαντήιον κτλ.

55 Hdt. 7.143. 56 Hdt. 1.68.

reality. The overall solution is not the outcome of a cumulative process, and not the sum total of the interpretation of the individual elements; but we are able to think about the most plausible interpretation of the individual elements once we have found the solution.

This, I think, is a possible, sympathetic way to picture the Derveni author's case. First comes the recognition about the general meaning of the text: this is what Orpheus must have meant by his *logos*. The detailed 'allegorical interpretation' comes in a second phase: the author can now put to work the items of his interpretative toolbox in order to match the individual elements, applying always the technique he finds the most suitable for the case. Here an etymology, there an argument based on the phonetic resemblance of two words, or another based on observations about dialectical differences. Also, one should not forget that all these techniques were generally, in the Derveni author's age, thought of as perfectly 'rational' and possessing a considerable persuasive potential.

My claim, of course, is not that this is the way the Derveni author actually arrived at the understanding of the poem he proposes in his text, but rather that this is how he pictures the correct hermeneutic attitude towards Orpheus' text. It has been argued by several scholars that the oracle stories presented by Herodotus are literary forgeries, sustaining political objectives.[57] Moreover, as the material collected and analysed by Parke and Wormell shows, the reality about oracles was generally far less refined.[58] Yet, these historical facts do not change that other fact that authoritative literary texts, notably Herodotus and Heraclitus, presented oracular responses as enigmatic and their interpretation as requiring a conscious hermeneutic effort. The Derveni author could well have found a model here for the method of reading Orpheus' text.

This presentation, once again, must remain a hypothetical construct. Nonetheless, approaching the author's exegetical attitude towards the Orphic poem by referring to the hermeneutics of oracles seems to me more promising than simply seeing it in terms of the traditional categories of natural and moral allegory. At all events, it cannot be objected that the comparison with oracles is completely far-fetched. One should bear in mind that in column 5 the author presents himself as a *mantis*, and describes how he goes into the oracular shrine questioning the god on behalf of those who consult the oracle. The reference to the knowledge of dreams in the same column presupposes a similar interpretative attitude. The argument of column 10 is no less interesting. In this passage, the author speaks about

[57] See, most recently, Struck (2002). [58] See Parke and Wormell (1956).

the oracular activity of Night in the poem, and explains how the oracular utterance is related to teaching.

For our author, then, salvation, the healing of the soul, does not merely depend on the mechanical enactment of certain ritual acts and the incantation of holy texts; all these practices are not worth much without understanding the situation which necessitates them, the underlying significance of the acts and texts applied as remedy, and the way they can change the given conditions for the better. Those who are involved should therefore understand the nature of the soul, the constitution and laws of the physical world, and the way the divinity governs both the soul and the world. Ultimately, salvation resides, to a great extent, in genuine knowledge about these factors. Irrespective of important doctrinal differences and philosophical complexity, this basic conviction was also shared by the Plato of the *Laws*. But the traditional practitioners of the *techne*, who think that what matters is only that the rites should be performed properly and all the ritual regulations be followed, must find this attitude just as ridiculous as the slave doctor finds it absurd that some of his colleagues want to teach the patients instead of simply ordering them the appropriate diet and drugs.

The Derveni author is naturally far more confident about the possibilities of explication and teaching than Heraclitus. For him, the quest for the truth is not a solitary individual endeavour; it is sufficient to find the right guide – the Derveni author, of course – who will give the necessary instruction. This proper instructor, then, unravels Orpheus' enigmatic language and makes explicit what is concealed. The task of the initiand is to understand what he or she is taught.

But this teaching position is not unproblematic either. Remarkably, the author seems somewhat apologetic about Orpheus' mode of expression, showing awareness of the fact that riddles can be debatable and misleading.[59] He urges in defence of the poet that Orpheus had good reasons to choose this style: he used this non-literal language in order to protect his deep message from the unworthy populace.[60] The assumption in itself is not absurd; for example Diogenes Laertius finds the very same motivation behind Heraclitus' riddling style.[61] Indeed, intentional obscurity as a didactic means in order to filter out unworthy disciples was a topos in the commentary tradition, and was often ascribed even to Aristotle. On this assumption Orpheus is writing a public text, but with a built-in defence

[59] See col. 7.5–7 and col. 23. [60] See esp. col. 7.9–10 and col. 25.12–13.

[61] D.L. 9.6: ἀνέθηκε δ' αὐτὸ εἰϲ τὸ τῆϲ Ἀρτέμιδοϲ ἱερόν, ὡϲ μέν τινεϲ, ἐπιτηδεύϲαϲ ἀϲαφέϲτερον γράψαι, ὅπωϲ οἱ δυνάμενοι <μόνοι> προϲίοιεν αὐτῷ καὶ μὴ ἐκ τοῦ δημώδουϲ εὐκαταφρόνητον ᾖ.

system, just like Heraclitus according to Diogenes, or Aristotle according to Simplicius.[62]

However, the Derveni author assumes that now this defence system can be switched off, because a sympathetic attitude towards the views expressed by Orpheus can be presupposed. The protective encoding can now be broken, the cryptic and allegorical can be made explicit. A number of interesting questions emerge at this point. First, how can the author claim to know Orpheus' intentions? Are they somehow encoded in the text itself? Or is the author relying on some independent piece of information which he has access to due to his privileged status or function in some religious or other institution? And why does he think himself entitled to reveal what, on his own interpretation, Orpheus wanted to hide in riddles? Moreover, does it make a difference that here he is explaining Orpheus' genuine meaning not only in an oral form, where he would have more control of the audience, but in writing? On the interpretation of Richard Janko, this move is equal to 'divulging Orpheus' mysteries'. But one may also speculate that he thought he was speaking to a select group of initiates, or at least initiands, or that the fire of the pyre was considered a protective means. Was there a religious community to which this text was specifically addressed? If so, how was it structured and how did it function? What role did secrecy play in it? To answer these questions we should need to know more about the religious, sociological and institutional context, and the *Sitz im Leben* of this particular text.

EMPEDOCLES

A brief comparison with Empedocles may highlight some further aspects of the Derveni author's attitude towards the relationship between Orpheus and cosmology. To begin with, as Christoph Riedweg has aptly shown, the figure of Empedocles, as it emerges from his poems and from the various biographic stories and legends recorded by later authors, shows features customarily associated with the activities of *orpheotelestai*.[63] I would go one step further. It seems to me that the social and cultural role assumed by Empedocles invites a comparison not only with the *orpheotelestai*, but also with the mythical figure of Orpheus: a poet, who sings about the gods and the world, who is more than a human and whose voice has magical

[62] Simpl. *in Cat.* 8.7.6. See Sluiter (1999) 179–80 with n. 31 for other references from Aristotle's commentators. One can also think of Plato's deliberate use of myth – as Lloyd (1990) 24 has put it – 'in order to insulate a viewpoint from a literalist-minded brand of criticism'.

[63] Riedweg (1995) 39–40.

power,[64] who proclaims oracles[65] and is capable of curing illnesses,[66] who is a *mantis* and an *iatromantis*, a propagator of cathartic and telestic rites,[67] a vegetarian, who condemns all bloodshed,[68] one who is able (or almost) to bring the dead back to life,[69] and who transmits his secret knowledge to his favourite disciple.[70] To be sure, none of these features is associated exclusively with Orpheus, but they are constitutive elements also in the figures of Musaeus, Pythagoras and Epimenides. Yet, I would maintain, the constellation of functions and abilities that make up the image of Empedocles finds its most immediate prefiguration in Orpheus. The core of the difference between the Derveni author and the Orphic initiators on the one hand, and Empedocles on the other is this: while the former try to sell their ideas and services under Orpheus' authority, the latter tries to elevate himself to Orpheus' religious and cultural standing by claiming the mythical poet's most important functions for himself.

Indeed, there are many links between Empedocles' text and that of the Derveni author, even if these are not so manifest at the level of specific doctrines. If the current trend in the interpretation of Empedocles – reinforced now by the Strasbourg papyrus – is correct, then Empedocles treated eschatological matters and cosmological theory within a unified framework. Even the overall structures of the Derveni text and Empedocles' *Physica* might show similarities. If we accept that B115 was originally part of the proem,[71] Empedocles started off with the demonological story that outlined his eschatological doctrine and thus created a specific context for his physical, cosmological theory. The relationship between the first six and the remaining columns of the Derveni papyrus shows a comparable

64 Orpheus: *OF* test. 46–55, 59, 83; Pl. *Prt.* 315a, etc. Empedocles: A15 DK with Wright (1981) 11.

65 Orpheus: *OF* test. 87–9, esp. Philochorus *apud* Clem. *Strom.* 1.21.134.4 in 87. Empedocles: e.g. B112.9–10 DK

66 Orpheus: Pl. *Phdr.* 244d5–245a1; Eur. *Alc.* 966–71, cf. also Eur. *Cyc.* 646–8. Empedocles: B111.1–9, B112.10–11 and Heracl. Pont. in D.L. 8.61; Celsus, proem. 2.11; Galen, *Meth. med.* 1.1.

67 Orpheus: *OF* test. 90–105. Empedocles: see all the fragments ranged under the title *Katharmoi* by Diels.

68 Orpheus: e.g. Eur. *Hip.* 952 = *OF* test. 213; Pl. *Leg.* 6 787c7; Ar. *Ra.* 1032 = *OF* test. 90. Empedocles: B115; B128; B130; B136; B139.

69 The first documented occurrence of Orpheus' descent to Hades for his wife is apparently in Eur. *Alc.* 357–62. It seems that the story circulated in two versions: in one Orpheus' attempt ultimately failed, while the other version had a happy ending. On this, see Linforth (1941) 16–21. For Empedocles, see B 111.9 DK and the story that he cured a woman believed to be dead by everybody (Heracl. Pont. *apud* D.L. 8.70 and Hermippus *apud* D.L. 8.69).

70 Riedweg (1995) 53–4 shows that Empedocles' teaching position towards Pausanias is similar to Orpheus' position towards Musaeus.

71 This has been proposed by Sedley (1989) and elaborated on in (1998a) ch. 1. The editors of the Strasbourg papyrus find that the evidence of the papyrus strongly supports this hypothesis. See Martin and Primavesi (1999) 112–13.

arrangement. Both authors start out by focusing on the fate of the souls and introducing the main eschatological factors, and turn in a second step towards a general physical, cosmological account. The ultimate message in both cases is that the best way to understand the status and current position of the individual souls and the possibilities of their redemption is to view them from the perspective of the history of the cosmos and its physical constituents.

Moreover, as I have argued in more detail elsewhere, not only did Empedocles assimilate certain religious taboos and cathartic methods from Orphic and kindred religious movements, but his cosmogonical theory may show a conscious and critical reflection on the overall structure and the structuring moments of the Orphic theogony.[72] And this is where the difference between two possible attitudes towards Orpheus becomes palpable. When the Derveni author explains his cosmological theory, he places it totally under the authority of Orpheus: this is what Orpheus actually meant by his poem. The Derveni author, as he presents himself, is no more than a *prophetes* of Orpheus' enigmatic teaching. It is his task, for example, to identify the physical forces Orpheus expressed by traditional divine names. Viewed from an external perspective, however, the author is trying to make Orpheus' teaching up to date by providing it with an allegorical interpretation involving the conceptual and explanatory framework of late Presocratic cosmological speculation. Empedocles, by contrast, and at least in the remaining fragments, does not mention Orpheus, but claims Orpheus' role and prestige for himself. He is not the *prophetes* of the divine poet, but he himself is the divine poet. And he is in no need of a *prophetes*, because he provides the physical interpretation in the same gesture: he uses traditional divine names, but he himself identifies them with the appropriate physical forces. The update does not come in a prose commentary attached to the poem, but is fully integrated into the poem itself, and Empedocles lays claim to all the rights of authorship.

[72] Betegh (2001).

Appendix: Diagoras and the Derveni author

Richard Janko has recently suggested that the Derveni papyrus is the work of Diagoras of Melos, probably his book called *Apopyrgizontes logoi*. Janko first developed this hypothesis in a longer paper about the authorship of the papyrus, and then further argued it in the discussion accompanying his new translation of the papyrus.[1] While I consider Janko's translation and textual suggestions an important advance, I find his hypothesis about the identity of the author unconvincing. I propose to check the suggestion about Diagoras' authorship in three steps. First, I shall compare Diagoras' general attitude towards religious matters to the image we have found in the papyrus. In a second step, I shall turn to the question of whether we can ascribe the more specific doctrines found in the papyrus to Diagoras. Finally, I shall examine the proposed identification of the text of the papyrus with the *Apopyrgizontes logoi*. I shall mainly rely on Marek Winiarczyk's collection of ancient evidence on Diagoras, accompanied by a circumspect discussion of this author, and the two other most important modern studies of Diagoras, one by Felix Jakoby and the other by Leonard Woodbury.[2]

Diagoras, a minor lyric poet, gained fame in antiquity as an 'atheist'. Indeed, he became a stock item on ancient lists of those who denied belief in traditional gods. The nature of his 'atheism' is debated, however. Two distinct views have been propounded by scholars. According to some, most notably Felix Jakoby, Diagoras was an outright atheist who uncompromisingly denied the existence of gods.[3] There is strong ancient support for this view. If this is the correct view, and barring a complex biographical hypothesis about Diagoras' intellectual development, totally unsupported by evidence, Diagoras cannot be the author of the papyrus. For the author of the papyrus clearly considers his cosmic principle a god (most conspicuously in col. 25.10).

[1] Janko (1997) and (2001). See also Janko (1998).
[2] Winiarczyk (1979), (1980) and (1981); Jakoby (1959); Woodbury (1965).
[3] Jakoby (1959) 26.

More promising is the milder view according to which Diagoras' 'atheism' is some kind of impious attitude. After a critical discussion of Jakoby's arguments and the ancient evidence, Winiarczyk arrives at this view, which I myself find more probable, too.[4] The different testimonies are rather confusing, and hence the case is complicated. Winiarczyk and Woodbury agree in that the only indisputable information we have is that Diagoras divulged and ridiculed the Eleusinian mysteries. This is what prompted Aristophanes' comic mention in the *Clouds* (v. 830), and this is why the Athenian jury condemned him to death some time later, in connection with the prosecution of those who mutilated the Herms and parodied the Mysteries. The Athenians issued a reward on his head when he was already in exile in Pellene, and out of their reach.[5] The Athenian decree condemning Diagoras and the Pellinians who refused to extradite him was inscribed on a bronze tablet. This we can know from Melanthius' work *On the Eleusinian Mysteries* and from Craterus, as reported by the scholiast on Aristophanes' *Birds* 1073. Indeed the information going back to Melanthius and Craterus is the most secure we have: Diagoras divulged the Mysteries, belittled them, and tried to deter those who wanted to become initiated.[6] This is, then, what most probably earned Diagoras the epithet 'the Atheist'. As Winiarczyk and Woodbury argue, all the rest of the evidence about Diagoras is forgery and false attribution. In particular, the anecdotes about the reasons for his becoming disrespectful towards the gods (e.g. that he was victim of plagiarism and perjury, which remained unpunished by the gods)[7] are fruits of the imagination of the Hellenistic biographers. Winiarczyk and Woodbury agree that the closest parallel to Diagoras' attitude is Cinesias, another lyric poet who belonged to the roistering group *Kakodaimonistai*, 'Fellowship of the Bad Daimon', at the meetings of which members constantly mocked the gods and their cults.

It might be claimed that the Derveni author is divulging the Orphic *logos*. Moreover, the argument in column 20, which objects to the activities of different types of initiators, may be interpreted as an attempt to deter those who wanted to be initiated. Yet, as I have tried to show in the last chapter, the criticism is not against initiation *tout court*, but rather against certain forms of it, and, more specifically, against the intellectual

[4] Winiarczyk (1980) 71–3.

[5] For the life of Diagoras, with a detailed discussion of the evidence, see Winiarczyk (1979).

[6] Schol. *Av.* 1073: ἐκηρύχθη δὲ τοῦτο διὰ τὸ ἀcεβὲc αὐτοῦ, ἐπεὶ τὰ μυcτήρια πᾶcι διηγεῖτο, κοινοποιῶν αὐτὰ καὶ μικρὰ ποιῶν καὶ τοὺc βουλομένουc μυεῖcθαι ἀποτρέπων, καθάπερ Κρατερὸc ἱcτορεῖ.

[7] Suda *s.v.* Diagoras = T 9A Winiarczyk; see also schol. Ar. *Nub.* 830b = T 6A2 Winiarczyk and Schol. in Clem. *Protr.* 1.304 Stählin = T 26 Winiarczyk.

and epistemological attitude of certain practitioners. The Derveni author's aim is not to deter people from initiation, but to convince them that the form he is propounding is the right way to get initiated. At any rate, the reference to the ritual activity of the initiates in column 6 is, in all probability, positive. This reinforces the feeling that the Derveni author did not want to ridicule or belittle the Orphic mysteries.[8]

Even more problematic is Diagoras' relation to the Orphic *logos*. As a matter of fact, in the extensive material collected by Winiarczyk only one testimony mentions anything related to Orpheus or Orphic lore. Athenagoras of Athens tells us that

> The Athenians reasonably accused Diagoras of atheism, a man who not only made the Orphic *logos* public, publicised the Eleusinian mysteries and those of the Cabiri, and chopped up Heracles' wooden statue so that he could cook his cress, but outrightly proclaimed that there is no god at all.[9]

Naturally, Janko gives special attention to Athenagoras' testimony because this is the only source that mentions not only the Eleusinian mysteries, but also the Orphic *logos* in relation to Diagoras. As Janko argues, 'Athenagoras used a well-informed Hellenistic source which knew *the same Orphic logos* as that in the papyrus; and that source claimed that Diagoras divulged not only the Eleusinian rites, but the Samothracian mysteries and an Orphic holy text as well.'[10] First of all, we should be more precise about the relationship between the Orphic *logos* known to Athenagoras' source and that in the papyrus. As Janko acknowledges in the sentence preceding the one I have just quoted, contrary to what this sentence claims, it is not the case that Athenagoras' source knew *the same Orphic logos* as that in the papyrus, but only a version of the Orphic theogony that West has supposed to be identical with the Hieronymus-Hellanicus version, which in turn he has taken to be an adaptation of the hypothesised 'Protogonos theogony', an abridged version of which he assumes to be quoted in the papyrus. Hence the connection between the theogony known to Athenagoras' source and the poem quoted in the papyrus, even if one accepts West's highly hypothetical and often criticised many-layered stemma, is loose at best.

On the other hand, Athenagoras in the very same sentence says also that Diagoras 'outrightly proclaimed that there is no god at all' (ἀντικρὺϲ δὲ ἀποφαινομένῳ μηδὲ ὅλωϲ, εἶναι θεόν). If Athenagoras' source was so well informed that among all our sources he is the only one to tell us

[8] So, most recently, Burkert (1999a) 104.

[9] Athenagoras, *Supplicatio pro Christianis* 4 = T 27 Winiarczyk.

[10] Janko (1997) 89 (my emphasis).

that Diagoras also divulged the Orphic *logos*, then we should give more credence also to his explicit assertion that Diagoras not only propounded some philosophically inspired cosmic god, but proclaimed overtly that there are no gods at all.[11] Athenagoras' report on Diagoras divulging the Orphic *logos*, then, is highly doubtful. Woodbury does not even mention it, while Winiarczyk rejects it.[12] Because none of the numerous other ancient discussions of Diagoras' 'atheism' mentions the Orphic *logos*, it seems to me that Athenagoras, who was indeed interested in the Orphic theogonic narrative, simply added the divulgation of the Orphic *logos* to the list of impious deeds committed by Diagoras. If so, we are left with no evidence linking Diagoras with the Orphic *logos*. So on the one hand, the evidence is less than meagre about Diagoras' treating the Orphic *logos*, while, on the other hand, there is hardly anything in the papyrus that could be described as ridiculing and divulging the Eleusinian mysteries with which Diagoras was accused.[13]

Did Diagoras have a physical system? As a matter of fact, many of the ancient sources call him not only a poet, but also a philosopher and even a natural philosopher (φυϲικόϲ).[14] The eleventh-century Arab author Al-Mubašir Ibn Fatik, in his *Life of Zeno of Elea*, says that the 'impious' (*al-māriq*) Diagoras lived at the same time as Zeno of Elea, Leucippus, Heraclitus, Empedocles, Melissus, Protagoras, Anaxagoras, Socrates, Democritus.[15] Pseudo-Hesychius even presents him as the slave and disciple of Democritus.[16] It seems significant, on the other hand, that our ancient sources merely affirm that Diagoras denied the existence of gods: they say nothing at all about his positive doctrines. We hear nothing about his arguments against the existence of (traditional) gods, or about a physical theory that he might have propagated to take the place of traditional piety. I tend to agree with those who assume that this is so because there was no such underlying theory. It seems to me probable that some of our sources group

[11] About Diagoras and the Cabiri, see the anecdotes about Diagoras mocking at votive offerings at Samothrace in T 36 and 37 Winiarczyk. For the story about chopping up the cult statue of Heracles, see T 27–33 Winiarczyk.

[12] Winiarczyk (1980) 71: 'Dieses vereinzelte Testimonium weckt jedoch kein Vertrauen.'

[13] It is true that Henrichs (1984) 266–8 raises the possibility that the author in col. 6 speaks about the Eleusinian rituals; but he also concludes that '[t]he proposed connection between col. 11 [=col. 6 according to the new numbering] of the Derveni papyrus is far from certain'. Note also that Henrichs was working on the very restricted transcript of the first columns available at that time. Tsantsanoglou (1997) 115–17 and Burkert (1999a) 106 agree that, although it cannot be excluded, the reference is probably not to the Eleusinian mysteries. West (1997) 84 thinks, I think rightly, that the ritual and the theogony belonged together, and that the initiates mentioned in the first columns 'are those of an Orphic-Bacchic cult society'.

[14] Cf. T 1 A, B, C; T2; T3; T9A Winiarczyk.

[15] T 10 Winiarczyk.

[16] Ps.-Hesychius, *De viris illustribus* 17 = T 9 B Winiarczyk.

Diagoras together with philosophers, or even claim that he was a natural philosopher, because they infer this from his reputation for being an 'atheist' – the assumption is that only people with philosophical education and physical theories deny the existence of (traditional) gods.[17]

There is an important detail, however, on which Janko lays much stress. In lines 826–31 of Aristophanes' *Clouds* (= T 38 Winiarczyk) Strepsiades explains to Pheidippides why there is no use in swearing by Zeus: Zeus has been dethroned by Dinos, the cosmic vortex. When Pheidippides asks who told Strepsiades this, the answer is 'Socrates of Melos, and Chaerephon, the expert on the footsteps of fleas.' As all the scholiasts hasten to remark, Socrates was of course Athenian, and therefore Aristophanes alludes here to Diagoras of Melos. Janko maintains on this basis that not only the Socrates of the *Clouds*, but also Diagoras followed the physical doctrines of Diogenes of Apollonia.[18] As against this view, Woodbury is right, I think, in pointing out that the doctrine of Dinos is attributed to Socrates and not to Diagoras.[19] Aristophanes evokes Diagoras because of his notoriously impious attitude towards the gods, and not because he shared the physical doctrines that Aristophanes attributes to Socrates. The numerous scholia on the passage speak about Diagoras' divulgation of the Eleusinian mysteries, and that he chopped up the cult statue of Heracles. None of them mentions anything about Diagoras' physical views.[20]

I do not think we would be much better off if we accepted Janko's suggestion that Diagoras shared the doctrine of a cosmic vortex. For even if we accepted that Diagoras agreed with Empedocles, Anaxagoras and the Atomists in explaining the birth of the cosmos by a whirl (hence the 'rule of Dinos'), we still have not arrived at the cosmogony of the Derveni author. It is true that, as Janko (2002) reports, Burkert has recently suggested supplementing the word τὴν δ[ίνην] in a lacuna in the first line of col. 18. It seems to me, however, that this supplement is not compelling. The resulting

[17] So Woodbury (1965) 206 and 209; Winiarczyk (1980) 67 and 73.

[18] As mentioned already in my short chapter on Archelaus, it seems to me that Diogenes' influence is sometimes overemphasised in the Socrates of the *Clouds*, whereas Archelaus of Athens gets less then his share. But this shift in emphasis would not substantially change the argument about Diagoras' assumed physical doctrines.

[19] Woodbury (1965) 186.

[20] A mention in Philodemus' *De pietate* Part II, PHerc. 1438, col. II (= p. 85 Gomperz = T 69 Winiarczyk) may parallel the passage in the *Clouds*. Philodemus says that the Stoics 'do not believe in anthropomorphic gods, but in airs, breaths and aithers, so I dare say that these people are more erring than Diagoras' (ἀνθρωπ̣[ο]ẹ[ι]δεῖϲ γὰρ ἐκεῖνοί γε <οὐ> νομίζουϲιν ἀλλὰ ἀẹρα[ϲ] καὶ πνẹύματα καὶ αἰ̣θέραϲ, ὥϲτ' ἔγωγε κἂν̣ τεθαρρηκότωϲ εἴπαιμι τούτουϲ Διαγόρου μᾶλλον πλημμελεῖν). Diagoras is not mentioned because he also held a doctrine involving 'airs, breaths and aithers', but because the Stoic view is just as, or even more, impious than the infamous Diagoras.

phrase, τὴν δ[ίνην] καὶ τἆλλα πάν[τ]α | εἶναι ἐν τῶι ἀέρι π̣[νε]ῦμα ἐόν, which is translated by Janko as 'that the vortex and all the other elements are in the Air, it being "breath"', seems to put the vortex on the same level as the other things (I see no clear reason to restrict καὶ τἆλλα πάν[τ]α to elements – the expression can refer to composites as well). Yet, if there is a mention of a cosmic vortex in the text, it must surely be a manifestation or function of the air, and not something 'in the air' together with the other beings. When there is a vortex in a cosmogony, then it is the vortex which produces the structure of the cosmos and triggers off the formation of the main constitutive elements of the cosmos. This is also why in Aristophanes' comic representation the vortex, Dinos, becomes the supreme divine ruler. Therefore, if there were a vortex in the Derveni text, it would not be something 'in the air' together with the other things, but rather a function of the divine Air/Mind itself. For this reason, I prefer Janko's own original supplement, τὴν δ[ὲ γῆν]: the earth (as an element, or rather as a composite, i.e. the earth we live on) is in the air together with all the other things. But without the supplement τὴν δ[ίνην] there is no trace of a vortex in the Derveni author's cosmology.[21] Thus, the attribution of a doctrine of a cosmic vortex is hypothetical both in the case of Diagoras and in the case of the Derveni author.

What can we say about the *Apopyrgizontes logoi* that the Suda ascribes to Diagoras? The title, the authorship and the contents of this work are heavily contested. The verb ἀποπυργίζω is a hapax the meaning of which is vexed. It might mean 'to fortify' or 'to destroy the towers', but numerous other possible senses have been suggested.[22] The most one can say is that this title recalls titles of works published by sophists, such as the *Kataballontes logoi* ('Knock-down Arguments') of Protagoras and the *Hyperballontes* of Thrasymachus.[23] Aristoxenus in his *Customs of the Mantineans*, as quoted by

[21] At the end of a paragraph developing a very interesting suggestion according to which Plato succeeded in veiling that Socrates followed Diogenes' theism in order to defend his master's image from the charge of irreligion, Janko (1997) 92 asks the following rhetorical question. 'Is it therefore coincidence that the Derveni papyrus professes a belief in *daimones* which uncannily resembles Socrates' *daimonion*?' It seems to me that the *daimones* of the papyrus can be explained with reference to the traditional eschatological conceptions of the *daimones* (for which see now Johnston (1999)), which were coloured, admittedly, by some measure of idiosyncrasy on the Derveni author's part. I certainly cannot see an uncanny resemblance with Socrates' *daimonion*.

[22] See the survey in Winiarczyk (1980) 54–8.

[23] So e.g. Woodbury (1965) 203 and Janko (2001) 7. The picture is further complicated by the fact that Tatian (*Or. ad Graec.* 27 = T 68 Winiarczyk) and Al-Mubašir (T 10 Winiarczyk) attribute to Diagoras a *Phrygioi logoi*. Winiarczyk suggests that this was an independent, euhemeristic work composed and attributed to Diagoras in Hellenistic times. Others have maintained that it is an alternative title

Philodemus, is the first to mention an 'atheist' book attributed to Diagoras (without telling its title), but he already claims that the attribution is false. Philodemus, for his part, adds that either Aristoxenus is right in doubting the authenticity of the work, or Diagoras was just jesting, for he showed himself pious as any other poet in his poems.[24] Al-Mubašir conspicuously reports that the 'atheist' book of Diagoras was discovered only after his death (T 10 Winiarczyk). Woodbury and Winiarczyk agree with Aristoxenus that the work is a forgery attributed to Diagoras exploiting his reputation as an 'atheist'. Winiarczyk maintains that the work is a sophistic forgery, and that this was the work that caused numerous ancient sources to attribute to Diagoras outright denial of the existence of gods. This characterisation is supported by the testimony of Epicurus, as quoted by Philodemus:

Epicurus criticised those who eliminate the divine from existing things for their total insanity, as in Book 12 (of *On Nature*) he criticises Prodicus, Diagoras and Critias among others, that they rave like lunatics, and he likens them to Bacchant revellers, admonishing them not to trouble or disturb us. For they explain the names of gods by changing letters, just as Antisthenes, substituting the most common (name for god), ascribes the particular ones to convention and even earlier through some act of deceit.[25] (trans. Obbink, modified by Janko)

This report clearly situates the book attributed to Diagoras within the intellectual anti-theology of the sophists. It is true that the linguistic or etymological method referred to here certainly recalls the linguistic explanations of the Derveni author. Yet I do not think that the reference is specific enough, and these methods were widely used as exegetical tools. Moreover, Epicurus, who provides us with this relatively explicit information about [Pseudo-?]Diagoras' method, treats Diagoras as someone who utterly denied the existence of the divine, which is, of course, not true of the Derveni author. What we have gained on one side, we have lost on the other.

for the *Apopyrgizontes logoi*. It is notable that there is a *Phrygios logos* attributed to Democritus (D.L. 9.48).

24 Philodemus' *De pietate* Part II, PHerc. 1438, coll. 11–12 (= p. 85 Gomperz = T 69 Winiarczyk), following the sentence quoted above in n. 20.

25 Philodemus, *De pietate* Part I col. 19. 518–41 Obbink = p. 112 Gomperz: [αὐτ]οῖς δὲ καὶ πᾶσαν μ[ανίαν Ἐ]πίκουρος ἐμ̣[έμψα]το τοῖς τὸ [θεῖον ἐ]κ τῶν ὄντων [ἀναι]ροῦσιν, ὡς κἀ[ν τῶι] δωδεκάτω[ι Προ]δίκωι καὶ Δια[γόραι] καὶ Κριτίαι κἄ[λλοις] μ̣έμφ[εται] φὰς πα[ρα]κόπτε̣ιν καὶ μ̣[αίνες]θαι, καὶ βακχεύουσιν αὐτοὺς [εἰ]κά[ζει, κε]λεύς[ας μ]ὴ πράγμα ἡμ{ε}ῖν παρέχειν οὐδ' ἐνοχλεῖν. κα[ὶ γὰρ] παραγραμ̣μίζ[ουσι] τὰ τ̣[ῶ]ν̣ θεῶν [ὀνό]ματα, [κα]θάπερ Ἀν̣[τισ]θέ[νης] τὸ κοινό[τατον] ὑποτ<ε>ίνων ἀν̣[αφέρει] τὰ κατὰ μέρος [τῆι θέ]σει καὶ διὰ τι̣[νος ἀπά]της ἔτι πρότερον.

What remains is the brief mention of the book in the Suda which implies that the book was about Diagoras' loss of faith and conversion to atheism. This description, yet again, is not easy to map on the Derveni text.[26]

In conclusion I would say that the evidence about Diagoras is astonishingly confusing and it is difficult to create a unified image from the divergent testimonies. But I do not think that the overall picture, however blurred, warrants his identification with the author of the Derveni papyrus. In particular, I do not think that there is sufficient ground to claim that Diagoras wrote on the Orphic *logos*, and that his 'atheism' was a theology based on a physical theory comparable to that of the Derveni author. Finally, I do not think Janko could demonstrate that Tsantsanoglou and others are mistaken in maintaining that the first columns of the papyrus present the author as a religious professional.

Even though I do not find Janko's hypothesis convincing, his discussion about Diagoras and the intellectual climate of the accusations of impiety in Athens raises an important issue. Janko rightly calls attention to the fact that a new type of theology, which preferred the conceptual scheme and explanatory methods of natural philosophy to the traditional cultic representations of gods, could easily be attacked by contemporaries as 'impious' and termed 'atheist' (not, of course, the same thing as we understand by the term). In such a climate the Derveni author could have been so regarded by his contemporaries. It is also conceivable, however, as I have tried to argue in more detail in chapter 10, that certain religious practitioners working on the basis of a *techne* could have arrived at the same intellectual stance. In my view Diagoras and the Derveni author represent two distinct attitudes towards traditional forms of religious practice in the age of Socrates.

[26] Suda *s.v.* πυργίσκοι = T67 B Winiarczyk (cf. *s.v.* Ἀποπυργίζοντας λόγους) . . . τοὺς καλουμένους Ἀποπυργίζοντας λόγους, ἀναχώρησιν αὐτοῦ καὶ ἔκπτωσιν ἔχοντας τῆς περὶ τὸ θεῖον δόξης· ἄθεος γὰρ <οὐκ Janko> ἦν τὸ πρότερον. It is also possible, however, as Winiarczyk (1980) 57 has suggested, that the mention in the Suda is based on the Hellenistic anecdotes about Diagoras' conversion.

Bibliography

Adoménas, M. (1999) 'Heraclitus on religion', *Phronesis* 44: 87–113.

Alderink, L. J. (1981) *Creation and Salvation in Ancient Orphism*, American Studies in Classical Philology 8. Ann Arbor.

Allen, R. E. and Furley, D. J. (eds.) (1975) *Studies in Presocratic Philosophy*. London.

Amersfoort, J. Van (1981) 'Traces of an Alexandrian Orphic theogony in the Pseudo-Clementines', in *Studies in Gnosticism and Hellenistic Religions Presented to Gilles Quispel on the Occasion of his 65th Birthday*, eds. R. van den Broek and M.-J. Vermaseren. *EPRO* 91: 13–30.

[Anon.] (1982) 'Der orphische Papyrus von Derveni', *Zeitschrift für Papyrologie und Epigraphik* 47: after p. 300, 1–12*.

Babut, D. (1975) 'Héraclite et la religion populaire: Fragments 14, 69, 68, 15 et 5 Diels-Kranz', *Revue des Études Anciennes* 77: 27–62.

Bargrave-Weaver, D. (1959) 'The cosmogony of Anaxagoras', *Phronesis* 4: 77–91.

Barnes, J. (1982) *The Presocratic Philosophers*. Revised edition in one volume. London and New York.

Barney, R. (1998) 'Socrates agonistes: the case of the Cratylus etymologies', *Oxford Studies in Ancient Philosophy* 16: 63–98.

Barrett, W. S. (1964) *Euripides, Hippolytus*, ed. with Introduction and Commentary. Oxford.

Baumgarten, R. (1998) *Heiliges Wort und Heilige Schrift bei den Griechen*. Tübingen.

Baxter, T. M. S. (1992) *The Cratylus. Plato's Critique of Naming*. Leiden, New York and Cologne.

Bernabé, A. (1989) 'Generaciones de dioses y sucesión interrumpida. El mito hitita de Kumarbi, la "Teogonía" de Hesíodo y la del "Papiro de Derveni"', *Aula Orientalis* 7: 159–79.

(1992) 'Una forma embrionaria de reflexión sobre el lenguaje: la etimología de nombres divinos en los órficos', *Revista española de linguística* 22: 26–54.

(2002) 'La théogonie orphique du papyrus de Derveni', *Kernos* 15: 1–37.

(forthcoming), *Poetarum epicorum Graecorum testimonia et fragmenta* II. *Orphicorum fragmenta*. Leipzig.

Bernabé, A. and Jiménez San Cristóbal, A. I. (2001) *Instrucciones para el más allá. Las laminillas órficas de oro*. Madrid.

Betegh G. (1993) *Le Papyrus de Derveni: un commentaire philosophique*. D.E.A. dissertation. EHESS, Paris.
(1994) 'A Nap és a láb: Hérakleitosz a Derveni papiruszban', *Magyar Filozófiai Szemle* 38: 70–84 (Hungarian with English summary).
(1996) review of Brisson (1995), *Ancient Philosophy* 16: 463–7.
(2002a) 'On Eudemus fr. 150 (Wehrli)', in *Eudemus of Rhodes*, eds. I. Bodnár and W. Fortenbaugh. New Brunswick: 321–41.
(2002b), 'Papyrus on the pyre', *Acta Antiqua* 42: 51–66.
Blank, D. (1991) 'The fate of the ignorant in Plato's Gorgias', *Hermes* 119: 22–36.
Böhme, R. (1988) 'Homer oder Orpheus?', *Zeitschrift für Papyrologie und Epigraphik* 71: 25–31.
(1989) 'Neue Orpheusverse auf dem Derveni-Papyrus', *Emerita* 57: 211–38.
Bottéro, J. and Kramer, S. N. (1989) *Lorsque les dieux faisaient l'homme. Mythologie mésopotamienne*. Paris.
Boyancé, P. (1952) 'La religion astrale de Platon à Cicéron', *Revue des Études Grecques* 65: 312–50.
(1974) 'Remarques sur le Papyrus de Derveni', *Revue des Études Grecques* 87: 90–110.
Bremer, J. (1981) 'Greek hymns', in *Faith, Hope and Worship*, eds. H. S. Versnel and F. T. Straten. Leiden: 193–215.
Bremmer, J. (1987) 'What is a Greek myth?', in *Interpretations of Greek Mythology*, ed. J. Bremmer. London: 1–9.
(1999) 'Rationalization and disenchantment in ancient Greece: Max Weber among the Pythagoreans and Orphics?', in *From Myth to Reason? Studies in the Development of Greek Thought*, ed. R. Buxton. Oxford: 71–83.
Breslin, J. (1977) *A Greek Prayer*. Pasadena, Calif.
Brisson, L. (1985a) 'Les théogonies Orphiques et le papyrus de Derveni: notes critiques', *Revue de l'histoire des Religions* 202: 389–420, reprinted in Brisson (1995).
(1985b) 'La Figure de Chronos dans la théogonie orphique et ses antécédents iraniens', in *Mythes et représentation du temps*, ed. D Tiffeneau. Paris: 37–55, reprinted in Brisson (1995).
(1987) 'Proclus et l'Orphisme', in *Proclus. Lecteur et interprète des Anciens*, Colloques internationaux du C.N.R.S. Paris: 43–103, reprinted in Brisson (1995).
(1990) 'Orphée et l'Orphisme à l'époque impériale', *Aufstieg und Niedergang der Römischen Welt* 2.36.4. Berlin and New York: 2877–82, reprinted in Brisson (1995).
(1991) 'Damascius et l'Orphisme', in *Orphisme et Orphée en l'honneur de Jean Rudhardt*, ed. Ph. Borgeaud. Recherches et Rencontres 3. Geneva: 157–209, reprinted in Brisson (1995).
(1992) 'Le corps "dionysiaque": l'anthropogonie décrite dans le *Commentaire sur le* Phédon *de Platon* (1.3–6), attribué à Olympiodore, est-elle orphique?', in *ΣΟΦΙΣΤΗΣ ΜΑΙΗΤΟΡΕΣ, Chercheurs de sagesse: Hommage à Jean Pépin*,

Collection des Études Augustiniennes, Série Antiquité 131. Paris: 481–99, reprinted in Brisson (1995).

(1995) *Orphée et l'Orphisme dans l'Antiquité gréco-romaine*, Variorum Collected Studies Series. Aldershot.

(1996) *Introduction à la philosophie du mythe* I, *Sauver les mythes*. Paris.

(1997) 'Chronos in column XII of the Derveni papyrus', in L&M: 149–66.

(2003) 'Sky, Sex and Sun: the meanings of αἰδοῖος / αἰδοῖον in the Derveni papyrus', *Zeitschrift für Papyzologie und Epigraphik* 142: 1–11.

Buffière, F. (1956) *Les Mythes d'Homère et la pensée grecque*. Paris.

Burkert W. (1961) 'Hellenistische Pseudopythagorica', *Philologus* 105: 16–43 and 226–46.

(1968) 'Orpheus und die Vorsokratiker. Bemerkungen zum Derveni-Papyrus und zum pythagoreischen Zahlenlehre', *Antike und Abenland* 14: 93–114.

(1970) 'La genèse des choses et des mots: le papyrus de Derveni entre Anaxagore et Cratyle', *Études Philosophiques* 25: 443–55.

(1972) *Lore and Science in Ancient Pythagoreanism*. Cambridge, Mass.

(1975) 'Le laminette auree: de Orfeo a Lampone', in *Orfismo in Magna Grecia. Atti del XIV convegno di studi sulla Magna Grecia*. Naples: 87–104.

(1977) 'Orphism and Bacchic mysteries: New evidence and old problems of interpretation', in *Protocol of the 28th Colloquy of the Center for Hermeneutical Studies in Hellenistic and Modern Culture*, ed. W. Wuellner. Berkeley: 1–32.

(1979) *Structure and History in Greek Mythology and Ritual*. Berkeley.

(1980) 'Neue Funde zur Orphik', *Informationen zum altsprachlichen Unterricht* 2: 27–42.

(1982) 'Craft versus sect: the problem of Orphics and Pythagoreans', in *Jewish and Christian Self-Definition* III, eds. B. F. Meyer and E. P. Sanders. Philadelphia: 1–22.

(1983) 'Eraklito nel Papyro di Derveni: due nuove testimonianze', in *Atti del symposium Heracliteum 1981*, ed. L. Rossetti. Rome: 37–42.

(1985a) *Greek Religion*. Cambridge, Mass.

(1985b) 'Herodot über die Namen der Götter: Polytheismus als historisches Problem', *Museum Helveticum* 42: 121–32.

(1986) 'Der Autor von Derveni', *Zeitschrift für Papyrologie und Epigraphik* 62: 1–5.

(1987a) *Ancient Mystery Cults*. Cambridge, Mass.

(1987b) 'Oriental and Greek mythology: the meeting of parallels', in *Interpretations of Greek Mythology*, ed. J. Bremmer. London: 10–40.

(1992) *The Orientalizing Revolution*. Cambridge, Mass.

(1994) 'Orientalische und griechische Weltmodelle von Assur bis Anaximandros', *Wiener Studien* 107: 179–86.

(1997) 'Star wars or one stable world? A problem of Presocratic cosmogony (PDerv. Col. XXV)', in L&M: 167–74.

(1999a) *Da Omero ai magi*. Venice.

(1999b) 'The logic of cosmogony', in *From Myth to Reason? Studies in the Development of Greek Thought*, ed. R. Buxton. Oxford: 87–106.

Cahn, H. A. (1994) Okeanos *s.v. LIMC* 7.1.

Calame, C. (1997) 'Figures of sexuality and initiatory transition in the Derveni theogony and its commentary', in L&M: 65–80.

Calvo, T. and Brisson, L. (eds.) (1997) *Interpreting the Timaeus-Critias. Proceedings of the IV Symposium Platonicum*, International Plato Studies vol. 9. Sankt Augustin.

Casadesús Bordoy, F. (1996) 'Metis, el nous, el aire y Zeus en el papiro de Derveni', *Faventia* 18: 75–88.

Casadio, G. (1986) 'Adversaria orphica et orientalia', *Studi e Materiali di Storia delle Religioni* 52: 291–322.

(1987) 'Adversaria orphica. A proposito di un libro recente sull' Orfismo' (review of Alderink (1981)), *Orpheus* 8: 381–95.

Cavallo, G. (1983) *Libri scritture scribi a Ercolano, Chronache Ercolanesi* suppl. 1. Naples.

Classen, C. J. (1965) 'Bemerkungen zu zwei griechischen Philosophiehistorikern', *Philologus* 109: 175–81.

Cleve, F. M. (1943) 'The god of Anaxagoras', *Review of Religion* 7: 124–6.

(1973) *The Philosophy of Anaxagoras*, 2nd edn. The Hague.

Cole, S. G. (1980) 'New evidence for the mysteries of Dionysos', *Greek, Roman and Byzantine Studies* 21: 228–31.

Collard, C., Cropp, M. J. and Lee, K. H. (1995) *Euripides. Selected Fragmentary Plays*, with Introductions, Translations and Commentaries vol. 1. Warminster.

Conche, M. (1986) *Héraclite. Fragments.* Paris.

Cook, A. B. (1914–40) *Zeus. A Study in Ancient Religion.* 3 vols. Cambridge.

Cornford, F. M. (1930) 'Anaxagoras' theory of matter', *Classical Quarterly* 24: 14–30 and 83–95, reprinted in *Studies in Presocratic Philosophy*, eds. R. E. Allen and D. J. Furley. London: 275–322.

(1950) 'A ritual basis for Hesiod's *Theogony*', in *The Unwritten Philosophy and Other Essays.* Cambridge: 95–116.

(1952) *Principium Sapientiae.* Cambridge.

Crahay, R. (1974) 'La bouche de la vérité', in *Divination et Rationalité*, ed. J.-P. Vernant. Paris: 201–19.

Dalimier, C. (1998) *Platon, Cratyle.* Paris.

Daux, G. (1962) *Bulletin de Correspondance Hellénique* 86: 792–3.

DeFilippo, J. G. (1993) 'Reply to André Laks on Anaxagoras' ΝΟΥΣ', *The Southern Journal of Philosophy* 31, Supplement (The Spindel Conference 1992): 39–48.

Deichgräber, K. (1933) 'Hymnische Elemente in der philosophischen Prosa der Vorsokratiker', *Philologus* 42: 347–61.

(1935) *Hippokrates über Entstehung und Aufbau des menschlichen Körpers.* Leipzig.

Delbrueck, R. and Volgraff, W. (1934) 'An Orphic bowl', *Journal of Hellenic Studies* 54: 129–39.

Detienne, M. (1964) 'Les origines religieuses de la notion de l'intellect: Hermotime et Anaxagore', *Revue Philosophique de la France et de l'étranger* 89: 167–78.

(1989) *L'Écriture d'Orphée.* Paris.

Dickie, M. W. (1995) 'The Dionysiac mysteries in Pella', *Zeitschrift für Papyrologie und Epigraphik* 109: 81–6.

Diels, H. (1902) 'Orphischer Demeterhymnus', *Festschrift für Th. Gomperz.* Vienna: 1–15.

Dieterich, A. (1891) *De hymnis orphicis capitula quinque*, Diss. Marburg, reprinted in Dieterich (1911): 69–110.

(1911) *Kleine Schriften.* Leipzig.

(1913) *Nekyia. Beiträge zur Erklärung der neuentdeckten Petruscapokalypse*, 2nd edn. Leipzig and Berlin.

Diggle, J. (1970) *Euripides: Phaethon.* Cambridge.

Diller, H. (1941) 'Die philosophiegeschichtliche Stellung des Diogenes von Apollonia', *Hermes* 76: 359–81, reprinted in Diller (1971): 162–86.

(1971) *Kleine Schriften zur antiken Literatur.* München.

Dodds, E. R. (1951) *The Greeks and the Irrational.* Berkeley and Los Angeles.

Du Bose, S. (1964) 'Anaxagoras' theory of mind', in *Studies in Philosophical Psychology.* Tulane Studies in Philosophy 13. The Hague: 50–4.

Dubois, L. (1996) *Inscriptions grecques dialectales d'Olbia du Pont.* Geneva.

Dunbar, N. (1997) *Aristophanes: Birds.* Oxford.

Edwards, M. J. (1991) 'Notes on the Derveni commentator', *Zeitschrift für Papyrologie und Epigraphik* 86: 203–11.

Evans-Pritchard, E. E. (1937) *Witchcraft, Oracles and Magic among the Azande.* Oxford.

Fehling, D. (1965) 'Zwei Untersuchungen zur griechischen Sprachphilosophie', *Rheinisches Museum* 108: 212–29.

Finkelberg, A. (1986) 'On the unity of Orphic and Milesian thought', *Harvard Theological Review* 79: 321–35.

Fränkel, H. (1960) 'Besprechung eines Buches über Anaxagoras' (review of Davide Ciurnelli, *La Filosofia di Anassagora*, Padua, 1947), in *Wege und Formen frühgriechischen Denkens*, 2nd edn. München: 284–93.

Frankfort, H. (1977) *The Intellectual Adventure of Ancient Man.* Chicago and London.

Frel, J. (1994) 'Una nuova laminella "orfica"', *Eirene* 30: 183–4.

Fritz, K. von (1943) 'Νοῦς and Νοεῖν in the Homeric poems', *Classical Philology* 38: 79–93.

(1945) 'Νοῦς, Νοεῖν and their derivatives in the Pre-Socratic philosophy (excluding Anaxagoras)' (Part I), *Classical Philology* 40: 223–42, reprinted in Mourelatos (1974): 23–85. References are to the reprint.

(1946) 'Νοῦς, Νοεῖν and their derivatives in the Pre-Socratic philosophy (excluding Anaxagoras)' (Part II) *Classical Philology* 41: 12–34, reprinted in Mourelatos (1974): 23–85. References are to the reprint.

(1971) 'Der Νοῦς des Anaxagoras', in von Fritz, *Grundprobleme der Geschichte der antiken Wissenschaft.* Berlin and New York: 576–93.

Funghi, M. S. (1979) 'Una cosmogonia orfica nel papiro di Derveni', *La Parola del Passato* 34: 17–30.

(1980) 'Phronesis nel papiro di Derveni', in *Miscellanea Papyrologica*, ed. R. Pintaudi, Papirologica Florentina 7. Florence: 79–86.

(1983) 'P. Bruxell. Inv. E 7162 e P. Med. Inv. 71.82. Due discussioni su "Olimpo"', *La Parola del Passato* 38: 17–19.

(1997) 'The Derveni Papyrus', in L&M: 25–37.

Furley, D. (1976) 'Anaxagoras in response to Parmenides', in *New Essays in Plato and the Presocratics, Canadian Journal of Philosophy*, suppl. vol. 2, eds. R. A. Shiner and J. King-Farlow: 61–85, reprinted in D. Furley (1989): chapter 6. References are to the reprint.

(1989) *Cosmic Problems: Essays on Greek and Roman Philosophy of Nature*. Cambridge.

Furley, D. and Allen, R. E. (eds.) (1970) *Studies in Presocratic Philosophy*, vol. 1. London.

Furley, W. D. (1995) 'Praise and persuasion in Greek hymns', *Journal of Hellenic Studies* 115: 29–46.

Garland, R. (1990) 'Priests and power in classical Athens', in *Pagan Priests: religion and power in the ancient world*, eds. M. Beard and J. North. Ithaca: 73–91.

Gemelli Marciano, M. L. (1991) 'L'atomismo e il corpuscularismo empedocleo: frammenti di interpretazioni nel mondo antico', *Elenchos* 12: 5–37.

Gerson, L. P. (1990) *God and Greek Philosophy. Studies in the Early History of Natural Theology*. London.

Ginouvès, R. (1994) *Macedonia from Philip II to the Roman Conquest*. Princeton.

Giouri, E. (1978) *Ὁ κρατήρας τοῦ Δερβενίου*. Athens.

Graf, F. (1974) *Eleusis und die orphische Dichtung Athens in vorhellenistischer Zeit*, Religionsgeschichtliche Versuche und Vorarbeiten 33. Berlin and New York.

(1980) 'Milch, Honig und Wein. Zum Verständnis der Libation im griechischen Ritual' in *Perennitas. Studi in onore di Angelo Brelich*. Rome: 209–21.

(1985), review of West (1983), *Gnomon* 57: 385–91.

(1991) 'Textes orphiques et rituel bacchique', in *Orphisme et Orphée en l'honneur de J. Rudhardt*, ed. P. Borgeaud, Recherches et Rencontres 3. Geneva: 87–102.

(1993a) *Greek Mythology. An Introduction*. Baltimore and London.

(1993b) 'Dionysian and Orphic eschatology: new texts and old questions', in *Masks of Dionysos*, eds. T. H. Carpenter and C. A. Faraone. Ithaca: 239–58.

(1994) *La magie dans l'antiquité gréco-romaine*. Paris.

Graham, D. (2002) 'La lumière de la lune dans la pensée grecque archaïque', in *Qu'est-ce que la philosophie présocratique?*, eds. A. Laks and C. Louguet. Lille: 351–80.

Gruppe, O. (1887) *Theogonien in Griechische Culte und Mythen*. Leipzig.

Guthrie, W. K. C. (1952a) 'The Pre-Socratic World-Picture', *Harvard Theological Review* 45: 87–104.

(1952b) *Orpheus and Greek Religion. A Study of the Orphic Movement*, 2nd revised edn. London.

(1955) *The Greeks and their Gods*. Boston.

Hahm, D. E. (1977) *The Origins of Stoic Cosmology*. Columbus.

Halleran, M. R. (1995) *Euripides, Hippolytus*. Warminster.

Harrison, J. E. (1922) *Prolegomena to the study of Greek religion*, 3rd edn. Cambridge.

Heinze, M. (1890) 'Über den Νοῦς des Anaxagoras', *Berichte der königlichen sächsischen Gesellschaft der Wissenschaften* 1: 1–45.

Held, D. (1972) *The Fragments of Anaxagoras: some problems textual and philosophical*, Diss. Brown Univ.

Henrichs, A. (1968) 'Die Erdmutter Demeter (P. Derveni und Eurip. *Bakch*. 275f.)', *Zeitschrift für Papyrologie und Epigraphik* 3: 111–12.

(1972) 'Towards a new edition of Philodemus' treatise On Piety', *Greek, Roman and Byzantine Studies* 13: 67–98.

(1974) 'Die Kritik der stoischen Theologie im PHerc. 1428', *Cronache Ercolanesi* 4: 5–32.

(1984) 'The Eumenides and the wineless libation in the Derveni Papyrus', in *Atti del XVII Congresso Internazionale de Papirologia, Napoli, 1983*, vol. 2. Naples: 255–68.

(1994) 'Antonimity and polarity: unknown gods and nameless altars at the Areopagus', *Illinois Classical Studies* 19: 27–58.

(1998) 'Dromena und Legomena. Zum rituellen Selbverständnis der Griechen', in *Ansichten griechischer Rituale. Geburtstag-Symposium für Walter Burkert*, ed. F. Graf. Stuttgart and Leipzig: 33–71.

Henry, M. (1986) 'The Derveni commentator as literary critic', *Transactions and Proceedings of the American Philological Association* 116: 149–64.

Holwerda, A. E. J. (1894) 'De theogonia Orphica', *Mnemosyne* 22: 286–329 and 361–85.

Hooke, S. H. (ed.) (1933) *Myth and Ritual: Essays on the Myth and Ritual of the Hebrews in Relation to the Culture Pattern of the Ancient East*. London.

Huffman, C. A. (1993) *Philolaus of Croton: Pythagorean and Presocratic*. Cambridge.

Hüffmeier, F. (1963) 'Teleologische Weltbetrachtung bei Diogenes von Apollonia?', *Philologus* 107: 131–8.

Hussey, E. (1972). *The Presocratics*. London.

(1999) 'The enigmas of Derveni', *Oxford Studies of Ancient Philosophy* 17: 303–24.

Inwood, B. (1986) 'Anaxagoras and infinite divisibility', *Illinois Classical Studies* 11: 17–33.

Irigoin, J. (1972) 'Les Manuscrits grecs 11', *Revue des Études Grecques* 85: 543–71.

Isler, H. P. (1970) *Acheloos: eine Monographie*. Bern.

(1981) Acheloos *s.v. LIMC* 1.1.

(1990) review of Weiss (1984), *Gnomon* 62: 661–3.

Jaeger, W. (1947) *The Theology of the Early Greek Philosophers*. Oxford.

Jakoby, F. (1949) *Atthis. The Local Chronicles of Ancient Athens*. Oxford.

(1954) *Die Fragmente der griechischen Historiker* vol. IIIb. Leiden.

(1959) *Diagoras ὁ ἄθεος*. Abhandlungen der deutschen Akademie der Wissenschaften zu Berlin, Klasse fur Sprachen, Literatur und Kunst, Abh. 3. Berlin.

Janko, R. (1986) review of West (1983), *Classical Philology* 81: 154–9.

(1997) 'The physicist as hierophant: Aristophanes, Socrates and the authorship of the Derveni papyrus', *Zeitschrift für Papyrologie und Epigraphik* 118: 61–94.

(1998) 'Plato's cover-up' (review of L&M), *The Times Literary Supplement*, No. 4958, April 10: 26.

(2001) 'The Derveni papyrus (Diagoras of Melos, Apopyrgizontes Logoi?): a new translation', *Classical Philology* 96: 1–32.

(2002) 'The Derveni papyrus: an interim text', *Zeitschrift für Papyrologie und Epigraphik* 141: 1–62.

Johnson, W. A. (1994) 'The Function of the paragraphos in Greek literary prose texts', *Zeitschrift für Papyrologie und Epigraphik* 100: 65–8.

Johnston, S. I. (1999) *Restless Dead. Encounters Between the Living and the Dead in Ancient Greece*. Berkeley, Los Angeles and London.

Jöhrens, O. (1939) *Die Fragmente des Anaxagoras*. Bochum and Langendreer.

Jouanna, J. (1965) 'Rapports entre Mélissos de Samos et Diogène d'Apollonie, à la lumière du traité hippocratique de natura hominis', *Revue des Études Grecques* 67: 306–23.

Kahn, C. H. (1973) 'Language and ontology in the Cratylus', in *Exegesis and Argument: Studies in Greek Philosophy Presented to G. Vlastos*, eds. E. N. Lee, A. P. D. Mourelatos and R. M. Rorty, *Phronesis* Supplement 1. Assen: 152–76.

(1979) *The Art and Thought of Heraclitus*. Cambridge.

(1986) 'Les Mots et les formes dans le "Cratyle" de Platon', in *Philosophie du langage et grammaire dans l'antiquité*. Brussels and Grenoble: 91–103.

(1994) *Anaximander and the Origins of Greek Cosmology*, 2nd edn. Indianapolis.

(1997) 'Was Euthyphro the author of the Derveni Papyrus?', in L&M: 55–63.

Kaibel, G. (1878) *Epigrammata Graeca ex lapidibus conlecta*. Berlin.

Kapsomenos, S. G. (1963) 'Der Papyrus von Derveni. Ein Kommentar zur orphischen Theogonie', *Gnomon* 35: 222–3.

(1964) "'Ο ὀρφικὸς πάπυρος τῆς Θεσσαλονίκης", *Archaiologicon Deltion* 19: 17–25.

(1964–65) 'The Orphic papyrus roll of Thessalonika', *Bulletin of the American Society of Papyrologists* 2: 3–12.

Kerferd, G. B. (1969) 'Anaxagoras and the concept of matter before Aristotle', *Bulletin of the John Rylands Library* 52: 129–43; reprinted in Mourelatos (1974): 489–503. References are to the reprint.

Kern, O. (1888) *De Orphei Epimenidis Pherecydis theogoniis questiones criticae*. Berlin.

Kingsley, P. (1994) 'Empedocles' sun', *Classical Quarterly* 44: 316–24.

(1995) *Ancient Philosophy, Mystery, and Magic. Empedocles and Pythagorean Tradition*. Oxford.

Kirk, G. S. (1970) *Heraclitus. The Cosmic Fragments*. Cambridge.

Komoróczy G. (1979) 'A kozmogónia mítoszai Mezopotámiában és a Kumarbi-eposzok', in *A šumer irodalmi hagyomány*. Budapest: 531–80.

Laks, A. (1983) *Diogène d'Apollonie. La dernière cosmologie présocratique*. Lille.

(1993) 'Mind's crisis. On Anaxagoras' ΝΟΥΣ', *The Southern Journal of Philosophy* 31 Supplement (The Spindel Conference 1992): 19–38.

(1997) 'Between religion and philosophy: the function of allegory in the Derveni papyrus', *Phronesis* 42: 121–42.

Laks, A. and Most, G. W. (1997) 'A provisional translation of the Derveni papyrus', in L&M: 9–22.

Lamedica, A. (1990) 'La terminologia critico-letteraria del Papiro di Derveni ai Corpora scoliografici', in *Lessici Tecnici Greci e Latini, Atti del I. Seminario di Studi, Accademica Peloritana dei Pericolanti, Cl. Lett. Filos.* 66, Suppl. 1: 83–91.

(1992) 'Il Papiro di Derveni come commentario. Problemi formali', in *Proceedings of the XIXth International Congress of Papyrology, Cairo 2–9 Sept. 1989*, vol. 1. Cairo: 325–33.

Lämmli, F. (1962) *Vom Chaos zum Kosmos*. Basel.

Lanza, D. (1966) *Anassagora: Testimonianze e Frammenti*. Firenze.

Lebedev, A. (1985) 'The cosmos as a stadium: agonistic metaphors in Heraclitus' cosmology', *Phronesis* 30: 131–50.

(1989) 'Heraclitus in P.Derveni', *Zeitschrift für Papyrologie und Epigraphik* 79: 39–47.

Lesher, J. H. (1995) 'Mind's knowledge and powers of control in Anaxagoras DK B12', *Phronesis* 40: 125–42.

Levin, S. B. (1995) 'What's in a name? A reconsideration of the *Cratylus*' historical sources and topics', *Ancient Philosophy* 15: 91–115.

Lewis, N. (1974) *Papyrus in Classical Antiquity*. Oxford.

Linforth, I. (1941) *The Arts of Orpheus*. Berkeley and Los Angeles.

Lloyd, G. E. R. (1966) *Polarity and Analogy*. Cambridge.

(1979) *Magic, Reason and Experience*. Cambridge.

(1987) *The Revolutions of Wisdom. Studies in the Claims and Practice of Ancient Greek Science*, Sather Classical Lectures, vol. 52. Berkeley, Los Angeles and London.

(1990) *Demystifying Mentalities*. Cambridge.

Lobeck, C. A. (1829) *Aglaophamus*. Königsberg.

Long, A. A. (1992) 'Stoic readings of Homer', in *Homer's Ancient Readers. The Hermeneutics of Greek Epic's Earliest Exegetes*, eds. R. Lamberton and J. J. Keaney. Princeton: 41–66.

Louguet, C. (2002) 'Note sur le fragment B4a d'Anaxagore: pourquoi les autres mondes doivent-ils être semblables au nôtre?', in *Qu'est-ce que la philosophie présocratique?*, eds. A. Laks and C. Louguet. Lille: 497–530.

Makaronas, C. I. (1963) *Archaiologicon Deltion* 18: 193–6.

Malinowski, B. (1954) *Magic, Science, and Religion and Other Essays*. New York.

Manetti, G. (1987) *Le teorie del segno nell'antichità classica*. Milan.

Mansfeld, J. (1971) *The Pseudo-Hippocratic Tract Peri hebdomadon, ch.1–11 and Greek Philosophy*. Assen.

(1985) 'Aristotle and others on Thales, or the beginnings of natural philosophy. With some remarks on Xenophanes', *Mnemosyne* 38: 109–29.

(1986) 'Aristotle, Plato, and the Preplatonic doxography and chronology', in *Storiografia e dossografia nella filosofia antica*, ed. G. Cambiano. Turin: 1–59.

(1987) *Die Vorsokratiker. Auswahl der Fragmente*. Stuttgart.

(1995) 'Critical note: Empedocles and his interpreters', *Phronesis* 40: 109–15.

Marazov, I. (ed.) (1998) *Ancient Gold: The Wealth of the Thracians*. New York.

Marcovich, M. (1967) *Heraclitus. Editio maior*. Merida.

Maróth K. (1951–2) 'Die Trennung von Himmel und Erde', *Acta Antiqua* 1: 35–66.

Martin, A. and Primavesi, O. (1999) *L'Empédocle de Strasbourg* (P. Strasb. gr. Inv. 1665–1666). Introduction, édition et commentaire. Strasbourg, Berlin and New York.

Mathewson, R. (1958) 'Aristotle and Anaxagoras: an examination of F. M. Cornford's interpretation', *Classical Quarterly* n.s. 8: 67–81.

Merkelbach, R. (1967) 'Der orphische Papyrus von Derveni', *Zeitschrift für Papyrologie und Epigraphik* 1: 21–32.

(1977) 'Ein neues "orphisches" Goldblättchen', *Zeitschrift für Papyrologie und Epigraphik* 25: 276.

Morris, I. (1992) *Death-Ritual and Social Structure in Classical Antiquity*. Cambridge.

Most, G. W. (1997) 'The Fire Next Time. Cosmology, Allegoresis, and Salvation in the Derveni Papyrus', *Journal of Hellenic Studies* 117: 117–35.

(1999a) *Commentaries-Kommentare*. Aporemata 4. Göttingen.

(1999b) 'Preface', in Most (1999a): v–xv.

(1999c) 'From Logos to Mythos', in *From Myth to Reason?*, ed. R. Buxton. Oxford: 25–47.

Mouraviev, S. N. (1985) 'The Heraclitean fragment of the Derveni papyrus', *Zeitschrift für Papyrologie und Epigraphik* 61: 131–2.

(1993) *Heraclitea II. A 1 Héraclite d'Ephèse. La tradition antique et médievale, I: D'Epicharme à Platon et Héraclide le Pontique*. Paris and Moscow.

Mourelatos, A. P. D. (ed.) (1974) *The Presocratics: a Collection of Critical Essays*. Garden City, New York.

Mugler, C. (1956) 'Le Problème d'Anaxagore', *Revue des Études Grecques* 69: 314–76.

Musgrave, J. (1990) '"Dust and damn'd oblivion": a study of cremation in ancient Greece', *Annual of the British School at Athens* 85: 271–99.

Nilsson, M. P. (1925) *A History of Greek Religion* (trans. from Swedish by F. J. Fielden). Oxford.

Obbink, D. (1994) 'A quotation of the Derveni papyrus in Philodemus' On Piety', *Cronache Ercolanesi* 24: 1–39.

(1997) 'Cosmology as initiation vs. the critique of Orphic mysteries', in L&M: 39–54.

Osborne, C. (2000) 'Rummaging in the recycling bins of Upper Egypt. A discussion of A. Martin and O. Primavesi, *L'Empédocle de Strasbourg*', *Oxford Studies in Ancient Philosophy* 18: 329–56.

Palmer, J. A. (1998) 'Xenophanes' Ouranian god in the fourth century', *Oxford Studies in Ancient Philosophy* 16: 1–34.

Panchenko, D. (1999), 'The shape of the earth in Archelaus, Democritus and Leucippus', *Hyperboreus* 5: 22–39.

(2002), 'Eudemus fr. 145 Wehrli and the ancient theories of lunar light', in *Eudemus of Rhodes*, eds. I. Bodnár and W. Fortenbaugh. New Brunswick: 323–36.

Parke, H. W. and Wormell, D. E. W. (1956) *The Delphic Oracle*, 2 vols. Oxford.

Parker, R. (1983) *Miasma. Pollution and Purification in Early Greek Religion*. Oxford.
(1995) 'Early Orphism', in *The Greek World*, ed. A. Powell. London: 483–510.
Patzer, A. (1986) *Der Sophist Hippias als Philosophiehistoriker*. Freiburg and Munich.
Peck, A. L. (1926) 'Anaxagoras and the parts', *Classical Quarterly* 20: 57–62.
(1931) 'Anaxagoras: Predication as a problem in physics', *Classical Quarterly* 25: 27–37 and 112–20.
Pépin, J. (1976) *Mythe et Allégorie. Les origines grecques et les contestations judéo-chrétiennes*, 2nd edn. Paris.
Pfeiffer, R. (1968) *History of Classical Scholarship. From the Beginning to the End of the Hellenistic Age*. Oxford.
Picard, C. (1966) 'Après la lecture – encore partielle – du Papyrus de Dherveni', in *Mélanges Carcopino*. Paris: 737–46.
Pippidi, D. M. (1967) 'In jurul papirilor de la Derveni si Callatis', *Studii Clasice* 9: 203–10.
Pirenne-Delforge, V. (1991) 'Le culte de la Persuasion. Peithô en Grèce ancienne', *Revue de l'Histoire des Religions* 208: 395–413.
(1994) *L'Aphrodite grecque. Contribution à l'étude de ses cultes et de sa personnalité dans le panthéon archaïque et classique, Kernos* suppl. 4. Athens and Liège.
Polito, R. (1997) 'Teofrasto e il "Bellissimo" disordine cosmico de Eraclito. A proposito de *Metafisica* 7a10–15', *ACME* (Annali della Facoltà de Lettere e Filosofia dell'Università degli Studi di Milano) 50: 31–63.
Pradeau, J.-F. (2002) *Héraclite, Fragments*. Paris.
Pugliese Carratelli, G. (1974) 'Un sepolcro di Hipponion e un nuovo testo orfico', *Parola del Passato* 29: 110–26.
(1993) *Le lamine d'oro 'orfiche'*. Milano.
Ramnoux, C. (1980) 'La récupération d'Anaxagore', *Archives de Philosophie* 43: 75–98 and 279–97.
Raven, J. E. (1954) 'The basis of Anaxagoras' cosmology', *Classical Quarterly* 48: 123–37.
Reale, G. (1974) *Aristotele. Sul Cosmo*. Naples.
Ricciardelli Apicella, G. (1980) 'Orfismo e interpretazione allegorica', *Bolletino dei Classici* 3: 116–30.
(1993) 'Le teogonie orfiche nell'ambito delle teogonie greche', in *Orfeo e l'Orfismo, Atti del Seminario Nazionale Roma-Perugia 1985–1991*, ed. A. Masaracchia. Rome: 27–51.
Richardson, N. J. (1975) 'Homeric professors in the age of the sophists', *Proceedings of the Cambridge Philological Society* n.s. 21: 65–81.
(1985) review of West (1983), *Classical Review* n.s. 35: 87–90.
Richir, M. (1994) 'Qu'est-ce qu'un dieu? Mythologie et question de la pensée.' Préface à la *Philosophie de la mythologie* de F. W. Schelling, trad. fr. par A. Pernet. Grenoble.
(1995) *La naissance des dieux*. Paris.
Riedweg, C. (1995) 'Orphisches bei Empedocles', *Antike und Abendland* 41: 34–59.
Robinson, R. (1969) 'A criticism of Plato's *Cratylus*', in *Essays in Greek Philosophy*. Oxford: 118–38.

Rohde, E. (1925) *Psyche*. London.

Rossi, L. (1996) 'Il testamento di Posidippo e le laminette auree di Pella', *Zeitschrift für Papyrologie und Epigraphik* 102: 59–65.

Rudhardt, J. (1986) *Le rôle d'Eros et d'Aphrodite dans les cosmogonies grecques*. Paris.

(1991) 'Quelques réflexions sur les hymnes orphiques', in *Orphisme et Orphée en l'honneur de Jean Rudhardt*, ed. Ph. Borgeaud, Recherches et Rencontres 3. Geneva: 263–88.

Rusjaeva, A. S. (1978) 'Orfism i kult Dionisa v Olvii', *Vestnik Drevnej Istorii* 1: 87–104.

Rusten, J. S. (1984) 'Phanes-Eros in the Theogony of "Orpheus" in P.Derveni', in *Atti del XVII Congresso Internazionale de Papirologia, Napoli, 1983*, vol. 2. Naples: 333–5.

(1985) 'Interim notes on the papyrus from Derveni', *Harvard Studies in Classical Philology* 89: 121–40.

Schibli, H. S. (1990) *Pherecydes of Syros*. Oxford.

Schmidt, M. (1975) 'Orfeo e orfismo nella pittura vascolare italiota', in *Orfismo in Magna Grecia. Atti del quattordicessimo Convegno di Studi sulla Magna Grecia*. Naples: 105–37.

Schofield, M. (1975) 'Doxographica Anaxagorea', *Hermes* 103: 1–24.

(1980) *An Essay on Anaxagoras*. Cambridge.

Schönbeck, L. (1993) 'Heraclitus revisited (Pap. Derveni col. I lines 7–11)', *Zeitschrift für Papyrologie und Epigraphik* 97: 7–22.

(1998a) *Sunbowl or Symbol: Models for the Interpretation of Heraclitus' Sun Notion*. Amsterdam.

(1998b) 'A note on Heraclitus', *Mnemosyne* 51: 576–8.

Schwabe, W. (1975) 'Welches sind die materiellen Elemente bei Anaxagoras?', *Phronesis* 20: 1–10.

Schwabl, H. (1978) Zeus *s.v. RE* Suppl. vol 15. Munich: 993–1481.

Scolnicov, S. (1997) 'Freedom and education in Plato's *Timaeus*', in Calvo and Brisson (1997): 363–74.

Seaford, R. (1986) 'Immortality, salvation and the elements', *Harvard Studies in Classical Philology* 90: 1–26.

(1993) 'Dionysos as destroyer of the household: Homer, tragedy and the polis', in *Masks of Dionysos*, ed. T. Carpenter and C. Faraone. Ithaca, N.Y.: 115–46.

Sedley, D. N. (1989) 'The proems of Empedocles and Lucretius', *Greek, Roman and Byzantine Studies* 30: 269–96.

(1997) '"Becoming like god" in the *Timaeus* and Aristotle', in Calvo and Brisson (1997): 327–39.

(1998a) *Lucretius and the Transformation of Greek Wisdom*. Cambridge.

(1998b) 'The etymologies in Plato's *Cratylus*', *Journal of Hellenic Studies* 118: 142–56.

(1999) 'The ideal of godlikeness', in *Plato* vol. II *Ethics, Politics, Religion, and the Soul*. ed. G. Fine. Oxford: 309–28.

Sider, D. (1981) *The Fragments of Anaxagoras*. Meisenheim am Glan.

(1987) 'Heraclitus B3 and 94 in the Derveni papyrus', *Zeitschrift für Papyrologie und Epigraphik* 69: 225–8.

(1997) 'Heraclitus in the Derveni papyrus', in L&M: 129–48.

Sideris, A. (2000) 'Les tombes de Derveni: quelques remarques sur la toreutique', *Revue Archéologique* fasc. 1: 3–36.

Silvestre, A. M. (1988) 'Significato e Ruolo del *Nous* nella Filosofia di Anassagora', *Il Contributo* 12: 29–52.

Sluiter, I. (1999) 'Commentaries and the didactic tradition', in Most (1999a) 173–205.

Snell, B. (1924) *Die Ausdrücke für den Begriff des Wissens in der vorplatonischen Philosophie*. Berlin.

(1944) 'Die Nachrichten über di Lehren des Thales und die Anfänge der griechischen Philosophie- und Literaturgeschichte', *Philologus* 96: 170–82, reprinted in Snell (1966): 119–28. References are to the reprint.

(1966) *Gesammelte Schriften*. Göttingen.

(1986) *Die Entdeckung des Geistes: Studien zur Entstehung des europäischen Denkens bei den Griechen*, 6th corrected edn. Göttingen.

Solmsen, F. (1958) 'Aristotle and Presocratic cosmogony' *Harvard Studies in Classical Philology* 63: 265–82; reprinted in Solmsen (1968): 356–73.

(1968) *Kleine Schriften*, vol. 1. Hildesheim.

Sourvinou-Inwood, C. (1995) *'Reading' Greek Death: To the End of the Classical Period*. Oxford.

Staudacher, W. (1942) *Die Trennung von Himmel und Erde*. Tübingen.

Stewart, D. J. (1968) 'Νοῦς in Aristophanes', *Classical Journal* 63: 253–5.

Stokes, M. C. (1962) 'Hesiodic and Milesian cosmogonies I', *Phronesis* 7: 1–37.

(1963) 'Hesiodic and Milesian cosmogonies II', *Phronesis* 8: 1–34.

(1965) 'On Anaxagoras', *Archiv für Geschichte der Philosophie* 47: 1–19 and 217–50.

(1971) *One and Many in Presocratic Philosophy*. Cambridge, Mass.

Strang, C. (1963) 'The physical theory of Anaxagoras', *Archiv für Geschichte der Philosophie* 45: 101–18, reprinted in Allen and Furley (1975): 361–80.

Struck, P. (2002) 'The ordeal of the divine sign: divination and manliness in archaic and classical Greece', in *ANDREIA. Studies in Manliness and Courage in the Classical World*, eds. R. M. Rosen and I. Sluiter. Leiden: 167–86.

Tambiah, S. J. (1973) 'Form and meaning of magical acts: a point of view', in *Modes of Thought*, eds. R. Horton and R. Finnegan. London: 199–229.

(1990) *Magic, Science, Religion, and the Scope of Rationality*. Cambridge.

Tarán, L. (1972) 'The creation myth in Plato's *Timaeus*', in *Essays in Ancient Greek Philosophy*, eds. J. P. Anton and G. L. Kustas. Albany: 372–407.

Teodorsson, S.-T. (1982) *Anaxagoras' Theory of Matter*. Arlöv.

Theiler, W. (1965) *Zur Geschichte der teleologischen Naturbetrachtung bis auf Aristoteles*, 2nd edn. Berlin.

Themelis, P. G. and Touratsoglou, I. P. (1997) *Οι ταφοι του Δερβενιου*. Athens.

Thomas, K. (1971) *Religion and the Decline of Magic*. New York.

Tilman, V. (2000) 'Archélaos d'Athènes', *Revue de Philosophie Ancienne* 18: 65–107.

Tinnefeld, F. (1980) 'Referat über zwei Russische Aufsätze', *Zeitschrift für Papyrologie und Epigraphik* 38: 68–71.

Tortorelli Ghidini, M. (1985) 'Euphrone: una divinità orfica?', *Parola del Passato* 40: 419–25.

(1989) 'Nephele: una metafora orfica arcaica', *Parola del Passato* 44: 29–36.

(1991) 'Due nuovi teonimi orfici nel papiro di Derveni', in *Orphisme et Orphée en l'honneur de Jean Rudhardt*, ed. Ph. Borgeaud, Recherches et Rencontres 3. Geneva: 246–61.

Tsantsanoglou, K. (1992) 'Heraclitus 1T', in *Corpus dei Papiri Filosofici Greci e Latini. Testi e lessico nei papiri di cultura greca e latina dal secolo IV a.C. al VII d.C.*, 1.1. Firenze: 221–6.

(1997) 'The first columns of the Derveni papyrus and their religious significance', in L&M: 93–128.

Tsantsanoglou, K. and Parrássoglou, G. M. (1987) 'Two gold lamellae from Thessaly', *Hellenica* 38: 3–16.

(1988) 'Heraclitus in the Derveni Papyrus', in *Aristoxenica, Menandrea, Fragmenta Philosophica*, ed. A. Branacci et al., *Studi e Testi per il Corpus dei papiri filosofici greci e latini 3*. Firenze: 125–33.

Turner, E. G. (1968) *Greek Papyri: An Introduction*. Oxford.

(1980a) *Greek Papyri: An Introduction*, 2nd edn. Oxford.

(1980b) 'Ptolemaic bookhands and Lille Stesichorus', *Scrittura e Civiltà* 4: 19–49.

Turner, E. G. and Parsons, P. (1987) *Greek Manuscripts of the Ancient World*, 2nd edn. *Bulletin of the Institute of Classical Studies*, Suppl. 46. London.

Turner, E. G., Tsantsanoglou, K. and Parássoglou, G. M. (1982) 'On the Derveni Papyrus', *Gnomon* 54: 855–6.

Tzanavari, K. (1997) 'Δερβένι, Μια νεκρόπολη της αρχαίας Λητής', *Makedonia kai Thraki* 10: 461–9.

Vander Waerdt, P. A. (1994) 'Socrates in the *Clouds*', in *The Socratic Movement*, ed. P. A. Vander Waerdt. Ithaca and London: 48–86.

Vernant, J. P. (1962) *Les origines de la pensée grecque*. Paris.

Versnel, H. S. (1987) 'Greek myth and ritual: the case of Kronos', in *Interpretations of Greek Mythology*, ed. J. Bremmer. London: 121–52.

(1993) *Inconsistencies in Greek and Roman Religion* II. *Transition & Reversal in Myth and Ritual*, Studies in Greek and Roman Religion, vol. 6, 2. Leiden, New York and Cologne.

Vian, F. (1984), review of West (1983), *Revue de Philologie* 58: 288.

Vinogradov, J. G. (1991) 'Zur sachlichen und geschichtlichen Deutung der Orphiker-Plättchen von Olbia', in *Orphisme et Orphée en l'honneur de J. Rudhardt*, ed. P. Borgeaud, Recherches et Rencontres 3. Geneva: 77–86.

Vlastos, G. (1950), 'The physical theory of Anaxagoras', *The Philosophical Review* 59: 31–57, reprinted in Mourelatos (1974): 459–88. References are to the reprint.

(1952) 'Theology and philosophy in early Greek thought', *Philosophical Quarterly* 2: 92–123, reprinted in Furley and Allen (1970): 92–129. References are to the reprint.

Vögelin, E. (1957) *Order and History*, vol. 2: *The World of the Polis*. Baton Rouge, Louis. and New York.
Weiss, C. (1984) *Griechische Flussgottheiten in vorhellenistischer Zeit*. Würzburg.
West, M. L. (1966) *Hesiod: Theogony*, ed. with Prolegomena and Commentary. Oxford.
(1982) 'The Orphics of Olbia', *Zeitschrift für Papyrologie und Epigraphik* 45: 17–29.
(1983) *The Orphic Poems*. Oxford.
(1987) *Euripides: Orestes*, ed. with translation and commentary. Warminster.
(1994) 'Ab Ovo. Orpheus, Sanchuniathon, and the origins of the Ionian world model', *Classical Quarterly* 44: 289–307.
(1997) 'Hocus-pocus in East and West. Theogony, ritual, and the tradition of esoteric commentary', in L&M: 81–92.
Willink, C. W. (1986) *Euripides, Orestes*. Oxford.
Winiarczyk, M. (1979) 'Diagoras von Melos – Wahrheit und Legende I', *Eos* 67: 191–213.
(1980) 'Diagoras von Melos – Wahrheit und Legende II', *Eos* 68: 51–75.
(1981) *Diagorae Melii et Theodori Cyrenaei reliquiae*. Leipzig.
Wöhrle, G. (1993) *Anaximenes aus Milet: Die Fragmente zu seiner Lehre*. Stuttgart.
Woodbury, L. (1965) 'The date and atheism of Diagoras of Melos', *Phoenix* 19: 178–211.
Wright, M. R. (1981) *Empedocles: The Extant Fragments*. New Haven.
Zhmud', L. (1992) 'Orphism and graffiti from Olbia', *Hermes* 120: 159–68.
Zuntz, G. (1971) *Persephone. Three essays on religion and thought in Magna Graecia*. Oxford.

Index verborum

An asterisk indicates words that are not in LSJ and its Supplements. A cross indicates words which appear only in the apparatus. The second column contains the forms as they occur in the papyrus, indicating sublinear dots and supplements. Phonetic transformations due to the position of the word (accents and consonant assimilations) have been changed, however. The definite article and καί are not listed. In those cases where a god's name is also a common noun, the occurrences of the name are listed under the common noun, but the upper case initial is retained.

Index of passages

P. Derv.

Other ancient works

Index of modern names

Index of subjects

For EU product safety concerns, contact us at Calle de José Abascal, 56–1°, 28003 Madrid, Spain or eugpsr@cambridge.org.

www.ingramcontent.com/pod-product-compliance
Lightning Source LLC
LaVergne TN
LVHW090258210926
848480LV00001B/60

* 9 7 8 0 5 2 1 0 4 7 3 9 5 *